International Symposium on Materials in the Space Environment (ISMSE16-ICPMSE14) 2024

IOP Conference Series: Materials Science and Engineering
Volume 1328

Saint Raphael, France
7-11 October 2024

ISBN: 979-8-3313-2245-8
ISSN: 1757-8981

Printed from e-media with permission by:

Curran Associates, Inc.
57 Morehouse Lane
Red Hook, NY 12571

Some format issues inherent in the e-media version may also appear in this print version.

Institute of Physics
Dirac House, Temple Back
Bristol BS1 6BE UK

Phone: 44 1 17 929 7481
Fax: 44 1 17 920 0979

techtracking@iop.org

Additional copies of this publication are available from:

Curran Associates, Inc.
57 Morehouse Lane
Red Hook, NY 12571 USA
Phone: 845-758-0400
Fax: 845-758-2633
Email: curran@proceedings.com
Web: www.proceedings.com

TABLE OF CONTENTS

IOP Conf. Series: Materials Science and Engineering **1328** (2025) 011001 doi:10.1088/1757-899X/1328/1/011001

ISMSE-16/ICPMSE-14 Conference Proceedings

IOP Conf. Series: Materials Science and Engineering **1328** (2025) 011001 doi:10.1088/1757-899X/1328/1/011001

The International Symposium on Materials in the Space Environment (ISMSE) and the International Conference on Protection of Materials and Structures from the Space Environment (ICPMSE) is jointly organized by ESA, CNES, ONERA and ITL. In 2024 (7-11 October) the main organizer was ONERA and the 16th ISMSE/ 14th ICPMSE took place in Saint Raphael, a beautiful city at the South of France. 155 participants representing 70 organisations from 20 countries joined this event.

At the conference 56 oral presentations (including 15 plenaries) and 58 posters were presented, along with round tables on such hot topics as Space Materials Outgassing, Environmental Testing and Sustainability.

These resulted in submissions of 22 papers which are herein provided in form of conference proceedings. The following relevant fields are represented: Space and Extreme environments (UV, charged particles, temperature...), Planetary environments (Lunar and Martian regolith...), Dust mitigation and related problems (degradation of materials performance, electrical build-up...), Micrometeoroids and debris in space environment and related effects, Atomic Oxygen effects (LEO/VLEO applications), Radiation-induced ageing, charging, outgassing, Contamination testing and modelling, thermo-optical properties of contaminated surfaces, species separation, Protection coatings (atomic oxygen, charging, contamination, radiation shielding), Innovative materials (self-healing, bio-inspired …) and processes in space sub-systems, Effect of space environment on "COTS" or "terrestrial" technologies (low budget, Cub-sat, short lifetime LEO missions), Smart / multifunctional / nano materials and coatings, Additive manufacturing, printed ink electronics…, Clean Space and sustainability (reusable materials, mitigation of debris), Long-term storage and materials obsolescence, Standards and Regulations, Ground testing (best practice, data), Test Facilities and platforms, Flight experiments: design, data and lesson learned.

All papers are published open access under the terms of the Creative Commons Attribution (CC BY) licence (creativecommons.org/licenses/by/4.0 or any newer version of the licence). Each submission was subjected to strict peer review process involving at least one independent Reviewer and many quality checks by the assigned Editorial Board Members. The following review criteria were regarded as priorities:

- novelty and relevance,
- suitability of methodology,
- completeness of provided information,
- completeness of references,
- editorial issues.

We hope the Readers will find this collection useful and inspiring for their engineering work and research.

ISMSE Proceedings Editorial Board:

Malgorzata Holynska (Chair, European Space Agency ESA/ European Space Research and Technology Center ESTEC, The Netherlands)
Sophie Duzellier (Main Organizer, Office National d'Études et de Recherches Aérospatiales, French National Office for Aerospace Studies and Research ONERA, France)
Adrian Tighe (European Space Agency ESA/ European Space Research and Technology Center ESTEC, The Netherlands)
Elisabeth Laurent (Centre national d'études spatiales, French Space Agency, CNES, France)
Jacob Kleiman (Integrity Testing Laboratory ITL INC, Canada)
Timothy Minton (University of Colorado Boulder, USA)

IOP Conf. Series: Materials Science and Engineering **1328** (2025) 011002 doi:10.1088/1757-899X/1328/1/011002

Peer Review Statement

All papers published in this volume have been reviewed through processes administered by the Editors. Reviews were conducted by expert referees to the professional and scientific standards expected of a proceedings journal published by IOP Publishing.

- **Type of peer review:** Single Anonymous
- **Conference submission management system:** Morressier
- **Number of submissions received:** 23
- **Number of submissions sent for review:** 22
- **Number of submissions accepted:** 22
- **Acceptance Rate (Submissions Accepted / Submissions Received × 100):** 95.7
- **Average number of reviews per paper:** 1.7
- **Total number of reviewers involved:** 20
- **Contact person for queries:**
  Name: Malgorzata Holynska
  Email: Malgorzata.Holynska@esa.int
  Affiliation: ESA ESTEC

IOP Conf. Series: Materials Science and Engineering **1328** (2025) 012001 doi:10.1088/1757-899X/1328/1/012001

Destination Moon : How "Luna" Dust affects Photovoltaics Modules ?

Philippe Voarino*, Timoté Lombard, Clément Jamin, and Eric Pilat

Univ. Grenoble Alpes, CEA, Liten, Campus INES, 73375 Le Bourget du Lac, France

*E-mail: philippe.voarino@cea.fr

Abstract. The possibility of generating electrical power using photovoltaic solar panels remains a challenge for future lunar bases. The photovoltaic solution chosen must be as resistant as possible to the extremely harsh lunar environment, particularly to lunar dust. The aim of this work is to develop an original, accurate method for depositing lunar dust on a large surface, combined with optical and electrical evaluation of the photovoltaic assemblies. By using lunar highland dust simulant, a first regime named "coating" associated with a linear decline, and a second regime named "overlapping" associated with an exponential decline were identified in accordance with what was suggested in the literature. According to the model established from experiments, if a semi-annual or an annual cleaning were planned, the associated power loss would be between ~0.5% - 2.5% or between ~1% - 5.5%, respectively.

1. Introduction

The Shackleton crater, with a diameter of 21 km and located at the South Pole, features areas with solar visibility exceeding 85% over a 20-year period [1]. This high visibility enables the nearly continuous generation of electricity through photovoltaic effects. The selected photovoltaic solution must be highly resistant to the extremely harsh lunar environment [2], particularly lunar dust. Lunar dust particles are irregular and elongated, with sharp edges and an average size of 70 µm [3]. Only 10 to 20% of lunar soil particles are spherical [4]. The lunar soil consists of various particle types, including basalts, mineral fragments (such as plagioclase, olivine, pyroxene, and feldspar), breccias, agglutinates, rock chips (lithic fragments), and glasses. The Apollo 16 site, situated in the highlands, exhibited high concentrations of Al_2O_3, SiO_2, and CaO, along with lower levels of FeO and MgO—a typical signature of highland rocks [5]. A crucial geotechnical parameter of lunar dust for engineering applications is its density, including specific gravity, bulk density, and relative density. However, due to the porosity (voids between particles) of lunar soil, interpreting density is complex [6]. Lunar soil exhibits three types of porosity: inter-granular (volume of space between individual grains), intra-granular (volume of re-entrant surfaces on the exterior of grains), and sub-granular (volume of enclosed voids within grains). Estimates of *in situ* lunar soil porosity (combining inter- and intra-granular porosity) range from 45% to 75% [6][7]. In addition to grain-to-grain cohesion, lunar soil is sticky, with small particles adhering to larger ones due to electrostatic charging and welding [8], as well as the soil's electrical insulation properties (very low conductivity) [9]. The transport mechanisms of these particles involve both natural processes (micrometeorite impacts, electrostatic levitation due to solar radiation) and anthropogenic activities (movement of astronauts and vehicles, mining, construction, and spacecraft landings and take-offs). Their effects on solar panels include obscuration leading to power loss, surface contamination, front surface abrasion, electrostatic discharge, and alteration of thermal properties. Table 1 provides an overview of the distance of dust deposition from the original site, the height, and the particle size distribution resulting from both natural and anthropogenic causes. Consequently, dust can impact locally or settle several kilometers away from its source.

Table 1: Overview of the distance of deposit from original site, height and Particle Size Distribution (PSD) of dust cloud originating from both natural and anthropogenic causes; data is gathered from the literature described in previous paragraphs. Question mark are considered as hypothesis and have to be consolidated with particular lunar missions.

	Natural causes		**Anthropogenic causes**		
	Radiation	Micrometeoroid	Human	Rover	Spacecraft
Distance of deposit from original site	Up to hundreds km (?)	Up to several km (?)	< 10 m	< 20 m	> 2 km
Height	From ground to ~100 km	From ground to several hundred km	< 5 m	< 5 m	> 60 m
Dust grain PSD	< 0.01– ~10 μm	< 1 mm (?)	< 1 mm (?)	< 1 mm (?)	< 5 mm

While anthropogenic causes are suspected to have a more significant impact on dust disturbance, and consequently on dust deposition and accumulation, they are often linked to human activities. Therefore, certain measures can be implemented to mitigate their effects on Photovoltaic Assemblies (PVAs), such as using well-designed fenders for rovers, selecting landing sites distant from power plants, elevating PVAs on poles, and generally keeping all activities away from essential components. In contrast, natural mechanisms are beyond human control and are geographically indiscriminate, although the amount of dust suspended and transported by natural processes is suspected to be less significant than that disturbed by anthropogenic activities. Thus, this study focuses solely on natural causes.

Two primary studies were conducted: one by Katzan *et al.* [4][10] on 2x4 cm^2 silicon solar cells, and another by Shahmoradi *et al.* [11] on photovoltaic (PV) modules, focusing on power evaluation (see Figure 1). Katzan *et al.* proposed a model based on simple overlapping to explain dust shading of solar panels. Experiments using "Minnesota Lunar Simulant-1" demonstrated that this exponential model effectively described light shading, with an approximation where the relative transmittance of light is proportional to the electrical power output. While this approximation holds true at the first order for current, it could be enhanced by conducting experiments on PVAs. Shahmoradi *et al.* investigated the effect of temperature on the module and showed that the emissivity of solar panels can change. However, the coating technique was not uniform, and the thickness was not well controlled.

In this paper, we propose using a method already developed for terrestrial applications to coat PVAs with lunar simulants. This method, based on experiences in the Atacama Desert conducted by CEA, combines optical and electrical evaluations of PV modules [12] [13]. Additionally, this method includes an estimation of cleaning frequency based on real data acquired from lunar missions. The paper is structured as follows: the second section describes the method used to coat PVAs with lunar simulant, and the third section presents an analysis of the electrical performance of PVAs in relation to real data from lunar missions, proposing a cleaning frequency based on electrical power losses.

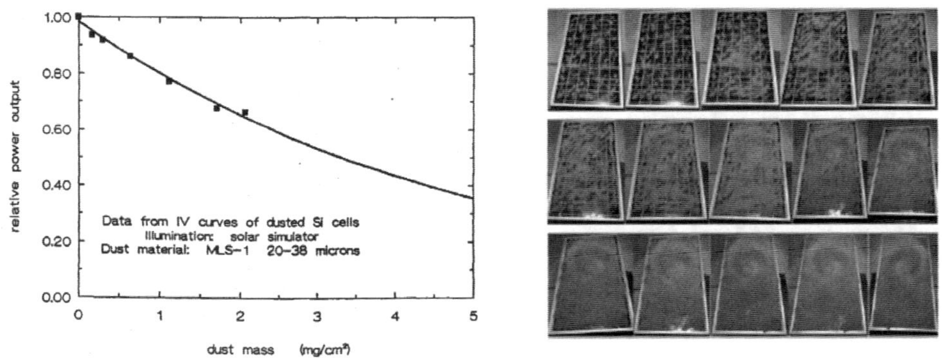

(a) (b)

Figure 1. (a) Evolution of PV performances with dust accumulation [14]. (b) Pictures of PV modules in function of different dust accumulations [11]

2. Method

Among the multitude factors that influence the performance of a PVA, one of the most significant is the accumulation and adhesion of dust onto the surface. The PVA's power output and then, its efficiency, can be substantially reduced when a significant amount of grains obstructs the sunlight reaching the PV cells. Just like lunar dust deposition rate, studies of degradation of PVA performances caused by lunar dust sedimentation are few in the literature. One of the objectives is to characterize the effect of a lunar dust simulant on a PV module through the Relative Maximum Power Output (RMPO) and the Relative Short-Circuit Current (RSCC) defined as:

$$RMPO = \frac{P_{max,\ \text{dust-covered PV module}}}{P_{max,\ \text{dust-free PV module}}} \in [0,1]$$

$$RSCC = \frac{I_{SC,\ \text{dust-covered PV module}}}{I_{SC,\ \text{dust-free PV module}}} \in [0,1]$$

The short-circuit current is also studied because it is closely linked to the optical properties of the PVA [15]. Then, by developing a model of the expected evolution of RMPO (or RSCC) as a function of surface dust density (mg/cm^2), the aim of this study is to evaluate the daily and annual RMPO (or RSCC) decline as a function of accumulated dust, using the annual lunar dust deposition rate reported in the literature. This approach will provide a projection of the operational lifetime of a future lunar PV power plant, considering only the dust issue and expected cleaning frequencies.

2.1 Manufacturing of PV modules

Three PVAs with 2x ½ cell Passivated Emitter Rear Contacted (PERC) type were manufactured. The PVA consists of a 150 µm cover glass as frontsheet, a 600 µm layer of ethylene vinyl acetate (EVA) as encapsulant, a 180 µm-thick solar cell string, another 600 µm layer of EVA and a composite substrate of 3 mm aluminum honeycomb. This sandwich was then laminated using a terrestrial PV process. Figure 2 presents a picture of one PVA with the associated electroluminescence image showing no crack and no defect . The dark lines are linked to the shadowning of the ribbons used to connect the two ½ cells (see figure 3).

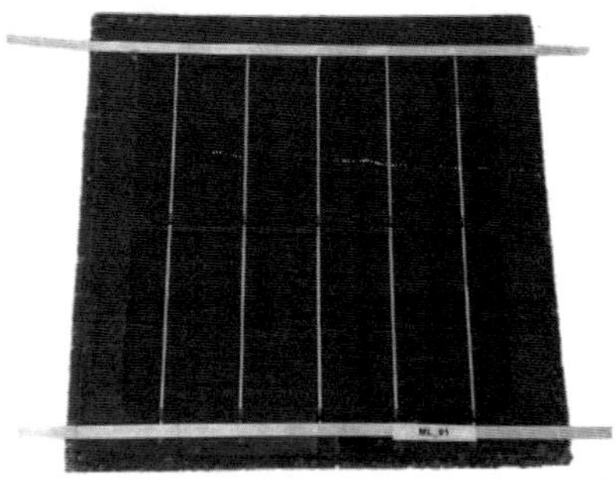

Figure 2. Picture of the PVA manufactured using as substrate for dust moon deposition.

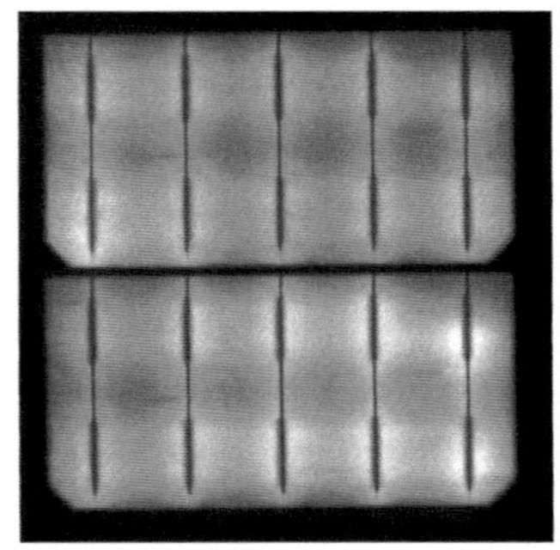

Figure 3. Electroluminescence of the same PV module.

2.2 Preparation of Simulant

The main objective is to assess the effect of lunar dust simulant accumulating on the PVA. The lunar dust simulant used in these experiments is the Lunar Highland Dust Simulant (LHS-1D) from Exolith Lab. These particles were selected because composition and shape are similar to values previously presented and characteristics of highlands of Schakleton Crater. A beaker filled with dust was beforehand stored for several days in dry condition in an oven at 60°C to prevent dust grains from sticking together due to laboratory humidity. Moreover, all necessary dishes and tools were stored in the oven as well, to avoid any issues with humidity. The dust and materials were also kept at 60°C when not in use in order to eliminate the clumping effects of moisture. Although being stored in a hot and dry environement, the dust was observed to cement strongly and was extremely adhesive despite being not exposed to any harsh radiation. All experiments were carried out at room temperature in a controlled laboratory atmosphere. From the baker (figure 4a), a given mass in mg of dust was taken with a metalic spoon (figure 4b) and weighed in a Petri dish (figure 4c) inside a high precision Sartorius Quintix® lab balance. The contents of the Petri dish were transferred into the tube through a plastic funnel.

Figure 4. (a) Beaker with dust, note the adhesion of some dust grains to beaker's wall. (b) Metallic spoon with dust, note the high angle of repose. (c) Petri dish with dust, note the high ability of dust to pack down and cement.

IOP Conf. Series: Materials Science and Engineering **1328** (2025) 012001 doi:10.1088/1757-899X/1328/1/012001

The injection was realized through a solid particle dispenser, the RGB 1000 from German company Palas GmbH [12]. Two injection tubes are available: the smaller has an inner diameter of 1.4 cm and the larger has an inner diameter of 2.8 cm which corresponds to a maximum filling quantity of 10.8 g and 43 g respectively. The dispenser can handle particle sizes ranging between <0.1 – 100 μm. Dust was inserted in the dispersing head through a piston that pushes dust inside the tube, at a precisely controlled feed rate, onto a rotative brush, so that dust was uniformly compressed. Adjustable compressed air flow streams at a very high speed and blows the particles out of the brush, toward the soiling chamber. Generally, the injection took a few minutes, the time it takes for the piston to push the dust inside the tube to feed the brush.

2.3 Dust deposition on soiling chamber

The soiling chamber has the following characteristics (see figure 5): the pressure is ambient (1 atm) with a terrestrial gravity of 9.81 $m.s^{-2}$; the humidity is less than 20% with a chamber temperature comprised between 40°C and 60°C identical to the temperature of the sample support. In this chamber, the size of PVA could be 200x200 mm^2. The deflector angle was fixed to 65°, coming from the laboratory scientific legacy, this angle was judged to be the best regarding dust injection homogeneity, and the PV module support was tilted to 30°, that matches the average slope of the Shackleton crater. At the end of injection, when the piston is at the end of its road, close to the brush, some residual dust remained in the tube, due to the injection device architecture and operation. This amount of dust remaining inside the tube was weighed to assess the dust density (mg/cm^3) that entered the chamber. PV module was exposed for 30 min in the chamber to achieve a suitable steady state, with a few dust particles potentially suspended inside the chamber.

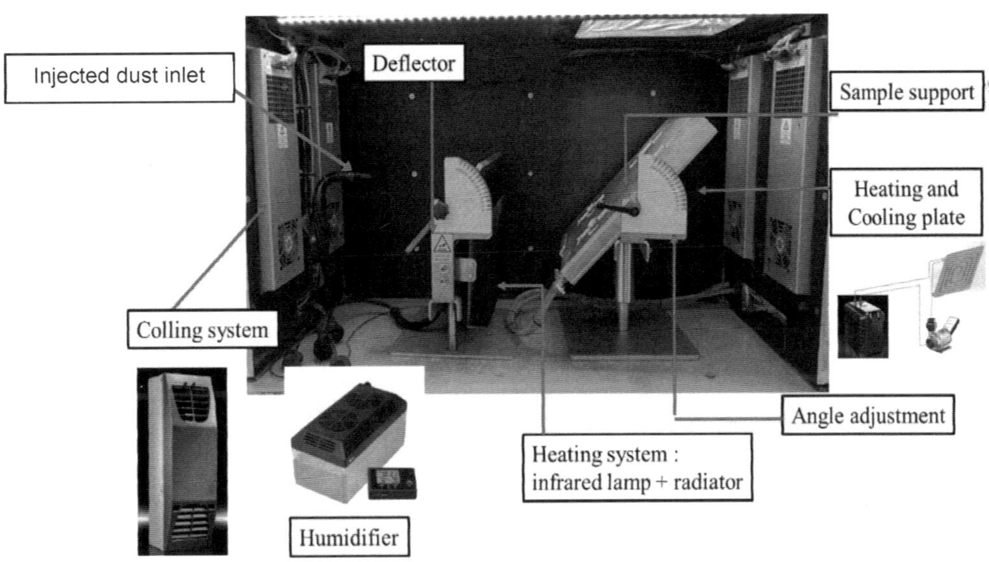

Figure 5. Soiling chamber developed for terrestrial photovoltaic purposes.

2.4. Electrical characterization

After the exposure time of 30 min, the soiled PV module was removed from the chamber and transported to the Solar simulator for flash testing. Solar simulator from PASAN were used to assess electrical performances of the PVA. In our case, the chosen spectrum is the Air Mass 1.5 G, and characterisation were carried out at 1000 W/m^2 with a standard temperature of 25°C. The sample were set vertically and normal to the irradiance. The figure 6 shows different pictures of PVA coated with different thicknesses of simulant. At left, the surface mass is around 0.2 mg/cm^2, and at right, it can reach 4.7 mg/cm^2. When the PVA is set vertical, no dust slide is visually observed. The PVA with the most of dust is of course the whiter (right one).

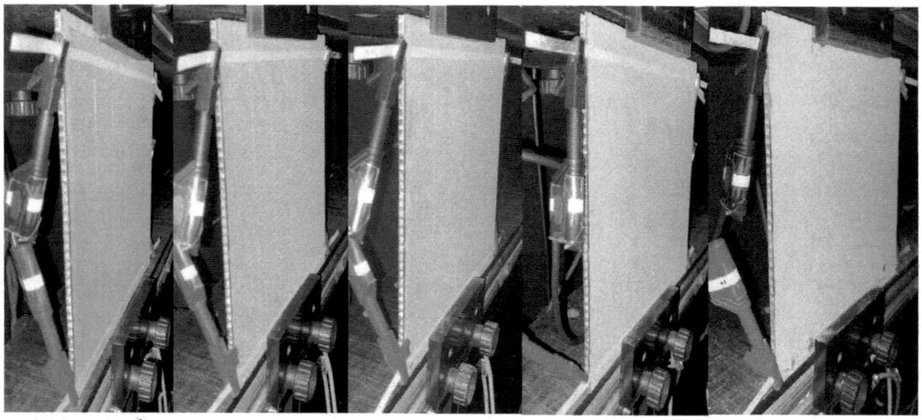

~0.20 mg/cm² *~0.30 mg/cm²* *~0.40 mg/cm²* *~1.40 mg/cm²* *~4.70 mg/cm²*

Figure 6. Pictures of PVA located into the solar simulator for different mass of dusts coated on the PVA (~0.2 mg/cm^2, ~0.3 mg/cm^2, ~0.4 mg/cm^2, ~1.4 mg/cm^2 and ~4.7 mg/cm^2).

The figure 7 presents the Current-Voltage (IV) curve of a dust free PV module (0 mg/cm^2) and IV curves for different dust accumulations and show a more dependent current degradation linked to optical path degradations.

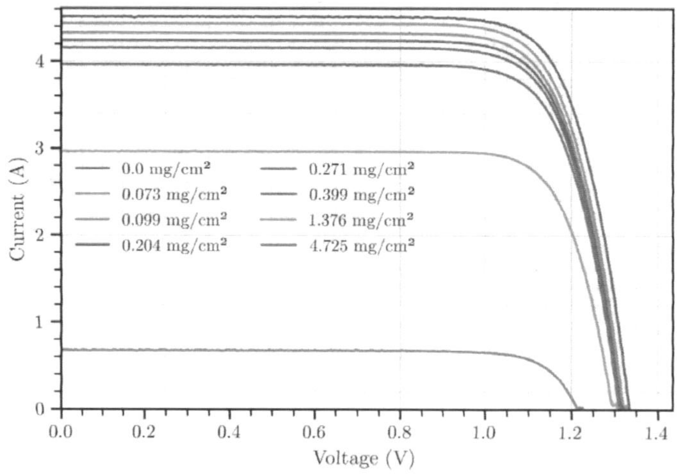

Figure 7. Evolution of IV curves for different dust surface density on the PV module for Air Mass 1.5 G at 1000 W/m^2 at a temperature of 25°C.

IOP Conf. Series: Materials Science and Engineering **1328** (2025) 012001 doi:10.1088/1757-899X/1328/1/012001

After the solar simulator flash test, the dust covering the PV module was collected using a classic paintbrush with synthetic bristles into a Petri dish or a small beaker. This dust was weighed to assess the surface density (mg/cm^2) of dust that had accumulated on the PV module. Afterward, the PV module was thoroughly cleaned with a water jet and air blast for the next measurement.

2.5 Conclusion

To ensure repeatability and reduce experiment's uncertainties, this process was performed five times. Discrepancies between measurements were inevitable due to the numerous handling steps and the need to move between different rooms. Additionally, experiments were conducted over a short period (less than 2 months) to ensure that only dust significantly affected the reduction of PV module performance, making other degradation factors such as UV or thermal cycling effects reasonably negligible.

3. Results and discussions

3.1 Assessment of deposition regimes

To study the optical degradation due to simulant deposition on PVA, the knowledge of RSCC is an interesting parameter (see figure 8). We can notice one exponential regime in function of the deposited dust thickness. If the quantity is very low, a linear law could be applied (see the zoom into the figure 8). The errors bars are linked to the five experiments carried out for each dust quantity.

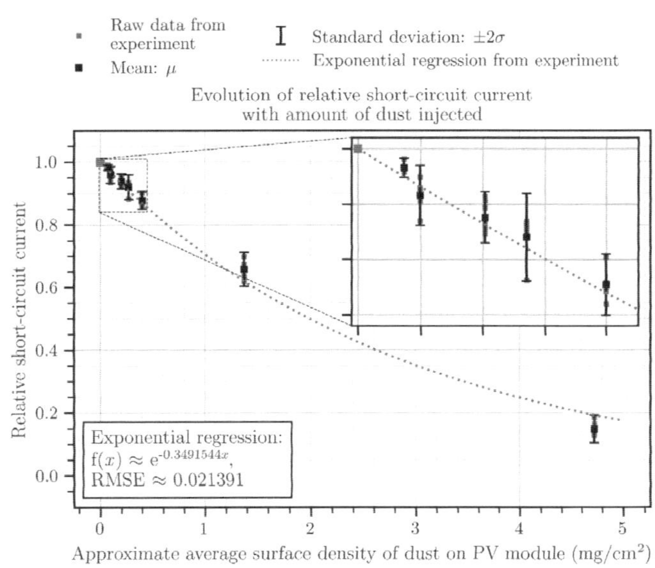

Figure 8. Evolution of RSCC with amount of dust deposited onto the PV module.

For comparison with the literature, Figure 9 shows the evolution of RMPO as a function of approximate and average surface density of dust on the PVA. The gray-square dots represent the raw measurement data (at least 5 points for each series). The black-square dots are the average point for each series and black error bars indicate two times the standard deviation for each series. Exponential regressions, as well as raw data, from Katzan *et al.* [4], and Shahmoradi *et al.*

[11], are plotted for reference, respectively, with green dotted line and green-square dots, and with red dotted line and red-square line. The gray dotted line is an exponential regression that fits the mean points from measurements. Results from experiments are not as good as expected from the literature. Even though, discrepancies with Katzan *et al.* may be due to the different PVA, different lunar dust simulant material with a different size range, different experimental conditions and different test facilities, and it may be roughly the same regarding Shahmoradi *et al.* As predicted by the literature and physics of dust accumulation, the decrease in RSCC and RMPO tends to follow an exponential pattern; here, a simple exponential function (f(x)=e^(-αx)) has been used.

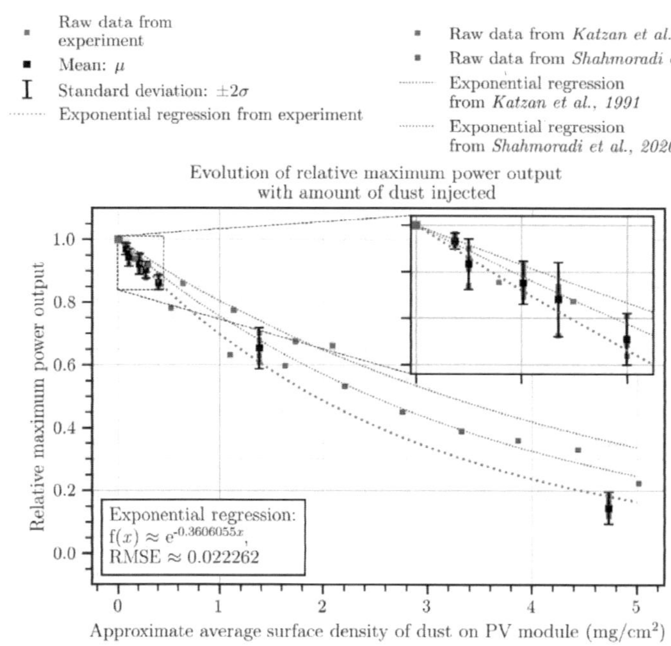

Figure 9. Evolution of RMPO with amount of dust accumulated on the PV module.

3.2 First estimation of time cleaning

With such lunar dust accumulation rates, the lifetime of a potential future lunar PV plant – considering only dust issues from natural sources and assuming the proposed model of exponential decline in power generation – can be estimated from figure 10 and figure 11. Only deposition rates measured by Chang'E 3 from Li et al. [16] [17] were estimated to be representative of *in-situ* natural dust deposition. Anthropogenic sources of dust are not taken into account as they might be avoidable with suitable behaviors and actions, but mostly because they are not discriminated from natural causes in other available annual deposition rates.

Figure 11 shows that after less than 4 years the 20%-loss may be reached. A future lunar base should be able to operate for more than 30 years. With the proposed model, the relative power output available after 30 years may vary between 20% and 80%, which is a quite wide interval, due to the natural dust deposition only (see figure 11). Radiations and thermal cycling will also

IOP Conf. Series: Materials Science and Engineering **1328** (2025) 012001 doi:10.1088/1757-899X/1328/1/012001

degrade the power generation from PVAs. So it's not so easy to perform an accurate prevision by taking into account only the dust simulant effect.

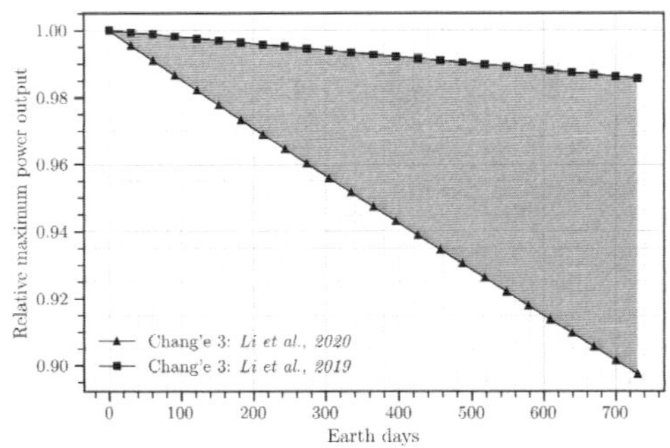

Figure 10. Expected evolution of relative maximum power output over Earth days (700 days), taking into account only the natural accumulation of dust.

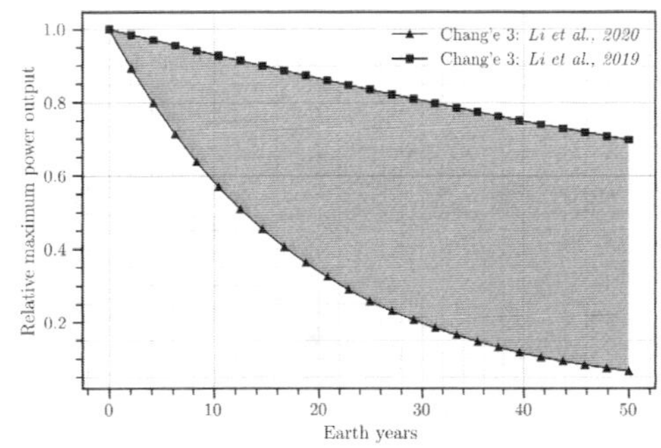

Figure 11. Expected evolution of relative maximum power output over Earth years (50 years), taking into account only the natural accumulation of dust.

Three power loss scenarios could be studied: 2%, 5% and 10% RMPO loss. Considering the minimum and the maximum range of the figure 10 and 11, the time between two PVA cleanings is respectively 5 months/2 years and 10 months, 11 months/ 7 years and 1 month, and 1 year and 11 months/ 14 years and 7 months. If a semi-annual or an annual cleaning were planned, the associated power loss would be between ~0.5% - 2.5% or between ~1% - 5.5%, respectively.

4. Conclusion

Lunar dust is ubiquitous and originates from several sources, both natural and anthropogenic. Dust simulant significantly affects PV performance, decreasing exponentially with the amount of dust. A new method was used to evaluate the lunar simulant impact onto the electrical performances of the PVA. This method is reproducible and allow us to have a good estimation of the RMPO and RSCC. However, this study deserves to be consolidated with a better understanding of interface properties between the PVA frontsheet and the dust simulant. Further experiments should more faithfully replicate the lunar environment (temperature and radiations). Moreover, improvements to the soiling chamber (inside IV measurements with the spatial Air Mass 0 spectrum, vacuum, humidity for instance) need to be carried out to better evaluated electrical performance under real conditions.

Acknowledgment

Authors would like to acknowledge Louis Chaput, Anthony Laraize for their support and advices. This work was supported by the French National Program "Programme d'Investissements d'Avenir - INES.2S" under Grant ANR-10-IEED-0014-01.

References

[1] P. Gläser, J. Oberst, G. A. Neumann, E. Mazarico, E. J. Speyerer, et M. S. Robinson, « Illumination conditions at the lunar poles: Implications for future exploration », *Planet. Space Sci.*, vol. 162, p. 170-178, nov. 2018, doi: 10.1016/j.pss.2017.07.006.
[2] A. Bermudez-Garcia, P. Voarino, et O. Raccurt, « Environments, needs and opportunities for future space photovoltaic power generation: A review », *Appl. Energy*, vol. 290, p. 116757, mai 2021, doi: 10.1016/j.apenergy.2021.116757.
[3] R. V. Morris, R. Score, C. Dardano, et G. Heiken, « Handbook of lunar soils ».
[4] C. M. Katzan et C. R. Stidham, « Lunar dust interactions with photovoltaic arrays », in *The Conference Record of the Twenty-Second IEEE Photovoltaic Specialists Conference - 1991*, Las Vegas, NV, USA: IEEE, 1991, p. 1548-1553. doi: 10.1109/PVSC.1991.169464.
[5] Taylor S.R., « Lunar science: a post-Apollo view: scientific results and insights from the lunar samples ». Pergamon Press, 1975.
[6] G. Heiken, D. Vaniman, B. M. French, and H. H. Schmitt, *Lunar Sourcebook: A User's Guide to the Moon*. Houston, TX, USA: Lunar and Planetary Institute, 2005.
[7] W. Fa, « Bulk Density of the Lunar Regolith at the Chang'E-3 Landing Site as Estimated From Lunar Penetrating Radar », *Earth Space Sci.*, vol. 7, nº 2, p. e2019EA000801, févr. 2020, doi: 10.1029/2019EA000801.
[8] S. Noble, « The Lunar Regolith », mars 2009. Consulté le: 17 mai 2023. [En ligne]. Disponible sur: https://www.semanticscholar.org/paper/The-Lunar-Regolith-Noble/6d1b78a5dff1c5bd01939398f3b6602e8a71125f
[9] C. M. Katzan et J. L. Edwards, « Lunar dust transport and potential interactions with power system components », nov. 1991. doi: 10.2172/10181067.
[10] C. M. Katzan et J. L. Edwards, « Lunar dust transport and potential interactions with power system components », NASA-CR--4404, 10181067, nov. 1991. doi: 10.2172/10181067.
[11] J. Shahmoradi *et al.*, « The Effects of Martian and Lunar Dust on Solar Panel Efficiency and a Proposed Solution », in *AIAA Scitech 2020 Forum*. doi: 10.2514/6.2020-1550.
[12] M.-Á. Muñoz-García, T. Fouris, et E. Pilat, « Analysis of the soiling effect under different conditions on different photovoltaic glasses and cells using an indoor soiling chamber », *Renew. Energy*, vol. 163, p. 1560-1568, janv. 2021, doi: 10.1016/j.renene.2020.10.027.
[13] S. Arbaretaz, E. Pilat, H. Colin, et E. Urrejola, « Quantifying the Energy Impact of Soiling—Thanks to the Tool SoilRatio », *Sol. RRL*, vol. 7, nº 8, p. 2200601, avr. 2023, doi: 10.1002/solr.202200601.
[14] C. M. Katzan, D. J. Brinker, et R. Kress, « The Effects of Lunar Dust Accumulation on the performance of PV array ». Space Photovoltaic Research and Technology Conference, 1991.
[15] Á. Fernández-Solas, L. Micheli, F. Almonacid, et E. F. Fernández, « Optical degradation impact on the spectral performance of photovoltaic technology », *Renew. Sustain. Energy Rev.*, vol. 141, p. 110782, mai 2021, doi: 10.1016/j.rser.2021.110782.
[16] D. Li *et al.*, « In Situ Investigations of Dust Above the Lunar Terminator at the Chang'E-3 Landing Site in the Mare Imbrium », *Geophys. Res. Lett.*, vol. 47, nº 17, p. e2020GL089433, sept. 2020, doi: 10.1029/2020GL089433.
[17] D. Li *et al.*, « In Situ Measurements of Lunar Dust at the Chang'E-3 Landing Site in the Northern Mare Imbrium », *J. Geophys. Res. Planets*, vol. 124, nº 8, p. 2168-2177, août 2019, doi: 10.1029/2019JE006054.

IOP Conf. Series: Materials Science and Engineering **1328** (2025) 012002 doi:10.1088/1757-899X/1328/1/012002

Lunar dust Contamination on habitable modules for exploration: impacts and mitigation approach for I-Hab

I. Locantore* [1], **A. Saverino**[2]

[1] Contamination Control, Thales Alenia Space Italia, Torino, Italy
[2] Contamination Control, Institution, City, Country

*E-mail: ilaria.locantore@thalesaleniaspace.com

Abstract. As humanity ventures towards prolonged lunar habitation and exploration, the management of lunar dust contamination within manned modules emerges as a critical concern. Lunar dust, characterized by its abrasive nature and electrostatic properties, poses multifaceted challenges to both human health and equipment integrity. This paper explores the strategy and challenges associated with controlling lunar dust contamination for I-Hab, the Gateway International habitat.
Key strategies for lunar dust contamination control encompass both proactive and reactive measures. Proactively, implementing dust-resistant materials and defining operational constraints can minimize dust entry into modules. Additionally, employing robust filtration systems and establishing strict decontamination protocols for spacesuits and equipment can mitigate internal dust accumulation. Reactively, developing efficient cleaning technologies, such as electrostatic dust removal systems and adhesive-based cleaning mechanisms, proves essential for maintaining habitable conditions within modules.
However, several challenges persist in the effective management of lunar dust contamination. The fine, abrasive nature of lunar dust necessitates the development of durable and resilient materials capable of withstanding its erosive effects over prolonged periods.
I-Hab is designed to properly work for 15 years of operations. This paper deals with the approach which was used to define criticality of internal and external sensitive items, together with possible mitigations and verification methods. Dust testing was proposed for materials characterization and design consolidation of non-maintainable items. TAS facilities are herein presented, as well as test definition approach and results.

1. Introduction

Apollo astronauts learned first-hand how dust can affect surface missions [1]. Indeed, lunar dust is a new contamination source, given by meteoroids impact, UV, solar wind over billions of years, so having properties and behaviour very different from what is known as classic molecular and particulate contamination.

In fact, lunar dust is made up of fine, abrasive particles that cling to surfaces due to their electrostatic properties. It has been defined as the biggest challenge for Moon exploration. It was found that the effects could be sorted into nine categories: vision obscuration, false instrument readings, dust coating and contamination, loss of traction, clogging of mechanisms, abrasion, thermal control problems, seal failures, and inhalation and irritation.

1.1 Gateway mission

Gateway is a multi-purpose, long-duration cis-lunar platform [2], including a variety of spacecraft, including the crewed Orion vehicle, the Human Landing System that will ferry astronauts to and from the lunar surface, logistics supply craft, and vehicles transiting further into deep space beyond the Earth-Moon system, such as to Mars. NASA is building an international cooperation with substantial cooperation from ESA and European companies to design, build, and launch Gateway's core elements to Near-Rectilinear Halo Orbit (NRHO) around the Moon, where it will operate for a minimum of 15 years. Gateway is humanity's next in-space science utilization platform, and its first in deep space, with three science payloads already selected to study solar and cosmic radiation [3]. As a lunar outpost, Gateway is a destination for deep space crew expeditions and science investigations, a port for deep space transportation, including landers transiting to the lunar surface or spacecraft embarking to deep space destinations beyond the Earth-Moon system [4].

The entire Gateway spacecraft is at preliminary design level of maturity, with some components at or near critical design review. In fact, Gateway (Figure 1) will be developed in various stages, including the Power and Propulsion Element (PPE), the Habitation and Logistics Outpost (HALO), Deep Space Logistics supply craft, the International Habitation module (I-Hab); Gateway External Robotics System, European ESPRIT Refuelling Module (ERM), and an Airlock (ALM).

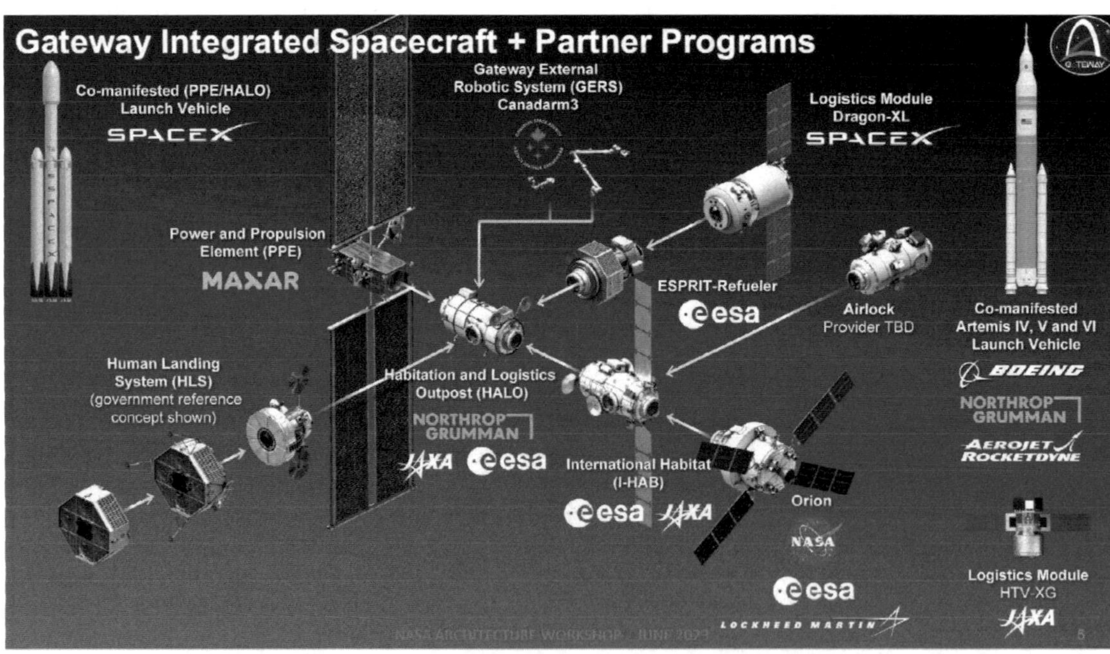

Figure 1. Gateway Integrated Spacecraft (NASA). Source: public domain.

1.2 The International Habitation module (I-Hab)

The European Space Agency (ESA) will provide habitation and refuelling to Gateway, with Thales Alenia Space as Prime Contractor for I-Hab and ERM, and as sub-contractor for HALO.

In particular, I-Hab will enhance Gateway capabilities for scientific research, life support systems and crew living quarters, enabling longer duration crewed Gateway missions [3].

The overall configuration of I-HAB includes four docking ports, two axial and two radial, adopting the IBDM docking system with adaptation to comply with the GDSS (Gateway International Docking System Standard Specification); out of the four ports present, three are Passive and only one is Active (the axial one that allows the initial docking with the Gateway). At each port, a hatch is installed, providing access (and isolation) of the cabin.

In order to thermally protect the ports during transit, specific covers are installed on the docking system. Once I-HAB reaches the Gateway, the Active axial port cover is opened (and left open) before the I-HAB docks to the Gateway; following I-HAB connection to the stack, the covers on the radial ports are opened, and kept open throughout the entire I-HAB mission, so to allow visiting vehicles docking.

Gateway, as an integral part of NASA's Artemis program, is expected to encounter lunar dust brought back by lunar landers returning from missions on the Moon's surface. In fact, for missions after Artemis III, the Human Landing System (HLS) is docked to the Gateway prior to each crewed HLS mission and returns to the Gateway after each HLS mission. Even though most of lunar dust largest fraction is expected to be removed or lost during transit, a certain amount will transfer onto Gateway exterior surfaces and brought inside the habitation volumes as landers dock after lunar operations. Characterization of the lunar dust induced environment around Gateway and contamination risk is accomplished with a comprehensive physics-based framework, the Gateway On-orbit Lunar Dust Modelling and Analysis Program (GOLDMAP), which has been developed by NASA to properly study and quantify potential dust transfer between HLS and Gateway and get results to define Gateway-level induced environments requirements for elements and subsystems [5].

Concerning I-Hab, external items which are sensitive to dust contamination are mainly the following (Figure 2):

- Deployable radiators, whose Optical Solar Reflectors (OSR) are reflecting surfaces. Obscuration of radiators radiative area would lead to a decrease in heat rejection capability. In addition, deployment and orientation mechanisms are sensitive as well: dust penetration would mean mechanisms' wear and blockage of moving parts and consequently a reduction of expected performance.
- Cameras: those items are designed for getting imagery data along the translation path. Both 360 cameras and 4k cameras are located outside the module. The latter ones will include lighting to support the imagery function. Lenses abrasion or area coverage by lunar dust would mean dealing with a strong performance loss.
- Batteries, whose increase of leakage current would imply malfunctioning and overheating.
- Robotic interfaces, which might be clogged by lunar dust particles over the lifetime or cause issues in mechanical connection with payloads.
- Docking mechanisms and targets, which might experience decrease of sealing tightness, torque increasing or jamming, or thermo-optical properties unwanted changes due to lunar dust deposition on active surfaces.

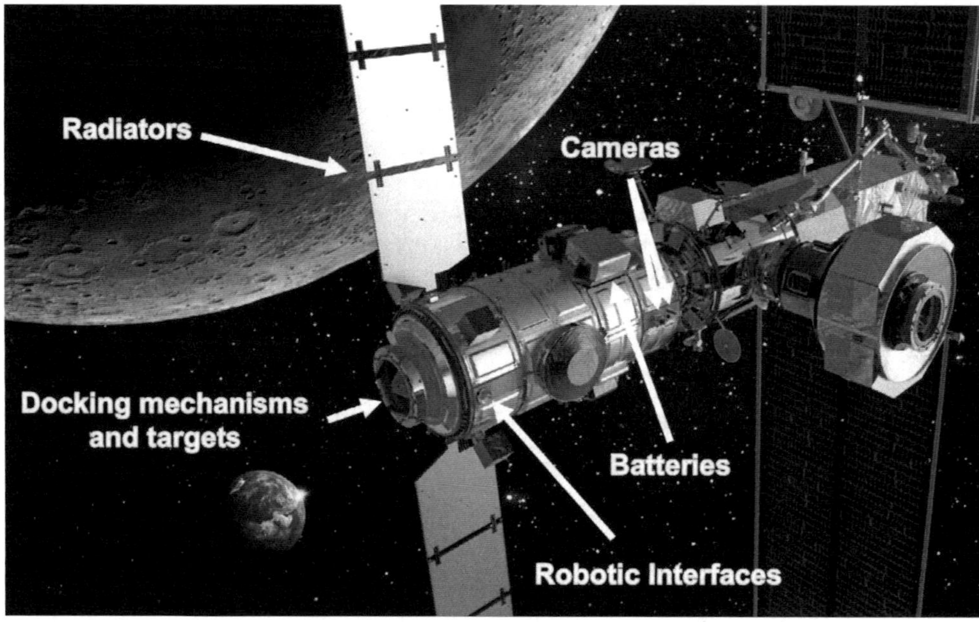

Figure 2. I-Hab sensitive items location. Credits: TAS.

2. Dust mitigation strategy

Baseline Gateway concept of operations currently foresee docking on HALO radial docking ports [6]. However, for contingency reasons, landers might dock on one of the I-Hab radial docking ports. In both those scenarios, lunar dust would be transferred to I-Hab module external and internal surfaces, with different loads.

Mitigation strategies fall broadly into three categories: Avoid, Remove, or Tolerate [7]. The most attractive approach, if it can be accomplished, is to simply avoid dust exposure. Whether to avoid, remove, or tolerate lunar dust exposure is a decision to be based on length of required service life, mass and complexity trades, and value (criticality) of the system.

A dedicated risk analysis was proposed, at Gateway community level, as efficient tool to define criticality of potential lunar dust contamination on sensitive items and to identify potential countermeasures.

2.1 Risk analysis

Most critical items were selected based on a risk factor (RF). It was obtained for each sensitive item by multiplying scores, from 1 (low level) to 5 (high level), of the following criteria:

A. Operability, which refers to operating cycles and exposure duration. This can vary between continuous exposure (score 5), which occurs consistently during crewed operations and leads to a high risk level, and intermittent exposure, which is linked to limited usage at certain intervals or even a single instance (score 1). For instance, radiator deployment mechanisms will not be operated repeatedly: once deployed after the first docking to Gateway, radiators will remain in the extended configuration throughout their entire lifetime.

IOP Conf. Series: Materials Science and Engineering **1328** (2025) 012002 doi:10.1088/1757-899X/1328/1/012002

B. Location, which defines its level of exposure and vulnerability by considering whether the item is directly exposed to the main cabin airflow (score 5) or positioned behind protective panels (score 1). This also ranges from having no protection at all (score 5) to implementing advanced measures such as IP-rated protective covers, filtration systems, or conformal coatings on printed circuit boards (PCB) (score 1).
C. Sensitivity, which addresses the system's vulnerability or resilience. Potential dust-induced failures were assessed for each relevant item, along with their criticality from a functional standpoint and in terms of higher-level consequences. Maintainable or cleanable items, as well as items with available flight spares, were considered less sensitive than others.

The above analysis was done for all sensitive items, both inside the pressurized volume and exposed to space, for a total of 180 items. The risk analysis purpose was to ensure appropriate testing is performed on critical hardware while avoiding unnecessary testing burdens for units that are not vulnerable or are protected.
Resulting scores allowed down-selection to more than 50 critical items, 20 of those were considered for dust testing since exceeding a risk factor threshold of 25 (Table 1 and Table 2, red-marked items). In fact, a risk factor of 25 implies at least 2 criteria above a score of 2, which was considered a reasonable condition to justify dust testing as necessary for improving knowledge on the item of concern.

Table 1. Risk factor analysis results for I-Hab. Critical items are listed for each I-Hab internal subsystem. Green-marked items are those whose risk is below threshold, whereas red-marked ones are those exceeding threshold, and therefore selected as potential candidates for a dedicated test campaign. Credits: TAS.

Dust-sensitive items (inside the pressurized volume)		A	B	C	RF
HATCH	Viewport	3	5	1	15
	Open mechanism rails	3	5	1	15
	Hand operated blind-gland	3	5	1	15
	Latching mechanisms	3	5	1	15
	Crank handle	3	5	1	15
	Hatch Seal	3	5	1	15
ECLS	HEPA filter	5	5	2	50
	Pressure valve	3	5	1	15
	Equalization valve	5	5	2	50
	Pressure relief valve	0	5	5	0
	Shut-off valve	0	5	5	0
	Fluidic connectors	5	1	1	5
SECONDARY STRUCTURE	Brackets	1	1	5	5
	Tilting Mechanism	2	5	2	20
	Portable Equipment Panel	5	1	3	15
	connectors	3	1	5	15
AVIONIC	connectors	5	1	2	10
	Mass Memory	5	1	2	10

	Flight Computer Module	5	1	2	10
	Interface Unit	5	1	2	10
	Switch Units	5	1	2	10
	Memory Storage	5	1	2	10
	Audio control unit	5	1	2	10
EPS	Battery Electronic Unit	5	1	2	10
	Power Distribution Assembly	5	1	2	10
	Docking avionic units	5	1	2	10

Table 2. Risk factor analysis results for I-Hab. Critical items are listed for each I-Hab external subsystem. Green-marked items are those whose risk is below threshold, whereas red-marked ones are those exceeding threshold, and therefore selected as potential candidates for a dedicated test campaign. Credits: TAS.

Dust-sensitive items (exposed to dust impingement)		A	B	C	RF
Deployable radiators	OSRs	5	5	5	125
	Orientation mechanism	5	5	2	50
Docking mechanism	Capture sensor	3	5	5	75
	Guide pin and receptacle	3	5	5	75
	Passive hook	3	5	5	75
SECONDARY STRUCTURE	Robotic interfaces and connectors	1	5	1	5
	External Cameras	5	5	5	125
	External Antennas	5	3	5	75
	Debris protection system	5	5	5	125
	Grapple fixtures	3	5	5	75
	Docking targets	3	5	5	75
	Batteries	5	5	5	125
	Docking camera	3	5	5	75
	Light detection and ranging sensor	3	5	5	75
	Docking light	3	5	5	75
	Docking light power module	3	5	5	75
ECLS	Intermodule ventilation valve	5	5	2	50
TCS	Mixing Valve	1	1	5	5
	Liquid On-Off Valve	1	1	5	5
	Cold plates	1	1	5	5
	Insulation blankets	5	3	1	15

The following paragraphs deal with design countermeasures and dust testing campaign which were proposed for some of the items under TAS-I design authority.

2.2 Mitigation solutions

Increasing the size of external large surfaces like radiators wings, in order to compensate for lunar dust potential effects, would have had relevant impacts to mass, cost, schedule and technical complexity. Therefore, performance loss due to partial wings OSRs obscuration was quantified, and impact on power budget at system level was assessed: total heat rejection capability was still in line with the target.

However, since radiators are a non-maintenable item, particular attention was paid to the orientation mechanism. This is made of many sensitive parts like bearings, a motor, an optical sensor, harness and flexible lines to allow wing rotation: any failure at this level could result in radiator malfunctions and potential loss of thermal control functionality, which is key for astronaut safety. The mitigation solution, designed by TAS France as responsible entity for radiator design, includes implementing labyrinths along potential dust paths to protect sensitive components (Figure 3). Additionally, a dust-tolerant V-seal was incorporated between the mobile and fixed sections to effectively prevent dust intrusion.

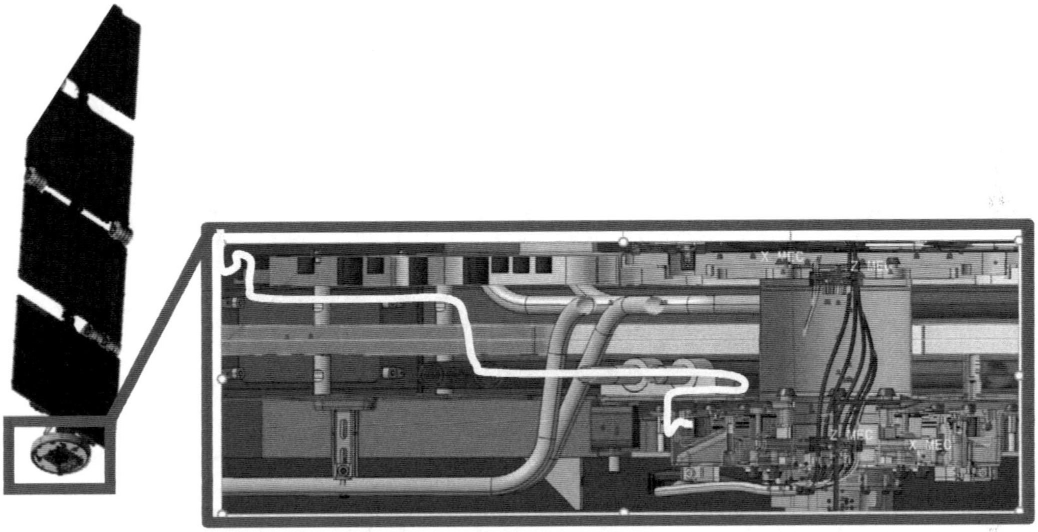

Figure 3. Radiators orientation mechanism with labyrinths. Dust path to sensitive parts is yellow-lined. Credits: TAS.

More in general, for all external items, dust mitigation solutions for I-Hab will include:

- selection of surface treatments and materials with low adhesion properties and optimal abrasion resistance;
- architectural solutions, such as debris protection panels exploitation for electronic units and EVA-replaceable connector savers to protect robotic interfaces from exposure;
- characterization of most critical equipment performance in presence of lunar dust simulant before CDR.

Inside habitats, prolonged exposure to lunar dust could lead to serious health effects (e.g., respiratory, cardiopulmonary, ocular, or dermal harm) to the crew or impact crew performance during missions, due to its unique properties [9]. Limits have been established based on detailed peer-reviewed studies completed by the Lunar Atmosphere Dust Toxicity Assessment Group (LADTAG) and are specific to the conditions relevant to the lunar surface.

Larger particles are unlikely to even reach cis-lunar orbit, as previously mentioned in paragraph 1.2. However, particles smaller than 100 microns are expected to be found and carried into the pressurized module by crew after extravehicular activities.

IOP Conf. Series: Materials Science and Engineering **1328** (2025) 012002 doi:10.1088/1757-899X/1328/1/012002

The environmental control strategy within the pressurized module will rely on HEPA filtration to capture airborne dust particles ranging from 0.02 to 10 microns, which are specified as posing risks to cabin crew. This type of filters are very effective against fine lunar dust, differently from lithium hydroxide filters used during Apollo missions which were notified as not cleaning cabin atmosphere [1].

The visible fraction (i.e., particles larger than 50 microns) will be removed from accessible surfaces by cleaning tools such as vacuums and wipes during regular housekeeping activities. Furthermore, internal components such as valves, directly injecting airborne dust, are designed to be maintainable, replaceable, or at least cleanable. With those controls in place, no design changes were deemed necessary for lunar dust mitigation inside the habitable volume.

Thanks to the availability of a full-scale 1:1 I-Hab mock-up (Figure 4 and 5), TAS-I was able to start checking actual cleanability of the module.

Figure 4. Luca Parmitano inside I-Hab mock-up during HITL test campaign. Source: public domain.

IOP Conf. Series: Materials Science and Engineering **1328** (2025) 012002 doi:10.1088/1757-899X/1328/1/012002

Figure 5. Internal view of I-Hab mock-up. Source: public domain.

Astronaut Luca Parmitano dedicated an entire week to verifying crew operations, including the evaluation of mechanisms and other contamination-sensitive components accessibility for vacuum cleaning. His valuable feedback provided critical insights for further improvements [8].

3. Dust testing

A risk mitigation test campaign was proposed alongside design countermeasures, aiming to identify potential non-compliances before the qualification phase. The campaign focuses on measuring actual performance losses and implementing design enhancements prior to the Critical Design Review, whenever compatible with program constraints.

Dust testing has been proposed in one of TAS-I test facilities, called PESCha (Planetary Environment Simulation Chamber), in thermal vacuum conditions and using simulants like LHS-1, which is representative of the landing site, particles size distribution and abrasiveness, or EAC-1 for thermal optical measurements, or others on a case by case basis.

These tests will play a critical role in enabling the detailed characterization and consolidation of the design, ensuring it performs effectively under the anticipated configurations and operating temperature ranges. Additionally, the testing process will contribute to advancing the Technology Readiness Level (TRL) of the technologies and components involved. By simulating operative conditions, these tests will provide valuable insights to validate performance and identify potential improvements in terms of mitigation actions and additional controls.

3.1 Mechanisms

One of the items which was selected for dust testing is the orientation mechanism of I-Hab deployable radiators. In particular, the V-seal, which serves as a barrier against dust intrusion, will be characterized for torque and wear resistance. Initial testing will take place under ambient conditions, followed by assessments in thermal-vacuum conditions, both with and without the presence of dust. These tests will enable measurements under both hot and cold scenarios, including post-lifecycle evaluations conducted from outside the chamber. The data obtained will be crucial for defining motorization margins and ensuring the seal's reliability in operational environments.

Following the same logic and objectives, docking mechanisms were also proposed for dust testing. Particularly, dedicated prototypes will be manufactured, focusing solely on the components directly involved in each mechanism functionality. These targeted tests aim to evaluate significant parameters, like the actuation force before and after dust exposure, to allow verification of proper performance and durability under the operative scenario.

2.1 TAS upcoming new facility

A new dust testing facility, named ALESS™ (Advanced Lunar Environment Simulator System), will come to TAS premises, in Turin, by summer 2025 (Figure 6). It will be used for materials, small items and mechanisms testing, during design phases to better address best solution definition.

The test chamber will enable automated dust sieving and distribution, in-situ simulant pre-treatment and charging, and operation across a wide temperature range (-150/+200 degC). It will also provide full solar spectrum exposure and, as a future upgrade, the capability to add molecular contamination, allowing evaluation of combined effects due to various contamination sources. Thermal-optical properties degradation and mechanical impacts will be measured directly within the chamber, complemented by high-resolution imaging to capture detailed observations during testing.

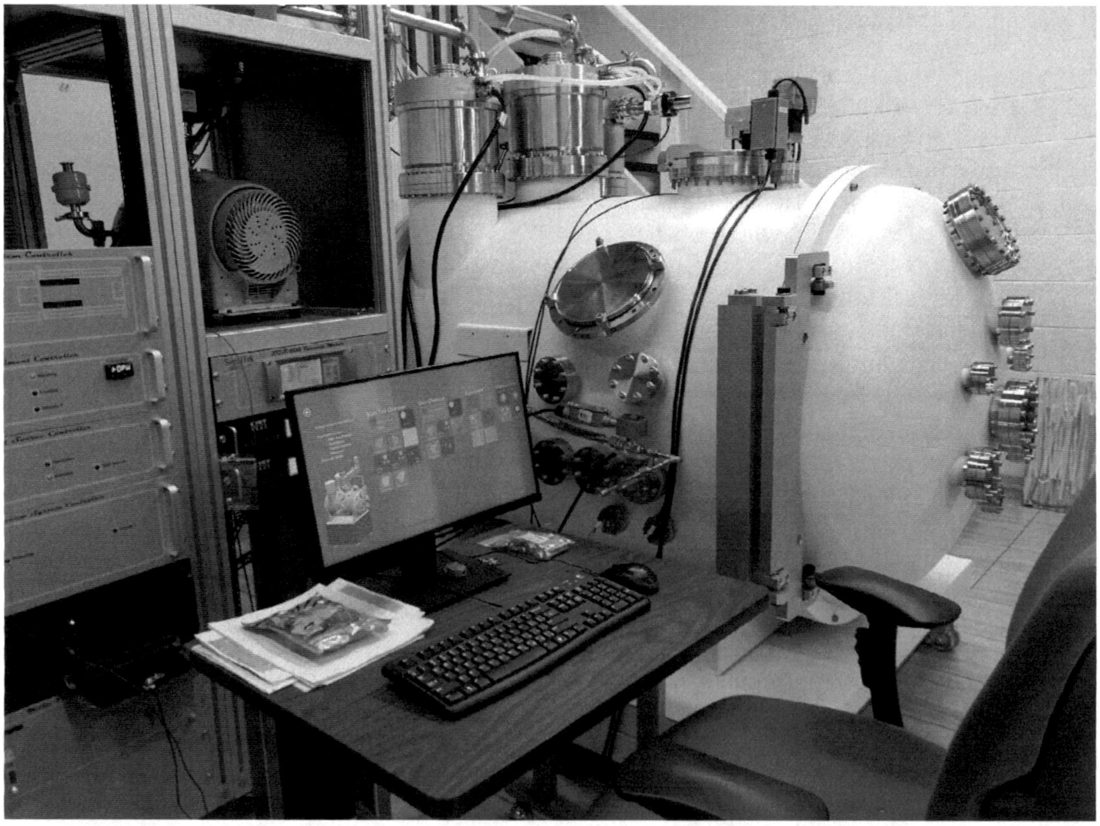

Figure 6. ALESS™: Advanced Lunar Environment Simulator System. Credits: TAS.

IOP Conf. Series: Materials Science and Engineering **1328** (2025) 012002 doi:10.1088/1757-899X/1328/1/012002

Conclusions

Apollo 17 Commander Gene Cernan stated, 'I think dust is probably one of our greatest inhibitors to a nominal operation on the Moon.' Lunar dust poses a significant challenge to both crew safety and equipment performance. Addressing and mitigating its effects is a critical step toward ensuring the long-term sustainability of lunar exploration.

The implementation of lunar dust mitigation solutions requires careful consideration, as it can significantly impact mass, cost, schedule, and operational constraints. These factors must be well-balanced against actual needs and benefits from an industrial perspective. Given the uncertainties in predicting the risks associated with dust exposure, and taking into account operational procedure and mechanical barriers already in place, no major redesigns were applied to units that had completed or were nearing the completion of their Critical Design Phase.

Nevertheless, TAS-I mitigation strategy incorporates targeted material selection, the introduction of design countermeasures, and the evaluation of impacts on vulnerable equipment and surfaces through dedicated development testing by in-house facilities. Therefore, positive outcomes are hoped and evisaged.

This approach ensures a robust and pragmatic response to the challenges posed by lunar dust in cis-lunar orbit.

References

[1] J R Gaier 2005 The Effects of Lunar Dust on EVA Systems During the Apollo Missions Nasa/TM-2005-213610

[2] Sean Fuller Emma Lehnhardt June 2023 White Paper: Gateway: The Cislunar Springboard for International and Sustainable Human Deep Space Exploration *NASA's Moon to Mars Architecture Workshop*

[3] Emma Lehnhardt Dylan Connella 2023 The Gateway Program as Part of NASA's Plans for Human Exploration Beyond Low Earth Orbit, *IEEE Aerospace Conference*

[4] Sean Fuller et all 2024 Gateway Program Development Progress

[5] Ronald G. Lee Gary L. Brown 2024 Overview of NASA Gateway Lunar Dust Mitigation and Contamination Modeling and Analysis

[6] NASA Gateway Concept of Operations 2022 GP 10027 REV B

[7] NASA/TP-20220018746 2021 Lunar Dust Mitigation: A Guide and Reference, First Edition

[8] https://www.esa.int/ESA_Multimedia/Images/2024/05/Luca_Parmitano_inside_Lunar_I-Hab

[9] https://www.nasa.gov/directorates/esdmd/hhp/risk-of-adverse-health-performance-effects-of-celestial-dust-exposure/

IOP Conf. Series: Materials Science and Engineering **1328** (2025) 012003 doi:10.1088/1757-899X/1328/1/012003

Interaction of Lunar Dust Simulants with Materials: Importance of Charging

J. Kleiman, S. Horodetsky, V. Issoupov, V. Verba, D. Artymowicz

Integrity Testing Laboratory, Markham, Ontario, Canada
jkleiman@itlinc.com

Abstract. Materials exposed to lunar regolith dust and other environmental factors on the Moon may suffer permanent damage, risking catastrophic failures. Lunar dust poses the greatest threat. Preventive measures are crucial, with principles emerging to deter dust accumulation in vacuum conditions. Charging of dust and surfaces significantly affects adhesion. Charging sources include photoemission, solar wind, and secondary electron emission from Earth's magnetosphere, resulting in positive charge on the dayside and negative charge on the nightside.

In framework of NASA's "Regolith Adherence Characterization (RAC) Payload" project, we initiated a program on conducting experiments in our Lunar Environment Simulator on interaction of lunar dust simulants with materials, similar to the RAC Payload experiment. As part of this program, we conducted a series of experiments to understand the effects of charges accumulating on dust simulants and the surfaces they interact with on the adhesion and mitigation of dust.

We adapted a number of methods to charge the dust, - tribological, vacuum ultraviolet and plasma, and used a nanocoulomb meter set-up to evaluate the dust charge. A rotating disk sample holder enhances dust flow uniformity. In the experiments we measured the dust charge acquired under different conditions, with the aim to understand interaction models. This paper presents initial findings and discusses relevant models.

1. Introduction

A number of unmanned [1-4] and manned [5,6] missions visited the Moon in the last 50-60 years creating an enormous pool of information that even today present a unique source of knowledge about the environment and conditions on the Moon. Among the Moon's inhospitable environmental conditions are the very thin exosphere that, practically, is a hard vacuum of 10E-14 atm, the space radiation environment that is close to GEO environment, and the temperature extremes that are in the range of -172 °C to + 125 °C, with 13 thermal cycles/year and the Lunar gravity being 6 times less than on the Earth [5,6]. Additionally, the Lunar surface is covered by a thick layer of sharp, multi-angular, highly erosive, very sticky dark-greyish or blackish (depending on the location on the Moon) regolith dust, that represents the planetary Lunar environment, developed for millions of years by meteors and meteorites striking the basalt Moon surface. Studies of Lunar regolith dust have been performed on dust samples returned from the Moon by USA, Soviet Union and, recently, China [7-9]. This dust has special chemical composition and particle size distribution, with a large percentage of very small fraction in it [1].

The interaction of Lunar dust and other environmental factors, like vacuum, temperatures, solar radiation, ultraviolet irradiation, and electron (e-) and proton (p+) irradiation with structures on the Moon and the outside of the future Gateway Lunar station may lead to permanent change or complete loss of thermal, optical, and other functionalities that could potentially lead to catastrophic failures. Among these factors, Lunar regolith dust is the most aggressive, causing the main problems. Charging of dust and surfaces significantly affects its adhesion to surfaces on the Moon. Charging sources include photoemission, solar wind, and secondary electron emission from Earth's magnetosphere, resulting in positive charge on the dayside and negative charge on the nightside [1].

On the illuminated side of the Moon, solar radiation, especially the ultraviolet part of its spectrum, and solar wind streams interact with the upper regolith layer, charging the regolith surface. The photoelectrons generated above the surface create, together with the charged regolith surface, a near-surface double layer. The electric field generated in this layer, as well as the particle charge fluctuations on the surface, create conditions under which electric forces may exceed the gravitational force and the van der Waals force of adhesion, resulting in micron- and submicron-sized regolith particles becoming detached from the surface and levitating above it [11,12].

Contact charging properties of two planetary analogue dust samples: lunar (JSC-1) and Martian (JSC-Mars-1) regolith simulants were investigated in a series of experiments by bringing dust particles into contact with various surfaces of known work functions (metals: Co, Ni, Au, Pt, and silica glass), and the resulting contact charges on the dust particles were measured. It was found that the measured contact charge of a dust particle increases with repeated agitation of the surface. The average contact charge also varies linearly with the work function of the contacting surface. The contact charging with oxidized metal surfaces was found to be independent of the metal's work function [13].

Theoretical models suggest that the dust grains on the lunar surface are charged by the solar ultraviolet (UV) radiation as well as the solar wind. Even without any physical activity, the dust grains are levitated by electrostatic fields and transported away from the surface in the near vacuum environment of the Moon. It is recognized that photoelectric emission induced by the solar UV radiation with photon energies higher than the work function (WF) of the grain materials to be the dominant process for charging of the lunar dust. To determine the charging and equilibrium potentials of individual dust grains, it is necessary to measure the photoelectric yields. Experiments were conducted in which the measurements of the photoelectric efficiencies and yields of individual sub-micron/micron size dust grains selected from sample returns of Apollo 17 and Luna-24 missions as well as similar size dust grains from the JSC-1 simulants were performed. The photoelectric emission measurements indicated grain size dependence with the yield increasing by an order of magnitude for grains of sub-micron to several micron size radii, at which it reaches asymptotic values. The yield for large size grains was found to be more than an order of magnitude higher than the bulk measurements on lunar fines reported in the literature [14].

In framework of NASA's "Regolith Adherence Characterization (RAC) Payload" project [15], we initiated a program on conducting experiments in our Lunar Environment Simulator on interaction of lunar dust simulants with materials, similar to the RAC Payload experiment [16] that was supported by CSA's LEAP Program [17]. To understand better, evaluate and explain the results from the RAC Payload experiment on the Moon, ITL conducted a series of experiments to understand the effects of charges accumulating on dust simulants and the surfaces they interact with on the adhesion and mitigation of dust.

We adapted a number of methods to charge the dust, - tribological, vacuum ultraviolet and plasma, and used a nanocoulomb meter set-up to evaluate the dust charge. A rotating sample holder enhanced dust flow uniformity. In the experiments we measured the dust charge acquired under different conditions, with the aim to understand interaction models. This paper presents initial findings and discusses relevant models.

2. Experimental

The advanced vacuum lunar environment space simulator developed by ITL [18,19] was used in present work (see Figure 1). A testing protocol was developed at that time [20] that is being used efficiently in all dust exposure experiments. The protocol is being critically assessed presently, in light of the new NASA Standard [21]. An overview of the early activities at ITL in this area can be found in [22].

To measure the dust charge, we used a nano-coulombmeter with either a Faraday cup supplied with the meter (Figure 2a) or a custom designed one (Figure 2b). In either case, the Faraday set-up consists of two cups that are electrically isolated from each other. To eliminate electrical interference, the outer cup was mounted on an insulating Teflon platform. The nano-coulombmeter cup was placed in a vacuum chamber with a vacuum level of 10E-6 Torr.

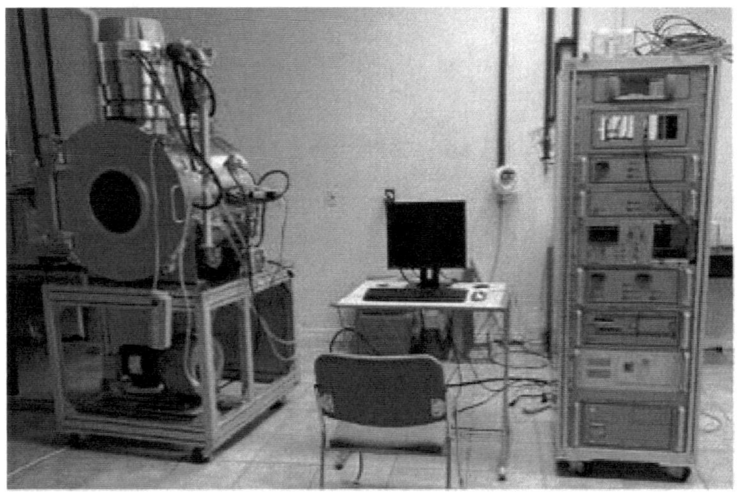

Figure 1 Image of Dusty Lunar Environment Vacuum Simulator designed and built by ITL and used in the present work.

It is important to note that during the distribution of the dust simulant from a vertically positioned source, dust accumulates at the bottom of the Faraday cup that affects subsequent measurements in both, the values and even the sign of the measured charge. Ideally, after each measurement, dust must be removed from the cup and each new measurement must start with an empty cup.

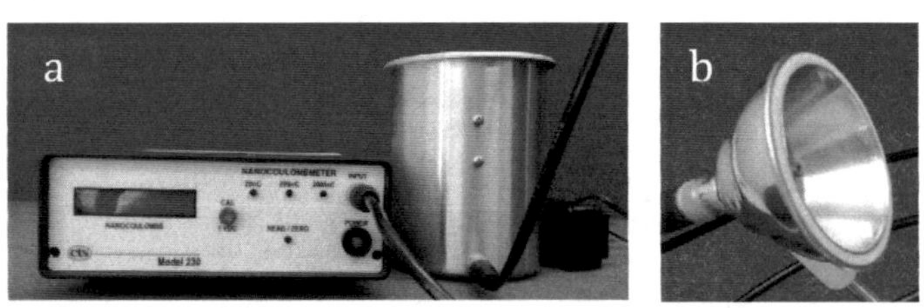

Figure 2 Nano-coulombmeter and Faraday cups used in the experiments.
a) ETS Model 230 – NanoCoulomb Meter with Model 231 – Small Faraday Cup supplied with the meter.
b) Faraday Cup designed at ITL for charge measurement when the dust is dispersed horizontally.

Designing experiments on exposing samples to dust simulants that are held vertically or horizontally with respect to the dust source in vacuum simulators may present, in many cases, a significant limitation. In real situations, the external sensitive hardware surfaces can be curved, and/or may be positioned at different angles to the incoming dust flux, either by original design or when moving, etc. We designed and used in our experiments a few sample holders, either stationary or rotating, with the samples placed at 45 degrees to the dust flux coming from a gravitational dust source from above (Figure 3). Such a set-up represents a general case and is accepted in the common materials dust testing standards in air, with the schematic of all forces, applied to the dust grains, presented in [23]. It is emphasized in [23], that both adhesion and dust cohesion forces influence the final results. This angle can be changed and we plan experiments with different angles that will allow investigating the angular dependence of dust adhesion to different surfaces. The LMS-1 Lunar dust simulant was used in the present testing program (Figure 4) [24]. It should be noted that particles with sub-millimetre size were found in the LMS-1 simulant and to prevent the dust source from clogging, we sieved the simulant through a 1mm sieve to remove the larger particles.

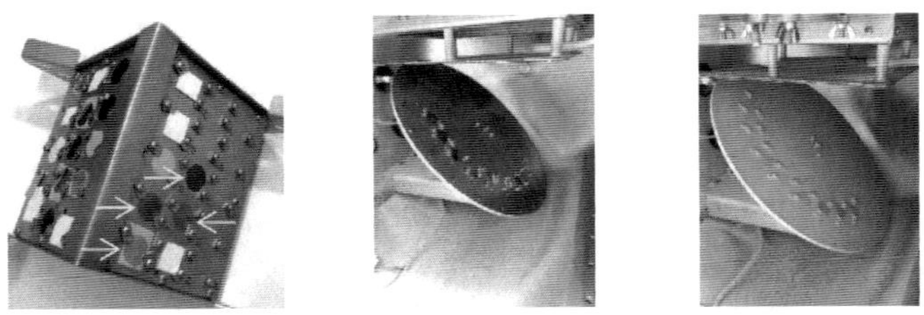

Figure 3 Images of the specially designed stationary (left) pyramid-like sample holder and rotating holder (canter and right) for holding the samples in the DLEVAS system that was used with a vertical gravitational dust source. Kapton samples on the pyramid-like holder are indicated with yellow arrows. The center and right images show Kapton samples before and after dust exposure respectively.

Figure 4 Optical (a) and scanning electron microscopy secondary electron (b, c) images of the LMS -1 used in present testing.

Kapton 100 HN, Al-metallized (aluminum-metallized polyimide film), is widely used in satellites and space structures due to its unique combination of thermal, mechanical, and electrical properties that make it ideal for the harsh environment of space. For these reasons we selected it as one of the base materials to apply our dust mitigation treatments.

We also were interested to study Kapton's behaviour in dusty environments because this material was used in our preliminary studies with dust mitigation technological processes, when a large number of different materials were exposed to dust simulants. The stationary pyramid-like holder in Figure 3 is shown with Kapton samples, in addition to white paint samples, also used in our studies. We used a number of charging methods that included tribological charging experiments that were conducted in air; charging from a short-wave vacuum ultraviolet (VUV) irradiation source; and by exposing the dust simulant to a low-energy plasma in a specially designed plasma source installed in the lunar simulator [25]. Table 1 presents a brief summary of the charging test methods used in this work. Only the VUV and Plasma charging results are discussed in this paper.

Table 1. Comparison of three methods of charging the Lunar Dust Simulant (LDS) used in the present work.

#	LDS charging methods	Advantages of the method	Disadvantages of the method
1	Tribological method (air test)	• Simplicity	• Not very effective • When shaking, an exchange of charges occurs between the particles of the LDS. This leads to an uncontrolled redistribution of this charge, up to its neutralization.
2	Short-wave ultraviolet ray (VUV) (vacuum test)	• Easy to install and use •Does not affect the experimental conditions • Does not change the composition and the structure of sample	• Limited cross-section of interaction of falling dust and VUV radiation. • Reduction in the intensity of VUV radiation due to the electrostatic interaction of the lamp glass and dust particles
3	Low Energy Plasma (vacuum test)	• Has high efficiency of charging the simulant • Does not require additional cleaning of the simulant and the sample surface	• Difficulties in selecting the operating mode • It is possible to change the structure and composition of the sample surface • Maximum allowable vacuum conditions for simulation (testing)

In addition to varying the charging methods, the influence of surface morphology of the samples was also evaluated in the conducted experiments. Two methods of surface preparation were employed – etching with atomic oxygen plasma (AOP) performed in the Denton Vacuum Etcher/Asher and etching with fast atomic oxygen (FAO) performed in a laser detonation fast atomic oxygen source. Etching parameters are listed in Table 2. Both, the fast atomic oxygen (FAO) and the plasma oxygen are affecting the surface morphology of Kapton, with the FAO having a much greater influence, due to much higher energy (4-5 eV for FAO versus 0.2-0.3 eV for the plasma oxygen). Also, because of the directionality of the FAO beams, sharp cones are produced in FAO treated surfaces versus much shallower bumpy structures in the case of plasma AO treatment.

No surface roughness measurements were conducted, but from the SEM images it can be suggested that the FAO treated samples had micron-sized roughness, while the AOP treated samples had sub-micron sized roughness.

Table 2. Etching parameters used to modify the surface of Kapton substrates and resulting Kapton erosion.

Etching method	Etching time	Oxygen Flux [atoms/cm2/s]	Total Erosion [mg/cm2]
AOP	3h	2.3E+20	1.5 E-1
AOP	6h	2.3E+20	3.0 E-1
FAO	24h	9.1e+15	2.2 E1

3. Results and Discussion

To evaluate the characteristics of the charging sources, we performed a number of tests with the tribological charging in air, and with the VUV charging source and with the Plasma charging source in vacuum. We estimated the specific charge value for each of the methods and the charge behaviour during the duration of the charging experiment. Table 3 summarizes the results from these experiments. As can be seen from Table 3, the dust simulant acquired a positive charge in the experiments with tribological charging in air and in VUV charging in vacuum. It attained, however, a negative charge in the experiments with plasma charging in vacuum. The specific charge value, calculated as nC per minute per gram (nC/min/gr) was the highest for the plasma charging (9.3 nC/min/g) and the lowest for the VUV charging (~0.25 nC/min/g).

Before the experiments, the samples were weighed, put in the vacuum chamber for 24 hours for degassing, removed for weighing again and returned to the chamber. After the experiments, the samples were removed from the vacuum chamber and weighed, taking exceptional care while venting the system and removing them from the holder that included very low venting rate, preventing the holder from any vibrational or sudden moves and removing each sample so that not to disturb the accumulated on them dust. At least 24h passed between removal of the dusted samples from the dusting chamber and (partial) removal of the dust by tapping and weighing again, with this weight used to establish the dust retention value expressed in mg/cm2. Dust retention was calculated as a difference between weight of samples after tapping and after degassing procedure. The samples were weighed again after 24 hours to account for the uptake of moisture. An OHAUS ANALYTICAL Plus microbalance was used, with readability of 0.00001 g. Figure 5 shows results of the dust retention values calculated for uncharged and charged dust for all samples with different surface morphology. VUV source was used for charging the dust simulant.

All samples were analyzed before and after dusting with the SEM with the most representative images shown in Figure 6 through Figure 8. Evaluating the surfaces after they were textured and before the dusting (Figure 6), it can be suggested that the pristine Kapton is basically featureless even at 10000x magnification. After 3h and 6h of AOP treatment, Kapton surface is covered with small, forked bumps of submicron height. The forking and distance between individual bumps increase with time of exposure to AOP. The estimated eroded mass is 0.15 mg/cm2 and 0.30 mg/cm2 for the 3h and 6h exposure respectively. On average, 22 mg/cm2 was eroded from Kapton substrates during FAO exposure. The surface of the FAO exposed samples was covered with cones several microns heigh and several microns apart. Cone sides were smooth and cone tops were 'decorated' with clumps of fibrous Kapton. SEM micrographs of dusted Kapton surfaces are shown in Figure 7 and Figure 8.

IOP Conf. Series: Materials Science and Engineering **1328** (2025) 012003 doi:10.1088/1757-899X/1328/1/012003

Table 3. Comparison of results for the three methods of charging the Lunar Dust Simulant (LDS) used in the present work.

<table>
<tr><th>#</th><th>LDS charging methods</th><th>Main Charging results</th><th>Charge vs. Time Graph</th></tr>
<tr><td>1</td><td>Tribological method (air test)</td><td>Specific charge value ~1.5 nC/min/g Polarity of charge - positive</td><td>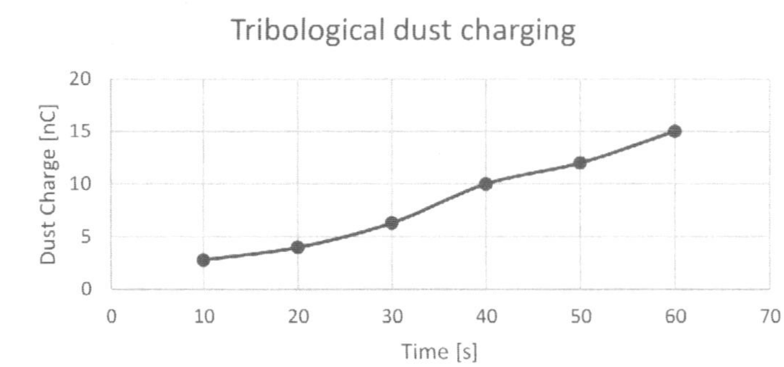
</td></tr>
<tr><td>2</td><td>Short-wave ultraviolet ray (VUV) (vacuum test)</td><td>Specific charge value ~0.25 nC/min/g Polarity of charge - positive</td><td>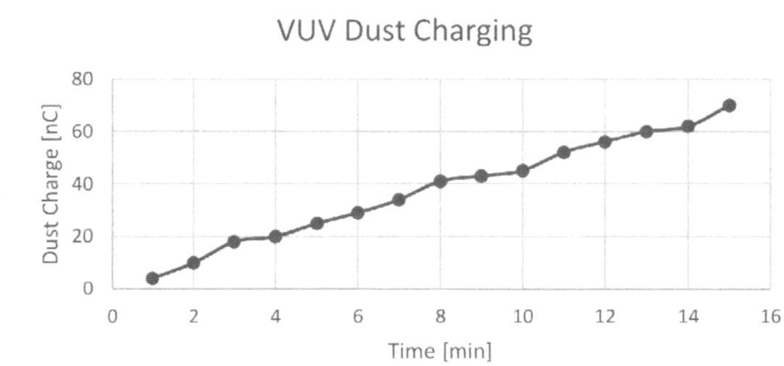
</td></tr>
<tr><td>3</td><td>Low Energy Plasma (vacuum test)</td><td>Specific charge value ~9.3 nC/min/g Polarity of charge - negative</td><td>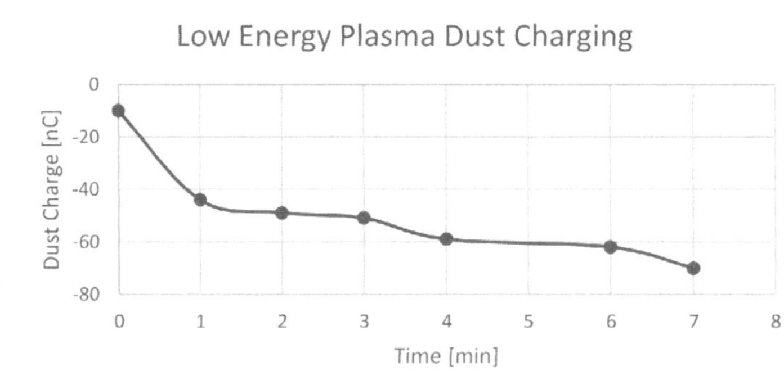
</td></tr>
</table>

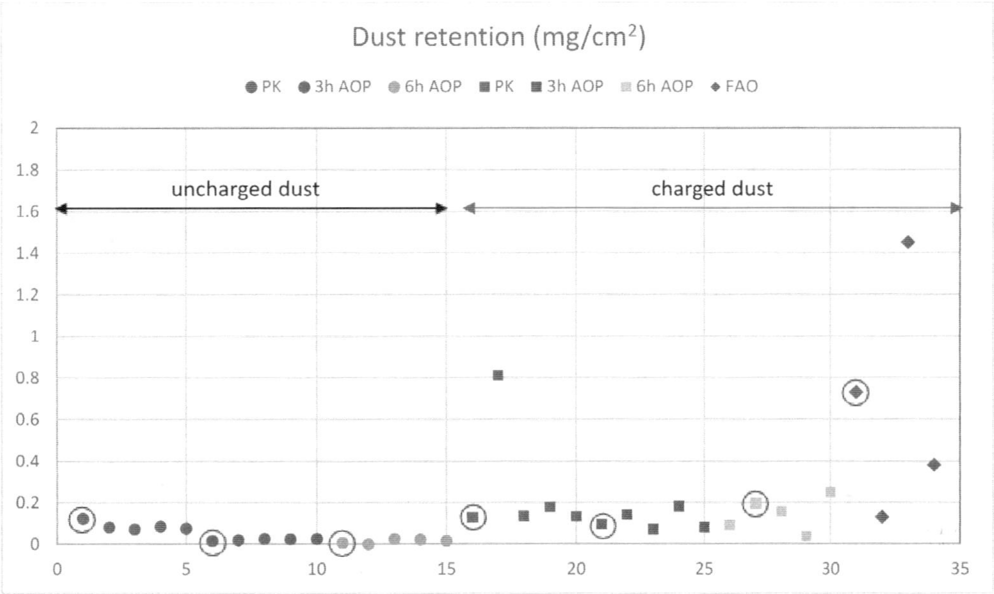

Figure 5. Dust retention for different substrates and different state of the dust simulant. Samples from 1 to 15 were exposed to uncharged dust. Samples from 16 to 34 were exposed to charged dust. Blue, brown and green data points correspond to Pristine Kapton (PK), 3h AOP and 6h AOP substrate respectively. Red diamonds are FAO substrates. Samples indicated with black circles were imaged with SEM. VUV source was used for charging the dust simulant.

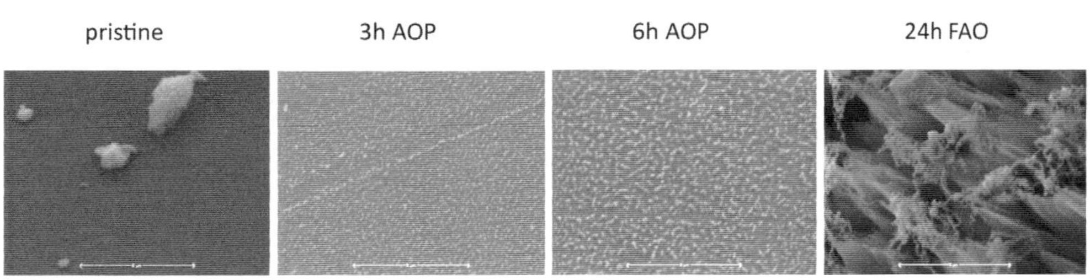

Figure 6. SEM micrographs of Kapton after different surface treatments. The objects on the surface of pristine Kapton are random dust particles. VUV source was used for charging the dust simulant. The scale bar in all pictures is 5 μm.

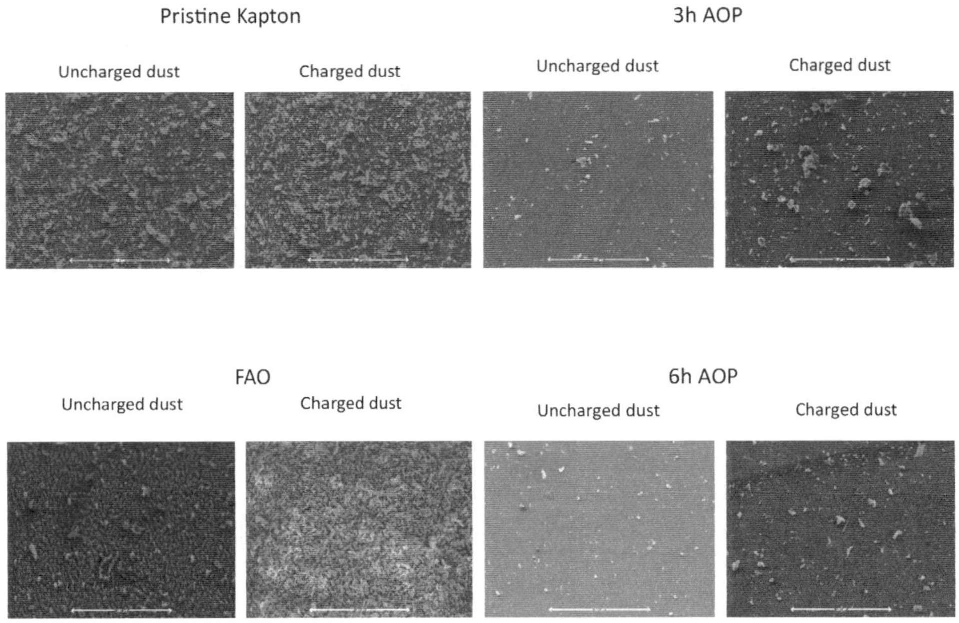

Figure 7. SEM micrographs of the dusted Kapton substrates after different surface etching procedures. The scale bar in all pictures is 100 μm. VUV source was used for charging the dust simulant.

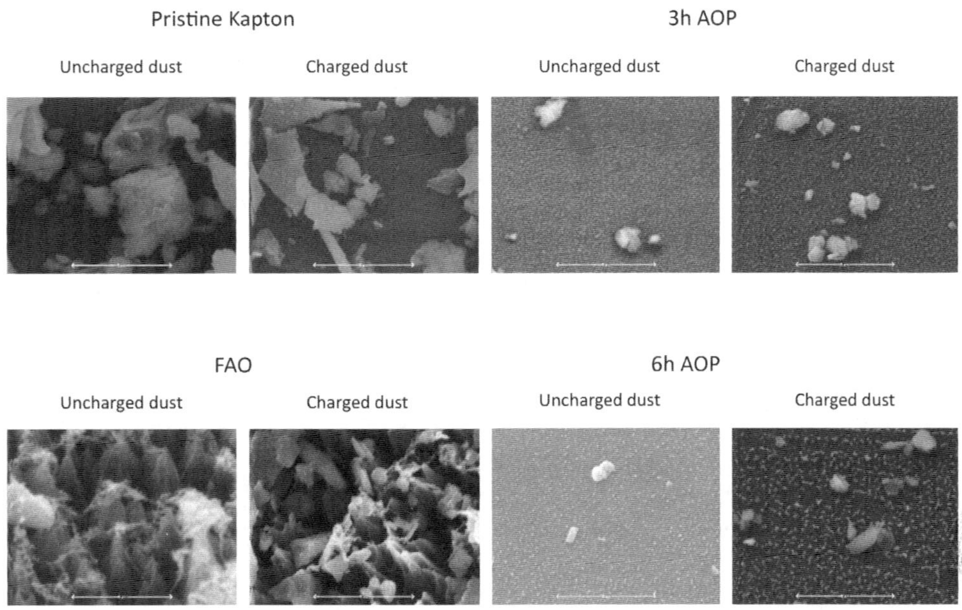

Figure 8. SEM micrographs of the dusted Kapton substrates after different surface etching procedures. VUV source was used for charging the dust simulant. The scale bar in all pictures is 100 μm.

Evaluating the results of the dusting experiments with plasma source charging, a number of observations can be made, as follows:

- The pristine Kapton and FAO etched Kapton show the highest density of retained dust.
- Dusting with a charged dust introduces significant scatter in mass-based data, calling for a significantly higher statistical pool of data to allow for valid conclusions.
- In general dust retention is higher in experiments with the charged dust than the uncharged one.
- Maximum dust retention was observed on FAO etched sample dusted with charged dust
- Minimum dust retention was observed on 6h AOP sample dusted with uncharged dust.
- SEM micrographs indicate that surface condition has much larger influence on dust retention rates than dust charging. However, an opposite conclusion can be drawn from numerical dust retention data.

In selected groups of samples weigh measurements were performed twice with 24h resting period to evaluate influence of moisture pick up. This was done on fully dusted as well as on tapped samples. Weight decreased in 24 out of 30 samples indicating that in the investigated cases the unforced dust shedding was more significant that moisture pick up. This type of measurements was done on the following three groups: 3h AOP, all substrates, fully dusted; 3h AOP, all substrates, tapped; 6hAOP, all substrates, fully dusted.

4. Conclusions

An evaluation of interaction of lunar dust simulant LMS-1 with different charging sources was conducted. We demonstrated that the acquired charges are positive in experiments with tribological land VUV charging sources. However, the induced charges were negative in experiments with the plasma charging source. The value of the accumulated charge was the biggest in the plasma experiments and considerably smaller in the case of tribocharging or VUV charging.

The initial results of a program to evaluate the importance of charging and surface morphology on the cohesive and adhesive behaviour of lunar dust simulant LMS-1 to Kapton 100HN indicate that the pristine Kapton and FAO etched Kapton had shown the highest density of retained dust. A significant scatter in mass-based data was noticed in dusting experiments with charging the dust. In general, it was noticed that the dust simulant retention is higher for the charged dust than the uncharged one, with the maximum dust retention observed in experiments with charging of FAO etched samples. The lowest dust retention rate was observed on 6 hrs AOP sample in experiment with uncharged dust. And, finally, the SEM analysis of the samples after the dusting experiments suggests that that surface conditions have a much larger influence on charge retention than dust charging.

The seemingly controversial conclusions from the dust retention rate experiments and SEM observations call for a significantly higher statistical pool of data to allow for valid conclusions. Also, in the present experiments we used the charging by VUV only. It would be interesting to investigate the dust interaction with materials with other sources of charging. New experiments are planned to confirm and to explain better the reported findings.

Acknowledgments

We acknowledge the financial support of Canadian Space Agency (CSA) [20STDP007] for certain aspects of this work.

References

[1] https://www.nasa.gov/mission_pages/LRO/multimedia/lroimages/lroc-20100318.html,
[2] Gromov V, Kemurdjian A et al 2003 Lunokhod 2 – A Retrospective Glance After 30 years, https://www.researchgate.net/publication/234403845_Lunokhod_2_-_A_retrospective_Glance_after_30_Years .
[3] https://www.planetary.org/space-missions/change-4
[4] https://www.space.com/india-chandrayaan-3-moon-landing-success .
[5] Gaier J R, 2005 The Effects of Lunar Dust on EVA Systems During the Apollo Missions Nasa/TM-2005- 213610.
[6] Creel R A 2011 Applying Thermal Control Experiences on Apollo Lunar Rover Project to Rovers for Future Space Exploration, Space Systems Engineer Member of the Apollo Lunar Roving Vehicle Team 3rd AIAA Atmospheric Space Environments Conference http://www.hq.nasa.gov/alsj/lrv_thermal_alsj.pdf .
[7] Vinogradov, A.P., Preliminary data on lunar soil collected by the Luna 20 unmanned spacecraft, Geochim. Cosmochim. Acta, 1973, vol. 37, pp. 721–729.
[8] McKay, D.S., Fruland, R.M., and Heiken, G.H., Grain size and the evolution of lunar soils, Proc. Fifth Lunar Sci. Conf., Houston: Lunar Sci. Inst., 1974, pp. 887–906.
[9] OUYANG ZiYuan et al 2010 Primary Scientific Results of Chang'E-1 Lunar Mission, SCIENCE CHINA, November 2010 Vol.53 No.11, pp 1565–1581.
[10] Colwell J E et al, 2007 Lunar Surface: Dust Dynamics and Regolith Mechanics, Reviews of Geophysics, 45, RG2006 / 2007, pp. 1 – 26, Paper # 2005RG000184.
[11] Zakharova A V, Zelenyia L M, and Popel S I 2020 Solar System Research, 2020, Vol. 54, No. 6, pp. 455–476, ISSN 0038-0946.
[12] Colwell J E, Robertson S R, Horányi M, Wang Xu, Poppe A, and Wheeler P 2009 Lunar Dust Levitation, J. Aerosp. Eng., 2009, 22(1), pp. 2-9.
[13] Sternovsky Z, Robertson S, Sickafoose A, Colwell J, and Hora´nyi M 2002 Contact Charging of Lunar and Martian Dust Simulants, J. Geophys. Res., 107(E11), 5105, doi:10.1029/2002JE001897.
[14] Abbas M M, Tankosic D, Craven P D, Spann J F, LeClair A, West E A 2007, Lunar Dust Charging by Photoelectric Emissions, Planetary and Space Science, v. 55, Issues 7–8, May, pp 953-965.
[15] https://news.satnews.com/2023/01/11/aegis-aerospace-racs-up-their-1st-lunar-payload-delivery/ - Aegis Aerospace RACs up their 1st lunar payload delivery January 11, 2023.
[16] Kleiman J, et al 2022 Regolith adherence characterization (RAC) experiment on the Moon and it's ground-based simulation: Materials Issues, Joint International Conference on Protection of Materials from Space Environment,

ISMSE15/ICPMSE13 , 18 -23 September, 2022 Leiden, The Netherlands, https://iopscience.iop.org/issue/1757-899X/1287/1
[17] Kleiman J and Iskanderova Z 2021 Innovative Lunar dust mitigation technologies: performance evaluation on the moon and in ground simulators Canada Reaches for the Moon - 2021: Canadian Lunar Workshop, June 14-16 (CLW-ALC 2021).
[18] Kleiman J, Horodetsky S, Sergeyev V, Issoupov V 2006 Critical Review of The Design of Space Environment Simulators: Lessons Learnt Proc. of the 10th Int. Symp. on Materials in a Space Environment / 8th Int. Conf. on Protection of Materials and Structure from the LEO Space Environment (Collioure France) June 19-23
[19] Kleiman J, Horodetsky S and Issoupov V 2012 A Planetary Environmental Simulator/test Facility Proc. of 12th European Conference on Spacecraft Structures, Materials & Environmental Testing, March 20-23 (Noordwjik The Netherlands).
[20] Materials samples Lunar dust simulants vacuum testing Protocol in Baseline Lunar Dust Simulant Vacuum Testing for Large Surface UUTS 2014 Integrity Testing Laboratory Report ITL-012914-14027-2M Submitted under Subcontract #26340 Dust Mitigation Technologies (Moon, Mars) Contract SC-26340/922250).
[21] NASA-STD-1008, Classifications and Requirements for Testing Systems and Hardware to Be Exposed to Dust in Planetary Environments, https://standards.nasa.gov/sites/default/files/standards/NASA/Baseline/0/2021-08-21_nasa-std-1008-approved.pdf.
[22] Kleiman J, Horodetsky S, Issoupov V and Verba V 2014 New Challenges in the Ground-based Technology for Simulated Planetary Environmental Testing Proc. 28th Space Simulation Conference, November 3-6, 2014 (Westin-BWI in Linthicum, Maryland) pp 213-229.
[23] Lee L-H 1995 Adhesion and cohesion mechanisms of Lunar dust on the moon's surface journal of Adhesion Science and Technology 9:8, pp 1103-24
[24] LMS-1 Lunar Mare Simulant | Fact Sheet, December, 202
https://sciences.ucf.edu/class/wp-content/uploads/sites/23/2019/02/Spec_LMS-1.pdf
[25] Horodetsky S, Kleiman J, Issoupov V, Verba V 2024 "New Plasma Source with Accelerator for creating a Dust Flow of a Lunar Dust Simulant" Proceedings of ISMSE-16/ICPMSE-14, Saint Raphael, France, 7-11 October, 2024

IOP Conf. Series: Materials Science and Engineering **1328** (2025) 012004 doi:10.1088/1757-899X/1328/1/012004

New Plasma Source with Accelerator for Creating a Dust Flow of a Lunar Dust Simulant

S. Horodetsky, J. Kleiman, V. Issoupov, V. Verba
Integrity Testing Laboratory, Markham, Ontario, Canada
jkleiman@itlinc.com

Abstract. In a framework of a program on interaction of lunar dust simulants with materials, we developed a new plasma source integrated into the lunar environment simulator and used in charging the dust simulants deposited onto various space materials. The designed source includes an electrically insulated dust container located at the bottom of the chamber, allowing to ground it or apply an electrical potential. Above the container a mesh electrode is placed, isolated electrically from the dust container. Magnets, installed in the system, allow forming and focusing of plasma. Samples used for testing are fixed on a holder located above the plasma dust source. It is also electrically insulated, and can be either grounded or supplied with an electrical potential. The plasma, ignited in the vacuum chamber, interacts with the dust in the container, charging it. The charged dust particles begin moving towards the sample holder, raised to a much higher potential. By selecting the electrical potentials applied to the sample holder or to the plasma source mesh, or to both, the dust particles will accelerate in the space between the dust container and the sample holder. This paper describes the idea, design and testing of this plasma source. It also describes the original studies of charging the dust simulator obtained both by this plasma source and by the VUV and tribological methods, carried out in the same vacuum chamber using the same measuring devices. A comparison of these original data was also made. A prototype of this source has been tested under various conditions, and its operation and possible advantages over other designs of dust distribution sources are presented and discussed in this paper.

1.Introduction.

A large effort has been devoted to studies of interaction of lunar dust simulants (LDS), usually having an electric charge, with various sample surfaces and materials [1-30]. Analysing the cited literature data allows summarizing the main directions in this important area of in research and development, as follows:

1. Selection of a suitable LDS that most accurately corresponds to a specific area of the Moon.
2. Creation of a method (source) for dust application to the surface of sample materials under study.
3. Electrostatic charging of the LDS.
4. Measurement of the parameters of the deposited dust, including its electric charge.

IOP Conf. Series: Materials Science and Engineering **1328** (2025) 012004 doi:10.1088/1757-899X/1328/1/012004

5. Study of the conditions influencing the processes of deposition (adhesion) of the LDS on sample surfaces, as well as the results of the simultaneous action of various physical factors (beams of electrons, ions and ultraviolet radiation, magnetic field) on adhesion.
6. The influence of the physicochemical state of the surface on the adhesion processes of the LDS.

Also, the analysis of these references [1-30] showed that the most suitable LDS for our research is JSC-1A.

This direction of research should be considered as particularly important, since it covers the extended research and development of modifying the surfaces to provide protection from and mitigate the effects of dust on the Moon. In this regard, the little-studied phenomenon as the levitation of lunar dust and its effect on adhesion processes is worth investigating.

This article is devoted to the second and third points from this list, i.e. the creation of a source with controlled parameters, allowing the application of charged LDS on the surfaces of the studied samples. The article will describe the analysis of literature data that allowed us to formulate the prerequisites and conduct a series of experiments on the basis of which this source was created.

1.1 Plasma-dust atmosphere of the Moon and its simulation.

The Moon is surrounded by plasma that contains dust, i.e. particles of solid matter. On the Moon, these dust particles become charged under the influence of UV-photons, electrons and protons. Their interaction with the positively charged surface of the Moon leads to the fact that they begin to repel, move and form dust plasma. In the process of dust plasma formation, an important role is played by the influence of the tail of the Earth's magnetosphere, which is formed as a result of the interaction of its magnetic field with charged particles from space [4]. Under the influence of a magnetic field, particles, for example, solar wind, deviate from their original trajectory and form a region around the Moon. This region has an asymmetric shape: on the day side it reaches a size of 8-14 Earth radii, and on the night side it is elongated and forms the Earth's magnetic tail with a length of several hundred Earth radii. For about a quarter of its orbit, the Moon is in the tail of the Earth's magnetosphere, which affects the movement of particles along the meridian: under the influence of the magnetic field, they begin to move from the polar region to the equator.

Consequently, the movements of dust particles are influenced by two opposing physical phenomena - the gravity of the Moon and the magnetic field of the Earth. The first force makes dust particles gradually fall to the surface of the Moon, and the second lifts them to a height of several meters from the surface of the Moon, provided that they have an electric charge [5-8]. The combined effect of both phenomena on lunar dust particles will cause them to either oscillate in space or levitate. In reality, lunar dust clouds most often levitate [9-13]. The existence of levitating dust clusters on the Moon is associated with two factors - the long duration of the lunar day, which lasts about 15 Earth days, as well as special interactions of charged dust particles with the "magnetic tail" of the Earth. Due to its action, clouds of levitating dust can exist in almost all regions of the visible hemisphere of the Moon. On the other hand, the key feature of the lunar regalite and lunar dust is the presence of a large amount of nanophase metallic iron. It is this feature that easily distinguishes lunar rocks from their terrestrial analogues. The appearance of nanophase metallic iron is a consequence of the constant impact of the solar wind and high-energy particles of solar and cosmic origin. The overall chemical composition of lunar dust varies across the lunar surface, but is about 50% SiO_2, 15% Al_2O_3, 10% CaO, 10% MgO, 5% TiO_2 and 5–15% iron. In some areas of the lunar surface, the content of ferromagnetic Ni and Co has also been detected, although in small concentrations. Due to the action of magnetic fields in the tail of the Earth's magnetosphere, charged dust particles can be transported over the lunar surface over long distances. Accordingly, dust plasma above the sunlit surface of the Moon can exist for the entire range of lunar latitudes [14–16]. The transfer of dust particles over long distances due to the uncompensated magnetic part of the Lorentz force is a new qualitative effect that does not exist in the absence of a magnetic field. The magnetic component of the Lorentz force acting on a dust particle on the surface of the Moon is comparable to the similar force for the magnetic fields of the Earth's magnetotail in the orbit of the Moon [17-18].

Based on the above, we can conclude that one of the main components of the lunar atmosphere is dust plasma - ionized gas containing dust particles (particles of solid matter of micron and submicron sizes), which either spontaneously form in the plasma as a result of various processes, or are brought into the plasma from outside.

Dust particles in the plasma acquire an electric charge and represent an additional charged component of the plasma. The sizes of dust particles in it are relatively large - from fractions to hundreds of micrometres. Dust particles are centres of recombination of electrons and plasma ions, and sometimes a source of electrons (thermo-, photo- and secondary electron emission). The charge of dust particles is not a fixed value, but is determined by the parameters of the surrounding plasma and can change both in time and space.

When in plasma, dust particles create an electric field around themselves that is significantly distorted due to the presence of other charged particles (ions, electrons, other dust particles) and is not constant in time. In this case, the field potential differs greatly from the classical Coulomb potential, and the screened potential theory is generally accepted for its description.

Simulation of real lunar environment conditions imposes very strict requirements on the experimental equipment used for these purposes. For sources of LDS flow, they are as follows:

- They must be compatible with high vacuum
- Allow cleaning of the LDS on site from contaminants (water and its vapor, organic compounds) and other gas atoms and molecules.
- Create uniform LDS flows with adjustable intensity and distribution
- Create charged flows of LDS.
- Have a long operating time, allowing testing for weeks.

Unfortunately, existing sources do not meet this set of requirements that significantly complicates their use. We have proposed a new type of LDS flow source. The principle of its operation is as follows: preliminary charging of the simulator is carried out using plasma. The energy value of this plasma is determined only by the geometry of the source and does not affect the output characteristics of the LDS flow. The criterion for selecting the plasma energy value is maximization of the number of charged dust particles. Then, electrostatic acceleration of the simulator particles towards the samples is performed. A number of advantages of this source over the previously used gravitational source (such as sieve-based sources) can be pointed out:

- No mechanical parts that significantly increases its service life.
- 100% dust charge inherent in its operating principle. An uncharged dust particle will not be able to leave the dust holder.

IOP Conf. Series: Materials Science and Engineering **1328** (2025) 012004 doi:10.1088/1757-899X/1328/1/012004

• No need for preliminary dust cleaning in vacuum conditions (only preliminary baking in the atmosphere in a muffle furnace is possible). Plasma treatment will reliably remove all possible organic and inorganic compounds from the surface of dust particles.
• Dust particles will move towards the samples with a certain acceleration. Its value can be changed by varying the energy (voltage applied between the dust holder and the sample holder). • It is possible to install an unlimited number of small identical plasma sources (honeycomb structure). In this case, the area treated with dust will depend only on the size of the vacuum chamber.

2. Experiments on the electrical charging of LDS particles.

A series of experiments on the deposition of charge and uncharged LDS on the surfaces of various samples have been conducted at the ITL. Much attention was paid to the study of the LDS charging processes. Our paper in these proceedings [3] will provide more details of these initial experiments. We shall concentrate in this paper on the obtained results that formed the basis of the proposed dust source. When creating a new source of charged lunar simulant dust particles, we had to overcome the following technological obstacles:
1. The lack of reliable data in the literature on the methods for obtaining charged flows of dust in a lunar dust simulator, required us to conduct a number of preliminary experiments related to the electric charging of dust simulant particles.
2. The next obstacle that had to be overcome was the features of plasma formation in vacuum chambers. Gaseous plasma has a maximum ignition efficiency and stability in the pressure range of 10^{-2} Torr. This value depends on the energy of the plasma and the gas in which it is ignited. The most favourable condition for plasma is its existence in a gas flow. However, in this case it is extremely difficult, and sometimes impossible, to maintain the required flow rate in the vacuum chamber of the lunar dust simulator, due to the limited pumping speed of the vacuum pumps used. In this case, even doubling the pumping speed will have virtually no effect.
3. Low vacuum level can lead to contamination of both the surface of the samples being studied and the dust particles themselves. This will lead to a discrepancy between the conditions of the simulated lunar environment and real conditions on the Moon that one tries to simulate.
4. In addition to charging dust particles, plasma can cause sputtering of the material from which some parts of the dust source are made, which in turn can lead to contamination of the surfaces of samples being studied and of the electrical insulators, used in construction of the source. In the latter case, this can lead to a loss of insulating properties of the insulators and damage to the source.
5. Non-uniformity of the electric field can cause non-uniform ionization of the gas (non-uniform plasma), which in turn should lead to non-uniform spatial distribution of dust particles in the simulator flow.

The need to develop new principles for charging dust particles of the lunar dust simulator arose due to the scarcity of traditional charging methods and their low efficiency. Among the most commonly used methods one can mention the tribological method and the irradiation of the LDS with a short wavelength ultraviolet (VUV) source. The tribological method, despite its simplicity, is extremely ineffective. This is explained by the fact that only the most superficial layer in the particles is involved in the friction process. Then, during mixing, an exchange of charges between dust particles occurs that, in turn, leads to an uncontrolled redistribution of this charge, up to its neutralization. To test this assumption, we conducted a series of experiments using an insulated

IOP Conf. Series: Materials Science and Engineering **1328** (2025) 012004 doi:10.1088/1757-899X/1328/1/012004

Teflon cup driven by a piezoelectric transducer. This simple device completely simulated tribological charging under our conditions and allowed us to fully confirm this conclusion. Another traditional method is VUV irradiation with a wavelength between 100 - 200 nm, centred around 176 nm. Along with a large number of advantages, this method of charging dust has a number of significant disadvantages, among which the limited cross section of interaction of falling dust and VUV radiation and the reduction in the intensity of VUV radiation due to loss of transparency of the output of the VUV lamp exit protective glass can be mentioned. Dust settling on the glass surface occurs quite intensively due to the electrostatic interaction between the glass and dust particles. Analysing all these shortcomings, we felt that a need exists in a more efficient source for dust charging that led us to the development of the new dust charge source based on the interaction of plasma with dust particles.

The idea of the new method of charging the LDS is the interaction of dust particles with plasma. Medium-energy plasma is understood as plasma having an energy of 1 keV - 5 keV. In our case, this energy was 3 keV. The principle of operation of the source is described in detail below and is shown in Figure 1. This source is compiled from shaking dust source we developed that has been used for several years and is described in [1] and [3]. A medium energy plasma is formed in a cavity between a shaking dust source and a stainless-steel mesh attached to the source via electrical insulators. A high negative voltage is applied between the mesh and the source. A pressure of 10^{-1} Torr - 10^{-2} Torr is required to create plasma.

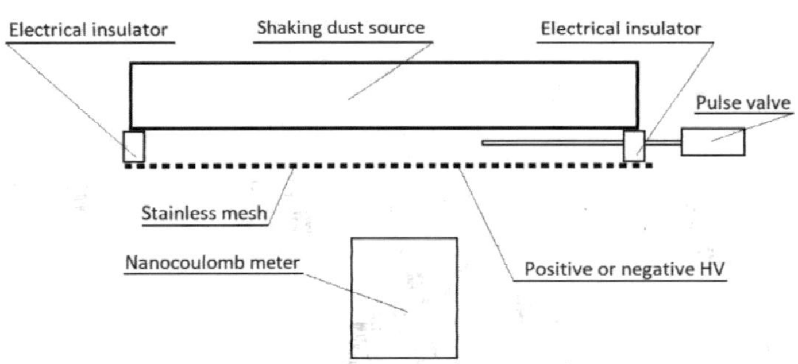

Figure 1. The idea of plasma charging the lunar dust simulant [Own work].

To locally increase the pressure in the plasma formation area (mesh - source), pulses of dry air (or any other gas) are used. They are formed using a pulse valve (see Figure 1). Selecting the operating mode of the pulse valve allows to obtain the pressure necessary for plasma ignition while maintaining a high vacuum in the chamber. The appearance of the dust charging system prototype is shown in Figure 2a, and the operation of the plasma source, with the plasma ignited, is shown in Figure 2b.

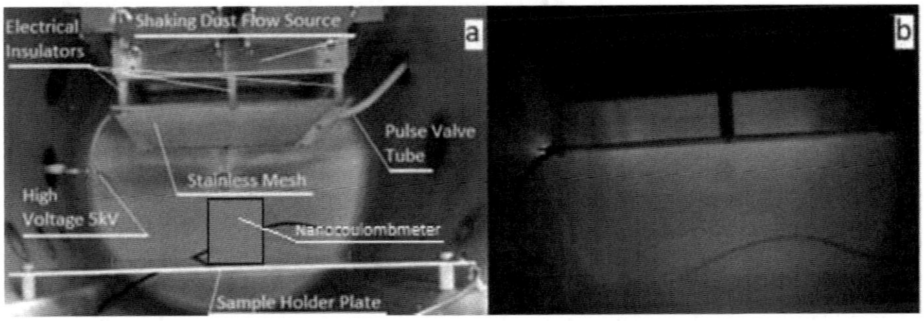

Figure 2. The appearance of the dust charging set-up in the vacuum chamber (a), and the operation of this plasma source (b) [Own work].

To measure the dust charge, we used a nanocoulombmeter, the design of which resembles a Faraday cup. It consists of two cylindrical cups that are electrically isolated from each other. To eliminate electrical interference, this outer cup was mounted on a grounded platform. The outer cup was grounded. The nanocoulombmeter cup was placed in a vacuum chamber with a vacuum level of 10^{-6} Torr. For reliable operation of the new dust charge source, the following conditions were chosen:

- The working pressure in the vacuum chamber was 10^{-6} Torr.
- The working pressure in the vacuum chamber when the source was turned on was 5x10-4 Torr.
- To obtain such conditions, the pulse valve described above operated with a frequency of 2 Hz and a gas pulse duration of 5 ms. Dry compressed air from a cylinder was used as the working gas. The pressure at the pulse valve inlet was 500 psi.
- The shaking dust source was set to a productivity of 0.5 ±0.1 g/min/cm^2.
- A negative voltage of -3 kV was applied to the charging grid.

Under these conditions, the plasma was stable and allowed us to obtain reproducible results. For a rough estimate of the plasma charge, a metal (aluminium) plate measuring 18 x 18 inches was placed on the bottom of the chamber and the current induced by the plasma in the plate was measured. It was 1–5 mA. The plate potential was changed between 1–23 V. It should also be noted that all plasma-related measurements were carried out with the nano coulombmeter measuring device always being turned off while the plasma was on, to avoid any damage to it.

The first series of experiments consisted of determining the charge of the Faraday cup itself in the plasma environment, by measuring it a number of times for 1 minute, with the meter, after each measurement, being reset to zero, and the cup discharged by grounding it. The cup charge varied from measurement to measurement in the range from -47.4 nC to -39.8 nC. The average value of this charge was -43.4 nC.

A number of assumptions were made in the conducted experiments:

- The plasma is stable over time.
- The dust source performance is constant over time.
- The dust does not charge before interacting with the plasma.

In the conducted experiments the main objective was to study the process of dust charge accumulation during minute-by-minute exposure to dust in the charge accumulation mode. The nano coulombmeter was zeroed and disconnected from the Faraday cup. Plasma was initiated and the dust flow was exposed to the plasma atmosphere, the nano coulombmeter was connected to the Faraday cup and the dust charge was measured. After that, the nano coulombmeter was switched off, without zeroing it, and an additional amount of dust, exposed to the plasma atmosphere, was collected in the Faraday cup for 1 minute. Then the measurements were repeated, etc. The results of these measurements are shown in Figure 3a, curve 2.

The next experiment used the same conditions, but 10 g of uncharged dust was added to the cup. The results of this experiment are shown in curve 1 (Figure 3a). This experiment was necessary for further development of the charging model of the dust particulars described below. This experiment with 10 g of uncharged dust allowed us to establish the exchange interaction between charged and uncharged dust particles. As a result of these two experiments, we found that the amount of dust in the nanocoulombmeter cup was 33.76 g (in the case of an empty cup, Figure 3a, curve 2) and 44.8 g (Figure 3a, curve 1). If we assume that there were already 10 g of dust in the cup during the experiment, we can assume that the productivity of the source was the same in both cases. The diameter of the inner cup is 8.5 cm. This means that its area is 56.7 cm2, and the specific productivity of the dust source during the experiment was 0.6 g/cm2. The total exposure time of the dust was 6 minutes; therefore, the charging capacity of the plasma source can be calculated. If the plasma source allowed charging 33.76 g to -102 nC (curve 2), then the charging capacity of the source is therefore 3 nC/g. In the case (curve 1) the dust charge was -70 nC, therefore its charging capacity was smaller and equal to 1.6 nC/g. The difference between the results shown on curve 2 and curve 1 is easily explained by the processes of dust charge loss for charging uncharged 10 g of dust. Therefore, curve 1 is smoother than curve 2.

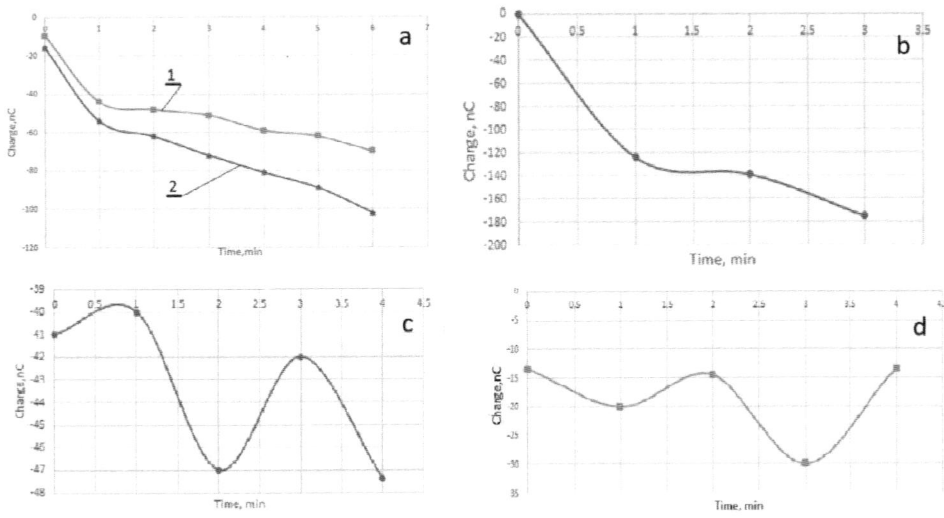

Figure 3. Study of SLP charging processes using low-energy plasma [Own work].

(a). Dependence of the dust charge on exposure time.
Curve 1. A cup contained 10g of uncharged sand. *Curve 2.* The cup was empty.
(b). Dependence of the empty cup charge on exposure time. The experiment was carried out in charge accumulation mode (without zeroing the device).
(c). Dependence of the dust charge on exposure time. The experiment was conducted with the device zeroed before each new measurement and the cup grounded to remove the charge.
(d). Dependence of the dust charge on exposure time.

An interesting result was obtained when an empty cup was exposed to plasma (Figure 3b). In this case, an intensive charging of the empty cup to -175 nC was noted in just three minutes. The charging rate of the empty cup was 58.3 nC/min. In the experiments with dust, it was 17 nC/min and 11.7 nC/min for the experiments shown in Figure 3b.

What is common to all three experiments is that the charge is measured in the accumulation mode, i.e. each subsequent result is obtained without zeroing the measuring device and grounding the cup to remove the charge. In these cases, the charge seemed to accumulate in the cup. The next experiment (Figure 3c) was conducted with zeroing the device before each new measurement and grounding the cup to remove the charge. The dust charge value during its exposure for one minute varied from -40 nC to -47 nC and, except for the spread caused, in our opinion, by instrumental factors, remained constant. This result is in excellent agreement with our first experiments on cup charging, where an average value of 43.4 nC was obtained. Similar results were obtained for the experiment with a cup filled with 10 g of dust (Fig.3.d). All other experimental conditions were similar to those in the previous case (Figure 3c). The dust charge value during its exposure for one minute varied from -13 nC to -30 nC and, except for the spread caused, in our opinion, by instrumental factors, remained constant.

At the end of this series of experiments, a final experiment was conducted (Figure 4), which completely repeated the conditions of the experiment described at the beginning (Figure 3, curve

IOP Conf. Series: Materials Science and Engineering **1328** (2025) 012004 doi:10.1088/1757-899X/1328/1/012004

1). These conditions are described in detail above. This experiment was necessary to check the correctness of the experiment and, accordingly, the obtained results. The complete identity of these curves once again proved the correctness of the experiments. All results were completely reproduced with an error not exceeding 5%.

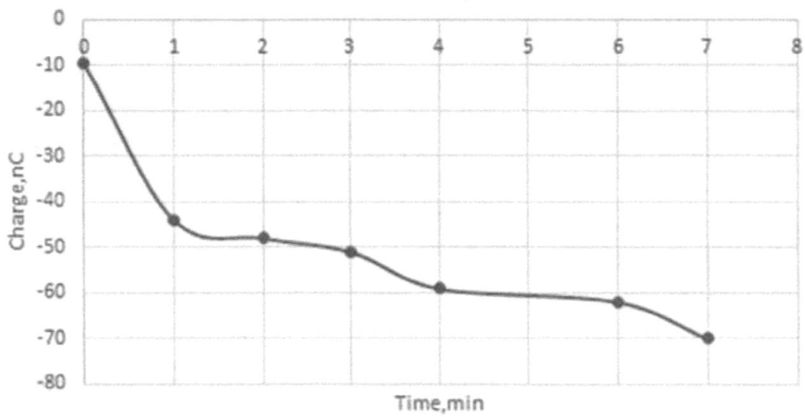

Figure 4. Dependence of the dust charge on exposure time [Own work].

To obtain a complete picture of the LDS charging process, we conducted several complementary experiments:

1. A series with charge accumulation, in which dust charged by a plasma source was poured into the nanocoulombmeter cup and the charge of this dust accumulated. In this case, the dust charge was measured as negative; over the same period of time (in our case, 1 minute), it increased by approximately the same amount. The monotonous, practically linea behaviour of this dependence proves the assumption about the accumulation of charge in the cup.

2. In the second series of experiments, the sequence of the first was practically repeated, with the difference that initially 10 g of uncharged dust was added to the cup. The total charge of the dust also increased, but more slowly compared to the first case. In both of these cases, the device was not reset before each exposure to dust, which meant that we worked exclusively in the accumulation mode.

3. And in the third series of experiments, the nanocoulombmeter was reset to zero before each exposure to dust, and the cup was grounded so that the charge flowed to the ground. In this case, the charge was approximately the same with a minute exposure to dust, but its value decreased slightly as the amount of dust in the cup increased.

4. In the fourth series of experiments, conditions similar to the third were observed; initially, only 10 g of uncharged dust was added to the cup. The charge measurement results showed similarity with the results of the third series; however, the charge obtained with each minute addition of dust was smaller.

5. In addition to the experiments with dust, the cup was exposed to plasma without dust. In this case, a very rapid negative charging of the cup was observed. After three minutes, its charge reached the practical limit of measurement of the nanocoulombmeter, and therefore further continuation of this experiment became impossible.

Two independent models can be proposed to explain these results. The first model does not take into account the instrumental factor. In it, the charge of dust that has undergone plasma treatment is NEGATIVE. It accumulates in the Faraday cup of the nanocoulombmeter. Its value is from 1 to 3 nC/g. This model does not take into account the interaction of charged dust with the nanocoulombmeter cup, as well as the redistribution of charge due to contact with other charged or uncharged dust particles. In the case of the second model, the interaction of charged dust passing through the plasma charger with uncharged dust or with dust charged to a different value located in the cup, as well as the charge of the cup itself, are taken into account. This charging of the nanocoulombmeter cup was measured in a series of experiments (Figure 3b).

Correctly taking into account all these parameters is a fairly complex mathematical problem, so we use a simplified calculation that allows us to obtain only qualitative results. If we take as a basis the initial curve shown in Figure 4 and the charge curve of the empty cup of the nanocoulombmeter (Figure 5) and subtract one curve from the other (taking into account the instrumental factor), then we get that the charge of dust is POSITIVE.

Unfortunately, this curve can only be plotted for a period of 3 minutes, for the reasons described above. If we assume that 33.7 g/6 min = 5.63 g of dust "falls" into the cup in one minute, then in three minutes there will be 16.9 g. Therefore, the plasma source has a charging efficiency of 6.1 nC/g. In conclusion of these considerations, it should be noted that all the presented results are the first and very approximate.

However, they clearly show that the plasma source charges the dust, and also give a complete picture of the fact that the process of measuring the dust charge is very important and taking this instrumental factor into account can even change the sign and magnitude of the results obtained, as was the case in our case. To obtain more reliable results on the dust charging processes, it is necessary to conduct a series of these experiments more carefully, with multiple repetitions, and, if necessary, add other experiments.

It is very important to use other devices measuring the dust charge, such as a Langmuir probe, instead of measurements using only one nanocoulombmeter. As mentioned at the beginning, in addition to a detailed study of the LDS charging processes using plasma, a systematic study of the charging processes of the simulant particles was carried out using other methods. These include charging using a VUV lamp and tribological charging. Various combinations of these three methods were also studied. These studies showed that different LDS charging methods produce not only different amounts of charge, but also different polarity. Therefore, the combination of these three LDS charging methods turned out to be ineffective. Analysis of the results described above allowed us to conclude that plasma charging of the simulant is the most effective.

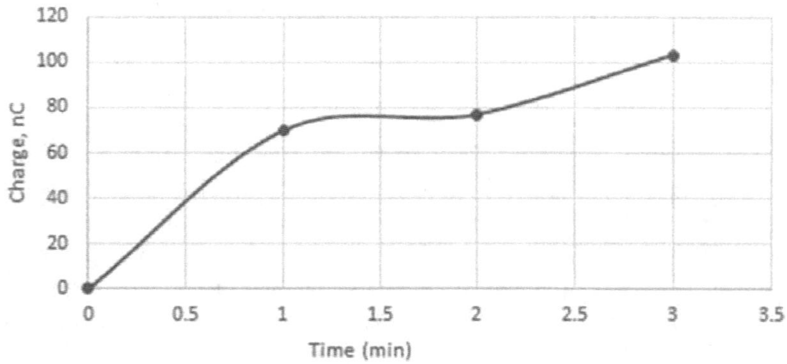

Figure 5. Dependence of the dust charge on exposure time with taking into account the instrumental factor [Own work].

3. Development of Dust Flow Simulant Plasma Source with Accelerator.

Tests of the plasma method of LDS charging allowed us to propose the idea of a fundamentally new source of a charged LDS flow (Figure 6). Its operating principle consists in charging a certain amount of dust with plasma with an energy of several kilovolts, followed by electrostatic acceleration in the direction of the samples. This source consists of a dust container that is filled with dust. Above it is a mesh electrode, isolated from the dust holder by electrically insulated posts.

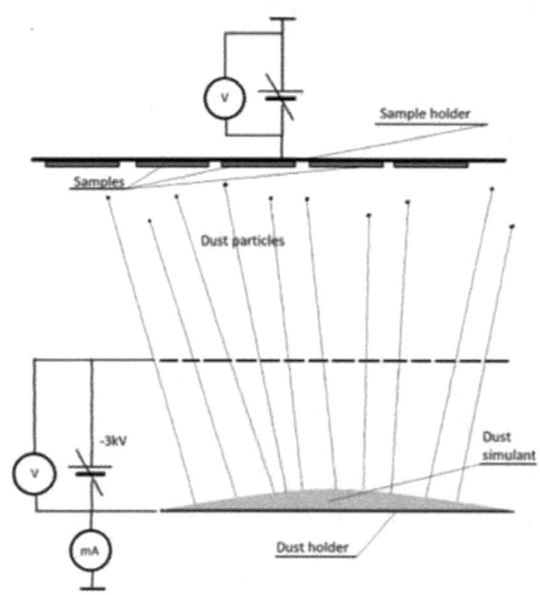

Figure 6. Promising plasma source of dust flow with accelerator [Own work].

IOP Conf. Series: Materials Science and Engineering **1328** (2025) 012004 doi:10.1088/1757-899X/1328/1/012004

The dust holder - mesh electrode system is fixed to an insulator, which allows it to rise to a high potential (up to 25 kV). The sample holder is located above the plasma dust source. It is also disconnected from the ground potential and can be supplied with a potential of up to 25 kV. The source works as follows. The dust container is filled with the required amount of dust and plasma is ignited in the dust source chamber. Interacting with the dust, the plasma charges it, and the charged dust particles begin to move towards the sample holder, razed to a much higher potential.

This idea of the dust source was tested by applying -3 kV to the mesh electrode and a voltage of about -5 kV across the entire source. In this case, the sample holder was grounded. Magnets installed in the system allow the plasma to be formed and focused. We implemented this idea on a simple device, which is a prototype of this source. The external appearance of this prototype is shown in Figure 7a, and its operation in a vacuum chamber is shown in Figure 7.b.

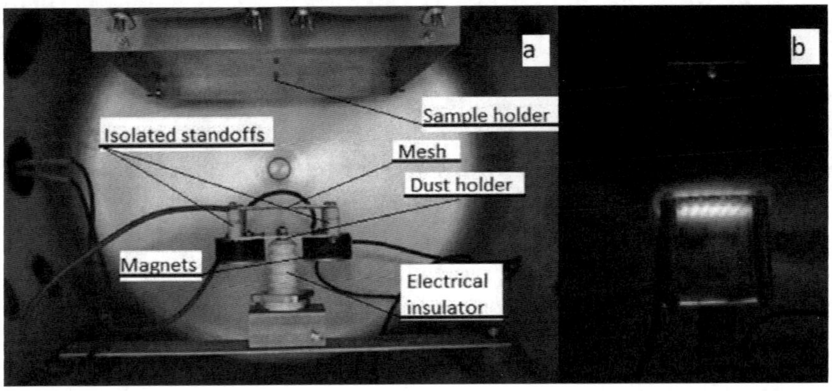

Figure 7. The appearance of the prototype of the plasma source of dust flow (a), and the operation of this prototype (b) [Own work].

During the testing of this prototype source, similar technical difficulties arose as with the plasma charging of dust described above several technical difficulties arose, described below: One of the main difficulties was associated with the discrepancy between the vacuum conditions of experiments with charged LDS and the optimal conditions for plasma ignition. As in the case of plasma charging of dust, this difficulty was associated with an obvious discrepancy between the optimal conditions for plasma ignition (in the pressure range of 10^{-2} Torr) and the necessary vacuum conditions for reliable simulation of the lunar environment. As in the previous case, to solve this problem, we used a pulsed system for pumping the working gas into the vacuum chamber. Such injection was carried out using a pulse valve, which made it possible to obtain short (up to several milliseconds) gas pulses. The repetition rate of these pulses was selected from the condition that the vacuum level in the chamber was not worse than 10^{-6} Torr, and on the other hand, so that the flow of charged dust was uniform. The frequency and duration of the pulses can vary within wide limits, which allows achieving the required experimental parameters. To reduce the contamination of the imitating particles and the surface of the sample under study, we use two methods: the first is to improve the vacuum conditions of the experiment, the second is to use inert gases to form plasma. It is the low adsorption capacity of inert gases that allows us to completely eliminate their contamination with atoms. In our experiments, we used Ar or He, although other inert gases can be used.

The next problem that arises during the operation of our plasma source could be the contamination of the surface of the test samples and electrical insulators caused by the plasma spraying of the material from which some parts of the dust source are made. In this case, these contaminants can lead to the loss of insulating properties of the insulators and damage to the source. To prevent contamination of samples, dust particles and electrically insulating parts of the plasma-dust source, its parts subject to sputtering were made of titanium. Titanium was chosen due to its lowest sputtering coefficient compared to other structural materials. As is known, the sputtering coefficient is defined as the number of emitted atoms per incident ion and depends on the mass of the incident particles, their energy and angle of incidence, as well as on the target material. For titanium, this value is from 0.3 atoms/ion to 0.45 atoms/ion, and for stainless steel - from 1.0 atoms/ion to 1.2 atoms/ion, i.e. almost three times more. - In addition to the above problems, the non-uniformity of the electric field can cause non-uniform ionization of the gas (non-uniform plasma), which in turn should lead to a non-uniform spatial distribution of dust particles in the simulator flow. To obtain a uniform dust flow, it is necessary to optimize the source parameters, such as the voltage supplied to the grids and the size of the cells of this grid. After numerous experiments, it was possible to select the optimal cell size. It was 5 µm.

4.Conclusions.

A new source for charging the dust using a low energy plasma was developed and described in this paper. Among the possible advantages of this source compared to the previously used gravitational source using a sieve-based dust distribution system we can mention the following:

- The absence of any mechanical parts, which significantly increases its service life.

- A 100% charging of dust, inherent in the principle of its operation. An uncharged dust particle will not be able to leave the dust container.

- There is no need for preliminary cleaning of dust in a vacuum (a preliminary heating in air atmosphere in a muffle furnace is enough to remove the moisture). Plasma treatment will reliably remove all possible organic and inorganic compounds from the surface of dust particles.
- Dust particles will move towards the samples with a certain acceleration, the value of which can be changed by varying the energy (voltage applied between the dust holder and the sample holder).

- It is possible to install an unlimited number of small identical plasma sources (honeycomb structure). In this case, the area treated with dust will depend only on the size of the vacuum chamber.

This source, in addition to all the advantages listed above, has another very important feature. Compared to others described in the literature, as well as sources developed by us, it allows simulating the lunar atmosphere with the greatest reliability. The fact is that dust in the surface layer above the Moon has its own characteristics. First of all, electrostatic and plasma-dust processes prevail there. As previously reported, the surface of the Moon is charged under the influence of electromagnetic radiation from the Sun, solar wind plasma and plasma from the tail of the Earth's magnetosphere. When interacting with solar radiation, both the lunar surface and dust particles emit electrons due to the photoelectric effect, thus forming a layer of

IOP Conf. Series: Materials Science and Engineering **1328** (2025) 012004 doi:10.1088/1757-899X/1328/1/012004

photoelectrons above the surface. But at the same time, both dust particles and the surface absorb photoelectrons, photons of solar radiation, electrons and ions of the solar wind, and if the Moon is in the tail of the Earth's magnetosphere, then electrons and ions of the magnetosphere plasma. All these processes lead to the charging of dust particles, their interaction with the charged surface of the Moon, and the lifting and movement of dust. All these conditions are modelled in the dust flow source we have developed.

References

[1] Kleiman J et al (2022) *Materials Issues, Joint International Conference on Protection of Materials from Space Environment ISMSE15/ICPMSE13* **18-23** September 2022 Leiden the Netherlands

[2] Joshua E Colwell Scott R Robertson et al *Journal of Aerospace Engineering* **22** Issue 1

[3] Kleiman J et al (2024) *Interaction of Lunar Dust Simulants with Materials: Importance of Charging, Joint International Conference on Protection of Materials from Space Environment, ISMSE16/ICPMSE14* **7 - 11** October 2024 St. Raphael France

[4] Horanyi M and Stern A 2018 *Dust Atmosphere and Plasma Environment of the Moon and Small Bodies* **156** 1

[5] Delory G et al 2009 *Lunar and Planetary Institute Science Conference Abstracts ser Lunar and Planetary Institute Science Conference Abstracts* **40** 2025

[6] Elphic R C 2012 *LPI Contributions* **1685** 3033

[7] Elphic R C et al 2013 *Lunar and Planetary Institute Science Conference Abstracts, ser. Lunar and Planetary Institute Science Conference Abstracts* **44** 3112.

[8] Grun E M et al 2011 *Planet. Space Sci.* **59** 1672

[9] Sternovsky et al 2008 *Journal of Geophysical Research (Space Physics)* **113** 12 10104

[10] Horanyi M et al 2009 *Lunar and Planetary Institute Science Conference Abstracts, ser. Lunar and Planetary Institute Science Conference Abstracts* **40** 1741

[11] Horanyi M *2012 Lunar and Planetary Institute Science Conference Abstracts Lunar and Planetary Institute Science Conference Abstracts* **43** 2635

[12] De B R and Criswell D R 1977 J. Geophys. Res. **82** 999

[13] Pelizzari M A and Criswell D R 1978 J. Geophys. Res. **83** 5233

[14] Randy I and Korotev 2024 *EARTH AND PLANETARY SCI* Department Earth, Environmental, and Planetary Sciences Washington University

[15] Heiken G H et al 1991 eds. *Cambridge University Press* New York NY 285

[16] Wentworth S J et al 1999 *Meteoritics and Planetary Science* **34** 593

[17] Gaier J Ret al 2005 *NASA STI Report Series,* NASA Technical Memorandum 213610

[18] Gaier J R et al 2009 *SAE Technical Paper* 2009-01-2420

[19] Popel S I et al 2018 *JETP Letters*

[20] Devaud G et al 2014, Proc. SPIE 9196, Systems Contamination: Prediction, Measurement, and Control doi: 10.1117/12.2062757

[21] Abbas M M 2007 *Planetary and Space Science* **55(7)** 953

[22] Walton O R 2007 *Adhesion of Lunar Dust,* NASA/CR-2007-214685

[23] Hong Gan et al 2023 Frontiers in Astronomy and Space Sciences 10.3389/fspas.2023.1213294 10

[24] Thoreson E J et al 2006 *Journal of colloid and interface science* **298(1)** 94

[25] Lin H et al 2013 *Langmuir* **29(9)** 3012

[26] Sternovsky Z et al 2001 *Journal of Vacuum Science & Technology A* **19(5)** 2533

[27] Baytekin H T et al 2011 *Science,* **333(6040)** 308

[28] William M et al 2023 *Reviews in Mineralogy and Geochemistry*, 10.2138/rmg.2023.89.1389 **1** 563

[29] Popel S I et al 2022 *Physics of Plasmas* 101063/5.0077732 **29** 1

[30] Li Yanwei et al 2023 *Applied Sciences* 10.3390/app13074441 **13** 4441

IOP Conf. Series: Materials Science and Engineering **1328** (2025) 012005 doi:10.1088/1757-899X/1328/1/012005

Characterization of thermo-optical properties of Power and Thermal Functional Surfaces Exposed to Lunar Dust Simulants

S. Duzellier[1*], JC. Mateo Velez[1], S. Soonckindt[1], A. Beyler[1], P.B. Hager[2]

[1] DPHY, ONERA, Toulouse, France
[2] ESTEC, ESA, Noordwijk, Netherlands

*E-mail: sophie.duzellier@onera.fr

Abstract. Measurements were conducted for a large set of samples and two lunar dust simulant (LDS) for estimating the impact of LDS contamination on thermo-optical properties of material surfaces. A systematic approach was used to get reproducible dust coverage rates on samples. The results indicate that the Rule-of-mixture approach is a conservative approximation for the change in solar absorptivity but is valid for predicting IR emissivity. For the coverglasses the reduction of transmissivity over dust coverage level was established. Finally, metrics could be proposed to link dust coverage to degradation of solar absorptance and IR emissivity for all tested material families.

1. Introduction

Moon surface exploration is facing many challenges, especially the dust coverage over time depending on both the lunar environment and mission operations. Plasma-matter models are useful tools to evaluate the mobilization of dust under different space plasma conditions including charging effects induced by the solar wind, VUV-induced photoemission and their collection by 3D objects [1][2].

By operating lunar surface assets, such as for example rovers, astronauts and robotic systems will necessarily generate additional contamination. Also, the landing systems (such as for example the ESA Argonaut [3]) will stir the lunar dust during landing.

What needs to be evaluated for future lunar missions is the change in the performance of a set of representative covering materials. It was previously demonstrated that "even a sub-mono layers of lunar dust (simulant material) increased the absorptivity of white paint and second surface mirrors significantly" [4][5].

The thermo-optical properties of "spacecraft outer surface" materials are critical for the thermal control and good functioning of power system of the spacecraft. Actually, an efficient thermal control of a spacecraft guarantees the performance and survival of the internal subsystems. In the same way, the performance of solar cells assemblies' materials are essential for the performance and reliability of the solar arrays. The coverglasses protect the solar cells while maintaining the required high transmission and still provide a radiation stabilized glass.

This paper reports on the characterization of the degradation of thermal and power properties of functional surfaces due to LHS-1 deposits (lunar dust simulant -LDS- from Space Resource Technologies).

The change in thermo-optical properties of a set of more than 20 different materials (Coverglasses CVG, Optical Solar Reflectors OSR, Second Surface Mirrors SSM, white and black paintings, MLI outer layers...) was measured with various levels of dust coverage. The large set of available data allows for highlighting the main response trends for IR emissivity and solar absorptance as a function of dust coverage.

2. Experimental details

2.1 Selection of materials

The selected material list is given in Table 1.

Table 1. Description of samples and corresponding with their ID used in the present study

Type	Description
CFRP	composite for structure, M18/55
AUCOAT	electroplated gold on aluminum
SURTEC	Surtec 650 with Al 6061 substrate, corrosion protection
OSRUVS	Optical Solar Reflector (silver/CMX 150μm)
OSRITO	with or without coating (ITO or UVS)
OSR	Radiator coating
SSMFEP	Second Surface Mirror (5 mils), Silver/FEP with or without coating (ITO)
SSMFEPITO	Flexible radiator material
SG121	White paintings
PSBN	Radiator coating for thermal and electrical control
PCBE	
AQ-PU1	Black paintings
PNC	Low outgassing coatings for thermal and electrical control
KAP	Polyimid insulation film, Kapton (200HN), 2 mils
VDAKAP	VDA on Kapton (1 mil)
KAPGE	Kapton with Germanium coating (1 mil)
VESPEL	Vespel SP1 for structural parts
BETACLO	Betacloth with VDA on one side, woven silica fiber used for shielding and thermal control
STAKAP	Black kapton with Stamet coating, thermal control RF transparency
TIT35	(Raw) Titanium T35 25 μm
NEXTEL	Nextel with VDA one one side Woven fabric, MLI outer layer
CVGARITO	coverglass with coating ITO/AR CMX 100μm
CVGAR	coverglass with coating AR CMG 100μm
CVG	coverglass un-coated CMX 100μm

IOP Conf. Series: Materials Science and Engineering **1328** (2025) 012005 doi:10.1088/1757-899X/1328/1/012005

This large set of test samples was selected to be representative of materials exposed to the dust environment in missions to the lunar surface:

- Radiator coatings (OSR, SSM, white paintings…),
- MLI outer surface coatings (Kapton films with or w/o coatings), Nextel, Betacloth and conversion coatings for thermal control,
- Coverglasses of solar cells (CMX standards with and w/o AR and ITO coatings),
- Structural material (like CFRP) and specific coatings (Surtec treatment and Gold electroplated), and Ti35 raw titanium foil candidate for HLCS/ERM/IHAB module.

2.2 Selection of lunar dust simulant

Currently, most mission concepts foresee landings in the Polar Regions, which are selenogical highland regions, and for the lunar simulant important parameters have been identified:

- Particle size range (0 – 50 micron) and distribution,
- Opaqueness (i.e. modal mineralogy),
- Appropriate maturity (how long a regolith sample may have been exposed at the surface and thus to "space weathering").

Absorptivity and IR emissivity properties tend to be lacking and are inconsistent within the literature for both authentic and simulant regolith samples. It was therefore chosen to go for a higher fidelity lunar highland simulant as an indirect way of better representing thermo-optical properties of regolith. One way of increasing fidelity is by simulating 'maturity'. Indicators of higher maturity, such as for example relatively longer exposure, can include: increased agglutinate modal proportion, nanophase iron enrichment, additions of rare gases, etc. Such properties may affect thermo-optical properties and therefore was considered as part of the simulant selection trade.

Two Lunar Dust Simulants have finally been selected, including:

- LHS-1 from Space Resource Technologies with custom sieve < 100 µm
- LHS-1-25A Lunar Highlands 25% Agglutinated Simulant from Space Resource Technologies with custom sieve < 100 µm. The simulant consists of 75% LHS-1 Lunar Highlands Simulant and 25% Anorthosite Agglutinates, by weight.

2.3 Test approach

This activity aims at performing a set of characterization tests with the purpose of linking the amount of dust coverage over time to degradation in performance.

The deposits and measurements have been performed at the ONERA's labs [6]. Requirements for testing systems to be exposed to lunar dust are described in NASA [7] guidelines. In the frame of this study, a methodology has been defined to perform test campaigns measuring the thermo-optical properties (alpha & epsilon parameters for radiators, and transmittance for coverglass), for two regolith simulants and varying quantities.

Reflectance curves in the [250-2500 nm] wavelength range (%R) is measured with the Perkin Elmer lambda 1050 spectrophotometer with a resolution of 1 nm allowing the calculation of solar absorptance of each sample, with a relative uncertainty of +/-1%. The (total) spectral transmittance %T is the ratio of the transmitted to the incident spectral flux (same wavelength range).

The AZTEMP 2000A reflectometer is used to determine room temperature emittance integrated over the spectral range of a 25°C blackbody in the [2.5-35 µm] wavelength range, with a relative uncertainty of +/-3%.

All measurements have been performed at ambient temperature.

A dust deposit chamber at DROP lab [6] has been calibrated to precisely control the quantity and homogeneity of dust particle coverage on a surface area of 3 cm². LHS-1 and LHS-1-25A lunar dust simulants have been sieved (by the supplier) to remove the fraction of dust particles with a size larger than 100 µm. The typical deposit in terms of Percent Area Coverage (PAC or %cov) is 5%, 15% and 30% for both LDS. Tests with a PAC larger than 50% have been performed on a subset of materials. In any conditions, the deposit at the sample surface is assumed to be a monolayer of LDS as confirmed by the mapping (for PAC <<100%).

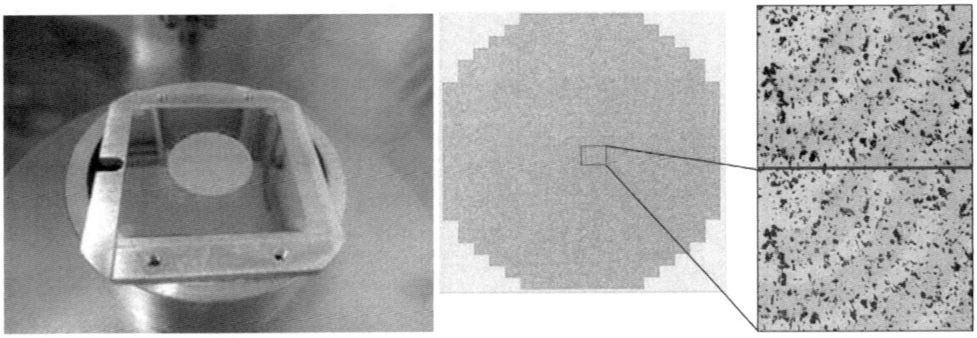

Figure 1. (left) LDS deposit onto OSR sample. (Right) mapping and counting of LDS grains (Credit: Onera)

A Nikon LV100 D microscope equipped with a motorized plate is used to observe dust grain on flat samples. It embeds a 2D/3D reconstruction mapping software and a particle counting software. The x5 optical objective allows scanning a 10 x 10 mm² surface with a resolution of 1 µm/px in about 10 minutes (see Figure 1 and [6]). Calibration and microscope settings allow for dust particle counting in the size range from 10 µm to 150 µm with absolute error on the size of 2 µm (twice the optical resolution) and relative error on the number of particles less than 10 %. Finally, the relative error on the percentage area coverage (PAC) is less than +/-15%.

Visual observation and mass measurement on witness samples is also used for crosschecking the LDS deposit properties.

The solar absorptance and IR emissivity of the lunar dust simulants have been measured using a thick puck of compacted dust. These properties are used as reference data for the dust particles. However, the compacted dust sample is not representative of a 100% PAC monolayer of LDS (same configuration as for contaminated samples). The validity of using this reference data will be discussed further in section 2. The solar absorptance, the IR emissivity and - in the case of transparent samples – the transmittance of the pristine material samples have been measured as well with increasing level of PAC.

One objective of the study was to establish whether some simple analytical modelling could fit the thermo-optical measurements. A Rule-of-Mixture model (RoM) assumes that the thermo-optical properties of a contaminated sample is proportional to that of the lunar dust simulants and to that of the material sample and to their relative surface area [4]. In the limit of a null percent area coverage, the measured property is that of the pristine sample material reference. In the limit of a fully covered surface, the property is that of the lunar dust simulant reference.

3. Main results

Results from measurement test campaigns before and after "contamination" provide a clear insight on the effect of %cov on the samples' thermo-optical response.

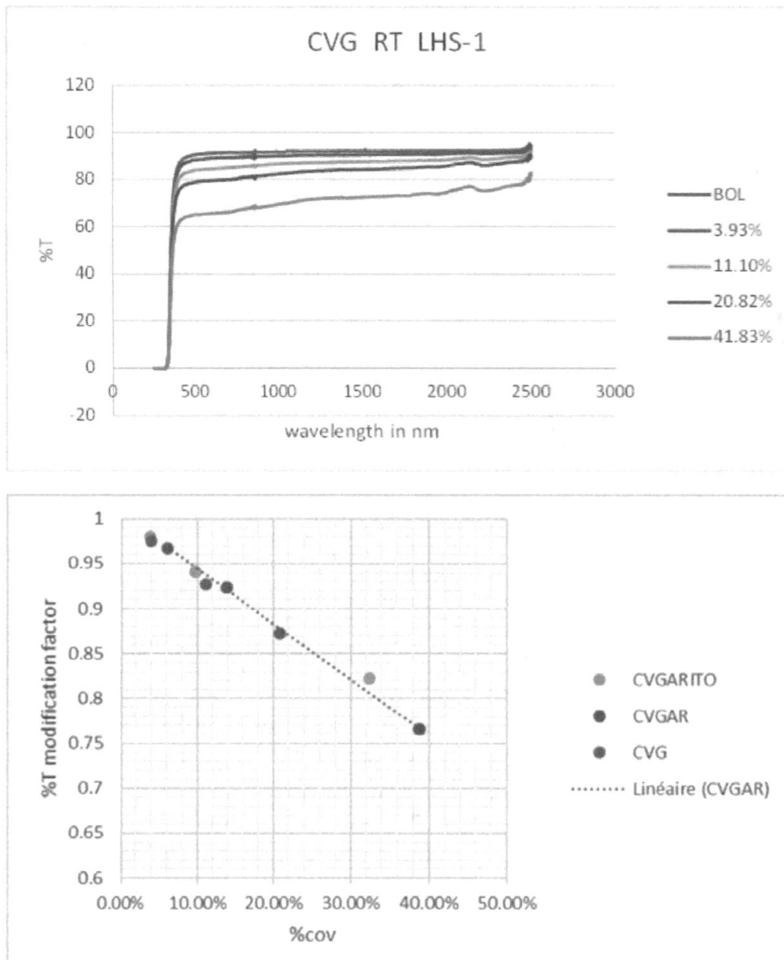

Figure 2. (Top) Transmittance curves for the set of CVG samples as a function of PAC. (Bottom) Evolution of transmittance modification factor (%T ratio of contaminated-to-reference) versus PAC at 500nm. (Credit: Onera)

For example, in Figure 2, the linear transmittance loss of all CVGs types with increased level of deposit show that LDS deposit acts as a "screening" layer (no deformation of the spectrum).

Regarding the evolution of solar absorptance, compliance with RoM is found with TiT35, KAP, KAPGE, VESPEL, and NEXTEL samples whereas ROM is conservative for most of the sample types as illustrated in Figure 3 (bottom) for white paintings.

One can also notice in Figure 3 (top) that including the reference LDS data (100% PAC) in the curve can lead to different fittings (or trend) than when considering sample data points alone.

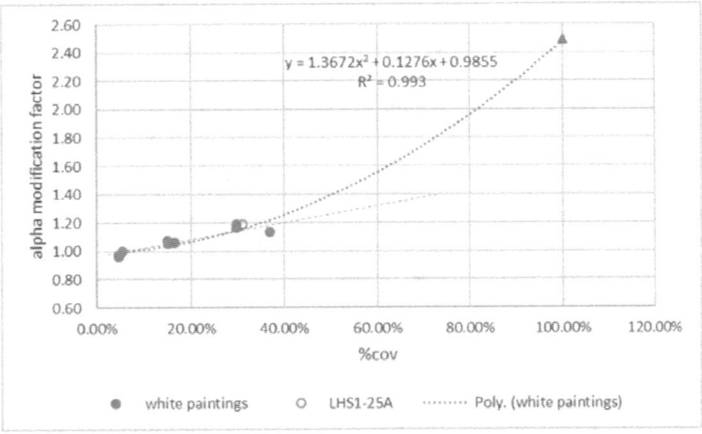

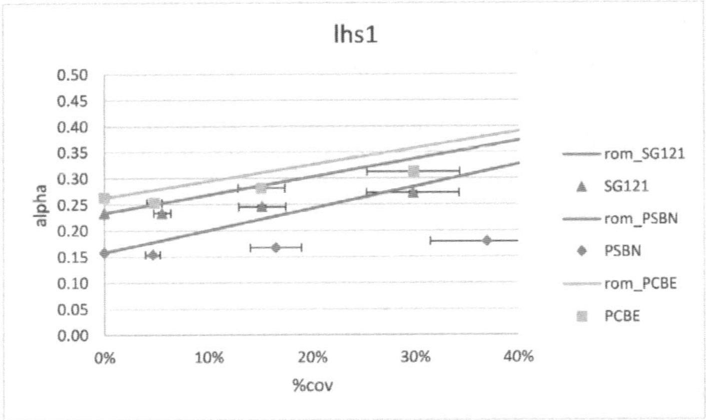

Figure 3. (Top) Evolution of solar absorptance modification factor of white paintings as a function of PAC (100% is the measured of the LDS itself - compacted). (Left) Comparison of experimental data for each paint type (symbols) with RoM model (lines). (Credit: Onera)

The use of these reference LDS data based on compacted sample shall thus be questioned as thermo-optical response seems different from single-layer deposit configuration.

Consequently, the range of investigated PAC here is not large enough and high PAC data measurement is missing to draw a clear conclusion.

More generally, the observed large range of the alpha modification factor indicates that the initial solar absorptance of samples is of prime importance to predict the overall trend. It is illustrated in Figure 4 where the modification factor (from 0.9 to up to 3.9) is plotted as a function of the relative solar absorptance data for all samples (i.e. the relative difference or ratio with the LDS solar absorptance) and for PAC ~30%.

The substrate feature and particles distribution in terms of transparency (composition and size) are also critical parameters leading to complex mechanisms when dealing with solar absorptance. Thus, fitting laws shall be defined specifically for each sample type (linear, exponential or polynomial).

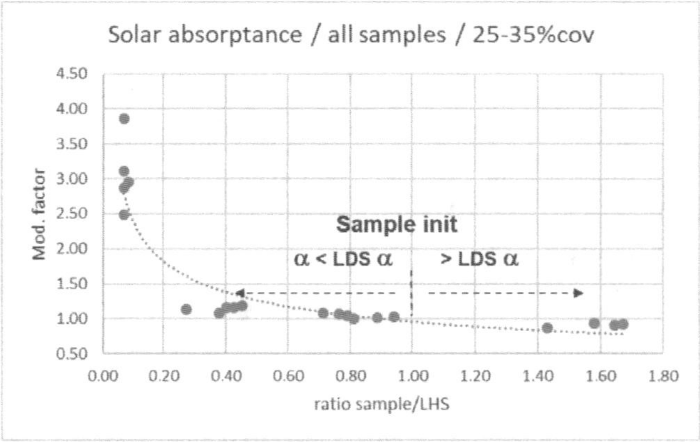

Figure 4. Solar absorptance modification factor as a function of the relative solar absorptance (relatively to LDS) for all samples at ~30% PAC. (Credit: Onera)

Contrary to solar absorptance, with emissivity, compliance with RoM is observed for all sample types except for SSM samples (model slightly conservative) (see Figure 5 for Kapton foils). It demonstrates that there is no interaction between the LDS deposit and the substrate: each one contributes to the overall response with proportionality (surface ratio).

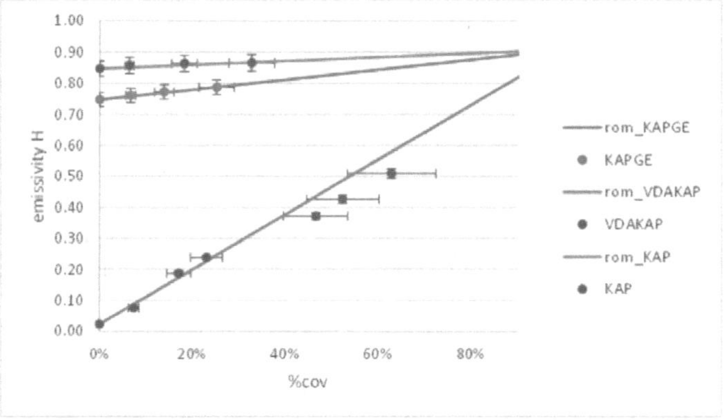

Figure 5. Comparison of emissivity experimental data for Kapton samples (symbols) with RoM model (lines). (Credit: Onera)

This outcome was expected with loose, low PAC and thus monolayer deposit. The thermal contact between a loose LDS deposit and substrate surface, or between LDS particles, is very poor and can be neglected. Therefore, radiative transfer dominates and temperature gradient across dust layer is driven by $\alpha_{substrate}$-to-α_{lds} ratio. The (semi) transparency of smaller LDS particles (<<100µm) to long IR wavelength is also consistent with this scheme. All these reasons are in favour of the observed validity of the ROM model in case of emissivity. Therefore, linear fits are appropriate for predicting the impact of LDS PAC on emissivity.

IOP Conf. Series: Materials Science and Engineering **1328** (2025) 012005 doi:10.1088/1757-899X/1328/1/012005

Finally, the type of fitting laws proposed for all sample families are disclosed in Table 2.

Table 2. Fittings laws per sample type

<table>
<tr><th>Sample</th><th>Alpha fitting laws</th><th>Epsilon fitting laws</th></tr>
<tr><td>SSM FEPITO</td><td rowspan="2">Exponential incl. 100%LDS</td><td rowspan="2">Linear, ROM conservative</td></tr>
<tr><td>SSM FEP</td></tr>
<tr><td>OSRUVS</td><td rowspan="3">Exponential incl. 100%LDS</td><td rowspan="19">Compliant with RoM</td></tr>
<tr><td>OSR</td></tr>
<tr><td>OSRITO</td></tr>
<tr><td>SURTEC</td><td>Polynomial compliant with ROM</td></tr>
<tr><td>SG121</td><td rowspan="3">Polynomial incl. 100%LDS</td></tr>
<tr><td>PSBN</td></tr>
<tr><td>PCBE</td></tr>
<tr><td>KAP</td><td>Polynomial incl. 100%LDS</td></tr>
<tr><td>KAPGE</td><td>Linear, compliant with ROM</td></tr>
<tr><td>VDAKAP</td><td>Exponential incl. 100%LDS</td></tr>
<tr><td>TIT35</td><td>Linear, compliant with ROM</td></tr>
<tr><td>CFRP</td><td>Linear</td></tr>
<tr><td>AUCOAT</td><td>Polynomial incl. 100%LDS</td></tr>
<tr><td>AQPU1</td><td rowspan="2">Linear incl. 100% LDS</td></tr>
<tr><td>PNC</td></tr>
<tr><td>VESPEL</td><td>Linear, compliant with ROM</td></tr>
<tr><td>BETACLO</td><td>Linear incl. 100% LDS</td></tr>
<tr><td>STAKAP</td><td>Linear incl. 100% LDS</td></tr>
<tr><td>Nextel</td><td>Linear incl. 100% LDS</td></tr>
</table>

4. Discussion

The α/ε evolution induced by dust contamination will affect the thermal response of surface materials as illustrated in Figure 6 where "hot" materials ($\alpha/\varepsilon>>1$) turn to "medium to cold" materials ($\alpha/\varepsilon\leq1$) for PAC as low as ~30% (or even less). The gold coating (AUCOAT) is particularly critical and constitutes a worst case here.

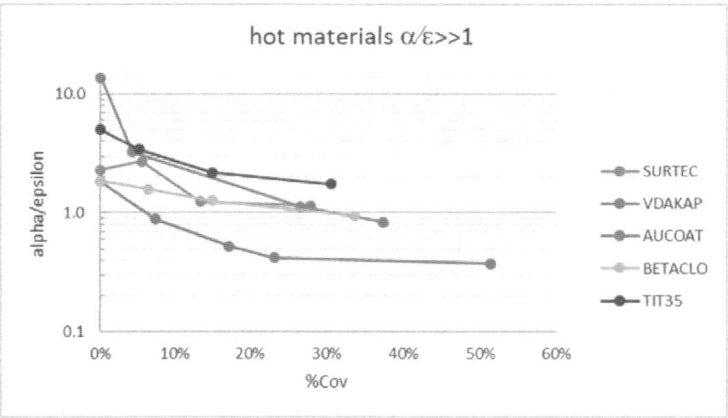

Figure 6. Evolution of α/ε as a function of PAC for "hot" materials (initial α/ε>>1). (Credit: Onera)

Considering a real lunar mission and dust properties different from the LDS selected for this study, the main conclusions drawn from this study can be expected to be still valid in case of mono-layer deposit (low PAC) and dust size <100µm for solar absorptance.

As already explained, the linear response of emissivity and compliance with RoM reveals the absence of interaction between LDS and sample. Changing the dust properties (epsilon, size or composition) should not modify this behaviour again in case of monolayer deposit (low PAC).

The similar trends observed with both LHS-1 and LHS1-25A shows the influence of dust composition is of second order here. However, contamination with LDS with much larger or greater reference alpha or epsilon values may lead to significant difference in adequate fitting laws.

It is difficult to conclude for lunar dust with very different size distribution, especially regarding the trends of solar absorptance, and for very high PAC (dust multi-layers). On Figure 7, one can observe the modification of the reflectance curve at UV cut-off due to the accumulation of LDS. It leads to strong variation in solar absorptance and shows that dealing with high PAC probably leads to response that is much more complex.

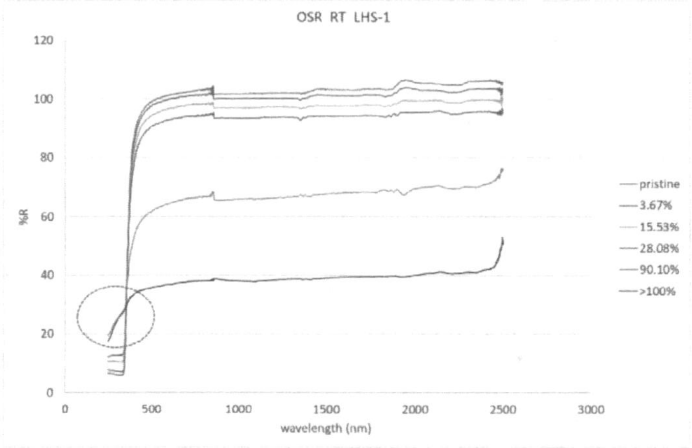

Figure 7. OSR reflectance curve for increasing levels of PAC up to accumulation (>100%). (Credit: Onera)

5. Conclusions

This study assessed the impact of contamination by lunar dust simulants (LHS-1) on the thermo-optical properties of various materials used in Argonaut space mission. The results indicate that solar absorptance and infrared emissivity of the tested surfaces may change significantly with increasing dust coverage.

The studied materials, including solar cell coverglasses, radiator coatings, and MLI outer layers, exhibited distinct response trends, requiring specific fitting laws (linear, exponential, or polynomial).

The Rule-of-Mixture (RoM) approach was found to be a conservative approximation for predicting solar absorptance but proved well suited for estimating IR emissivity.

The study also highlighted the importance of accumulation of LDS in altering optical properties, particularly for solar absorptance.

Therefore, for low to moderate contamination levels (single-layer dust deposits), the authors believe the conclusions remain applicable to real lunar conditions, although uncertainties persist for high coverage levels (multi-layer deposits).

Finally, these findings emphasize the need to carefully consider lunar dust contamination effects in the design of materials for lunar missions to ensure the performance and longevity of spacecraft thermal and power systems.

6. Acknowledgements & Disclaimer

This activity was funded by the European Space Agency (contract 4000139649/22/NL/AT). Special thanks to Thales Alenia Space, Airbus DS, Excelitas, Dunmore, CNES and ESA for providing the samples and materials.

References

[1] Hess S *et al* 2015 New SPIS capabilities to simulate dust electrostatic charging, transport and contamination of lunar probes *IEEE Trans. on Plasma Science* **43** pp 2799-2807

[2] Oudayer P *et al* 2019 Multiscale Modeling of Dust Charging in Simulated Lunar Environment Conditions *IEEE Transactions on Plasma Science* **47** pp 3710-3716

[3] Cropp A *et al* 2024 Argonaut LDE – European Access to the Moon *75th International Astronautical Congress*

[4] Gaier J R *et al* 2008 The effect of Lunar Simulated Dust on the Absorptivity, Emissivity and Operating temperature on AZ-93 and Ag/FEP Thermal Control Surfaces *NASA-TM* **215492**.

[5] Katzan C M *et al* 1991 The Effect of Lunar Dust Accumulation on the Performance of Photovoltaic arrays, *NASA technical Report* ***4404***

[6] Mateo-Velez J C *et al* 2024 Testing with Dust at the DROP lab of ONERA: facilities and approach *ISMSE16-ICPMSE14 symposium*

[7] Classifications and requirements for testing systems and hardware to be exposed to dust in planetary environments *NASA Technical Standard NASA-STD-1008* 2021-09-21.

IOP Conf. Series: Materials Science and Engineering **1328** (2025) 012006 doi:10.1088/1757-899X/1328/1/012006

Testing with Dust at the DROP lab of ONERA: facilities and approach

R. Pacaud[1*], S. Duzellier[1], JC. Mateo Velez[1], G. Murat[1], A. Jankowiak[2], A. Suarez Kahan[1], L. Moisset[1], S. Soonckindt[1], A. Beyler[1]

[1] DPHY, ONERA, Université de Toulouse, 31000, Toulouse, France
[2] DMAS, ONERA, Université Paris-Saclay, 92320, Châtillon, France

*Corresponding author: remi.pacaud@onera.fr

Abstract. This paper describes the DROP laboratory facilities and test approaches developed to investigate the topic of lunar dust in the context of future lunar missions. It presents a set-up developed to investigate the effect of dust on thermo-optical properties of space grade materials and the facility designed to study the physical mechanisms at the root of dust adhesion.

1. Introduction

In the context of future lunar missions, the evaluation of the effect of particulate contamination on materials' performance is required. One important question that needs investigating is the evolution in thermo-optical properties of representative materials as a function of dust coverage. The final objective is to predict the change in performance over time of surfaces exposed to natural or man-induced particulate contamination. This complicated task requires mastering the production of lunar dust simulant (LDS) deposits in terms of quantity (percentage area coverage or PAC), uniformity (monolayer, accumulation of LDS) and reproducibility. The fine, sharp particles are also known to damage mechanisms and cause wear.

The understanding of the mechanisms of lunar dust adhesion to surfaces is critical, however, this is not a well understood phenomenon. Lunar dust particles adhere through several mechanisms, including electrostatic forces due to the radiative environment.

ONERA has developed characterization means and testing facilities in order to address these problems. The (Dust Regolith Or Particle) DROP platform, located in Toulouse, hosts a dust deposition chamber, microscopy capabilities for visual inspection and mapping of dust deposits, and a vacuum chamber that reproduces the lunar radiative environment (VUV/electron irradiation).

This paper describes how this platform is useful to investigate the contamination-induced degradation of surface properties of materials and the charging behaviour of lunar dust simulants under representative conditions.

2. Test facilities and capabilities

The test capabilities at DROP lab primarily aim to research the behaviour and effects of dust on surfaces, requiring a well-controlled dust deposition system (see Figure 1). Good reproducibility in producing the deposition is critical, as is the accurate characterisation of deposit characteristics.

IOP Conf. Series: Materials Science and Engineering **1328** (2025) 012006 doi:10.1088/1757-899X/1328/1/012006

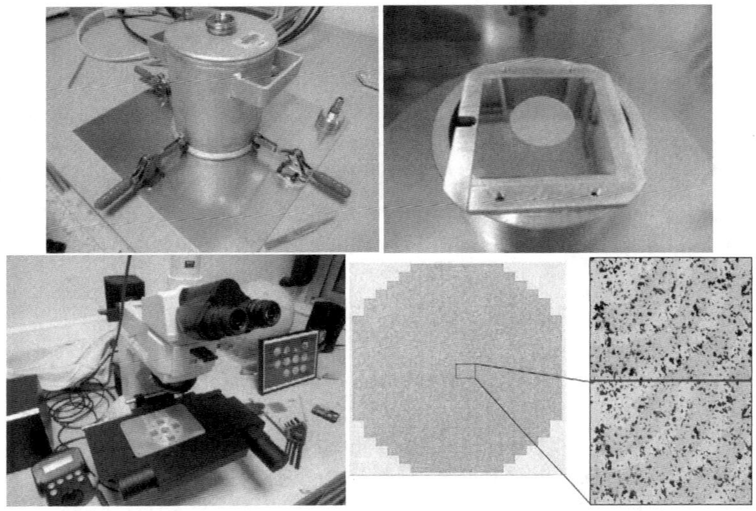

Figure 1. Dust Deposit Chamber DDC (top left), OSR (Optical Solar Reflector) sample with deposit (top right). Microscope for mapping of deposit (bottom left). Mapping and counting of dust particles (bottom right). Credit: Onera

2.1 Dust deposition and contamination of samples

Dust depositions is performed in the dust chamber (DDC). It is equipped with a dust dispenser, a primary vacuum pump and a compressed air inlet controlled by an electrovalve (Figure 2). The pressurized air is cleaned by appropriate oil, particle and humidity separators. A given amount of dust is deposited into the dispenser. Once the vacuum level inside the chamber is sufficient, the electrovalve is turned on. The combined effect of the depressurized chamber and pulsed air inlet disperse the dust inside the chamber. The characteristics of the deposit depend on the amount and type of dust injected and on the pulsed jet duration.

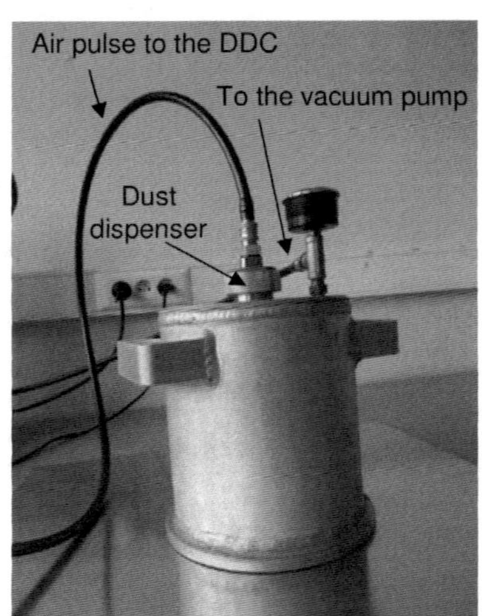

Figure 2. Description of the Dust Deposition Chamber (DDC). The dust dispenser is mounted on top of the DDC where dust is previously added. Vacuum is done through a vacuum pump and an air pulse is generated in order to have dust deposited. Credit: Onera

2.2 Characterization of Lunar Dust Simulants (LDS)

The characterization of LDS is of great importance in order to better understand their electrical and thermal properties. Both the size distribution and composition of a dust layer have an impact on its conductivity, hence charging levels and adhesion forces. For these reasons, various instruments are available to the lab for the characterisation of simulants used in testing. Some example measurements are described below in which the particle distribution size, microstructure and composition of LDS were characterized for 50-µm sieved JSC1-A and 100-µm sieved LHS-1 powders.

The LDS particle size distribution was measured using a laser granulometer (Horiba LA-950v2) allowing measurement in the 0.01 to 3000µm range. Before the measurements, the LDS were suspended in water and subjected to ultra-sound for 20 minutes. The particle distribution measurements were performed in volume mode using the refraction index of SiO_2 and then recalculated to compare with data obtained through other techniques (Figure 3). JSC-1A exhibits a monomodal particles size distribution after sieving (50µm) whereas the particle size distribution is multi-modal for LHS-1.

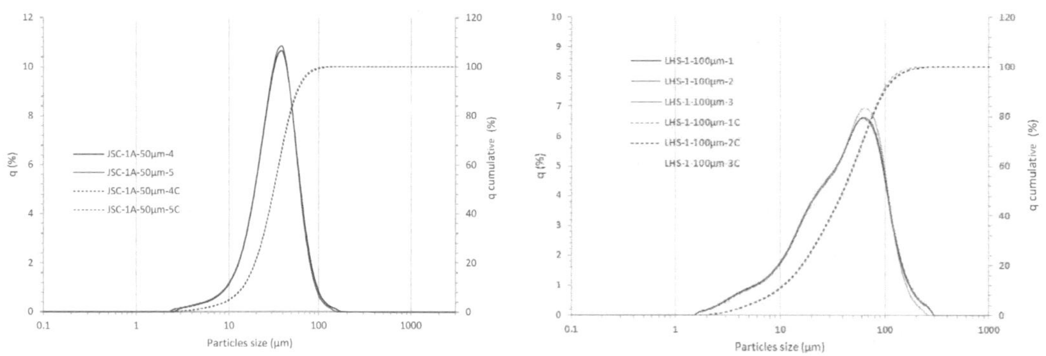

Figure 3. Volume distribution of 50-µm sieved JSC-1A raw (a) and 100-µm sieved LHS-1(b) powders. The lines represent differential distributions. The dashed lines represent cumulative distribution of dust particles. Credit: Onera

Microstructure and composition of LDS powders were determined using scanning electron microscopy (SEM) (Tescan Mira3 Schottky FEG-SEM) equipped with an EDAX Octan Elite EDS (Energy Dispersive Spectroscopy). Before SEM/EDS imaging and chemical analysis, powders are embedded in a resin and polished down to ¼ µm roughness. Backscattered electron images (SEM, Figure 4) are in good agreement with the granulometric measurements for both powders. Particles are angular and one particle can contain several mineral phases. After elemental quantification with EDS, several mineral phases were identified and reported in Figure 4. In the case of JSC-1a, Olivine (Mg_2SiO_4) and plagioclase (Na, Ca, K, Al, Si and O elements) phases are detected. For the LHS-1 power, Olivine (Mg_2SiO_4) and the anorthosite rock, which is mainly composed of the anorthite phase (plagioclase $CaAl_2Si_2O_8$), are detected as well. The analyses are representative of the LDS powers and are confirmed with XRD measurements and datasheets. The main objective was to locate the phases in the grains.

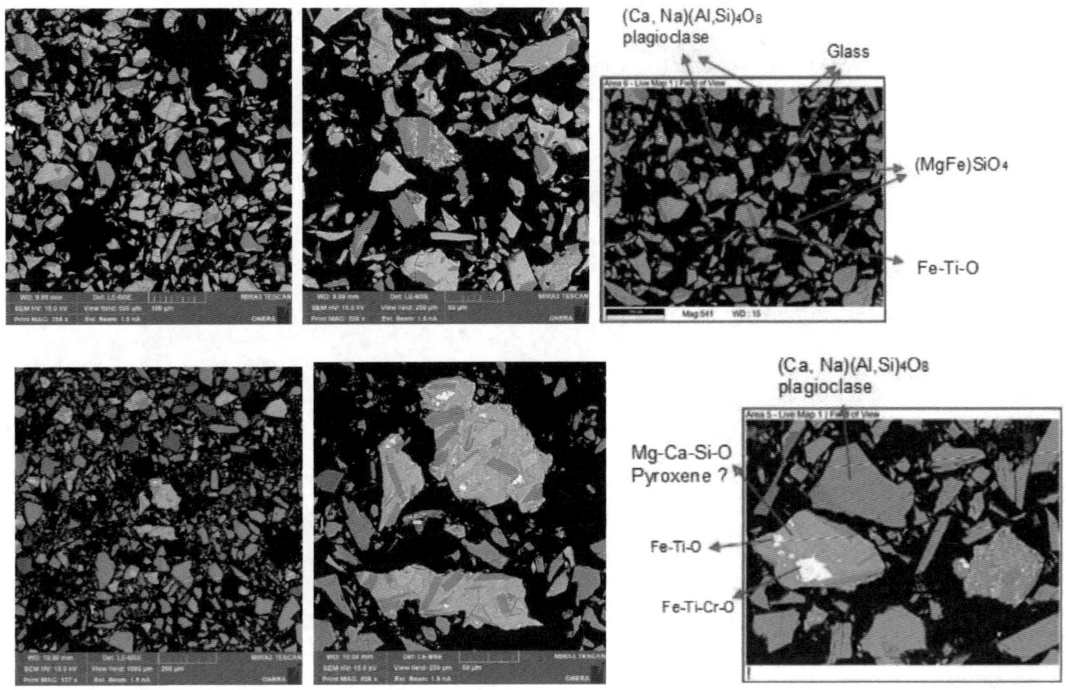

Figure 4. Microstructures and local compositions of 50µm-sieved JSC-1A (top) and 100µm-sieved LHS-1 (bottom) powders. Credit: Onera

LHS-1 can be found here: https://cdn.shopify.com/s/files/1/0398/9268/0862/files/lhs-1-spec-sheet-Dec2023.pdf?v=1703001782. JSC-1a has already been well documented in previous studies [1], [2] but datasheets are not available anymore.

2.3 Characterization of dust deposits

A Nikon LV100 D microscope equipped with a motorized stage is used to observe dust grains on flat samples. It embeds a 2D/3D reconstruction mapping software and a particle counting software. The x5 optical objective allows scanning a 10 x 10 mm^2 surface with a resolution of 1 µm/px in about 10 minutes. x10 and x20 objectives are also available.

Calibration and microscope settings allow for dust particle counting in the size range from 1 µm to 150 µm with absolute error on the size of 2 µm (twice the optical resolution) and relative error on the number of particles less than 10 % (Figure 5). For very low concentration deposits below <10% PAC, we assume that particles are mostly separated, but the accuracy of the microscope software particle separation for the case where particles do touch has not yet been quantified. Finally the relative error on the percentage area coverage (PAC) is less than +/-15%.

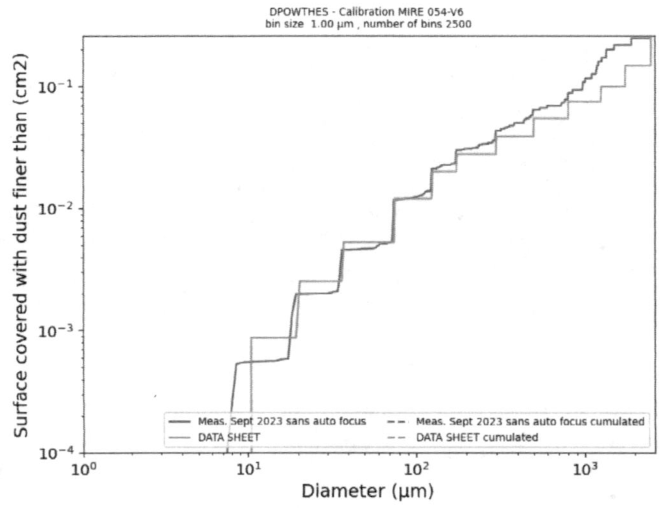

Figure 5. Cumulative area of disks finer than a given size. Comparison with manufacturer data at calibration. Credit: Onera

Several limits have been identified regarding the capacity to carry out PAC measurements. The main limitations come from the nature of the technical surfaces, especially when the sample is opaque and its surface is rough. In these cases, light scattering reduces the ability of the optical microscope to separate dust particles from substrate features.

2.4 Degradation of thermo-optical properties due to dust deposit

The main objective when investigating the degradation of thermo-optical properties of contaminated samples is to link the percentage area coverage (PAC) with the performance loss

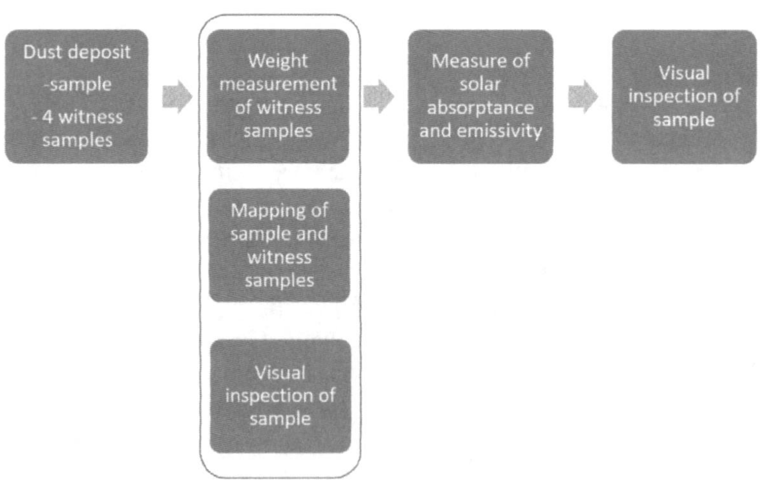

Figure 6. Main steps of the test procedure for the analysis of the degradation of thermo-optical properties due to dust deposit. Credit: Onera

(modification factor). It is therefore critical to ensure a good reproducibility and uniformity of the dust deposit, accurate measurement of PAC at the position of measurement, and the integrity of dust deposit at any step of the test campaign.

The approach developed to ensure the reliability of the data includes cross-checking tasks at critical steps of the process (see Figure 6), the use of witness samples (transparent glass) to allow for monitoring of PAC when no mapping is possible on a test sample and for relating PAC with deposited mass of dust (Figure 7).

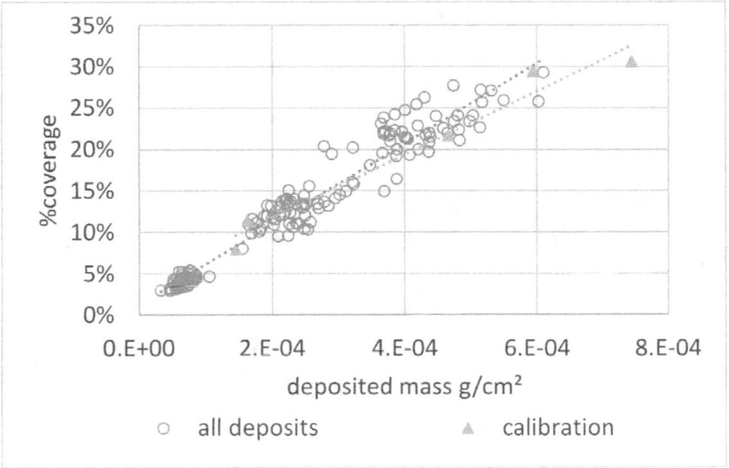

Figure 7. PAC versus dust deposited mass for all deposits and a calibration step. Credit: Onera

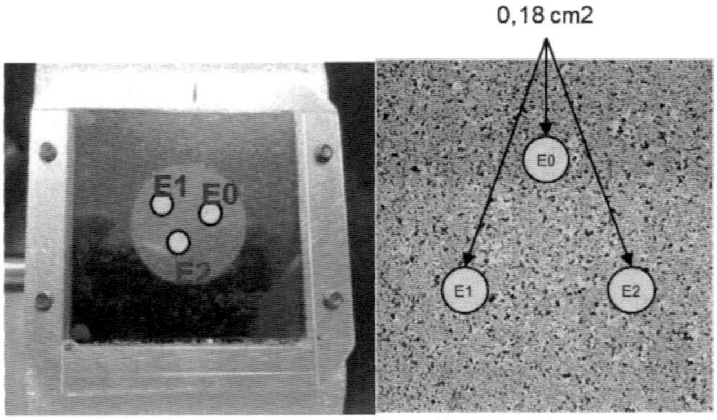

Figure 8. Local measurement and mapping at 3 different zones of the deposit. Credit: Onera

Particle size distribution (PSD) and PAC can be obtained through mapping for the whole dusty surface, but local mapping is also carried out at three different location points defined by the set-up for measuring reflectance (solar absorptance) and emissivity.

3. The DROP chamber

The DROP dusty vacuum chamber is described in Figure 9 and is used to carry out studies on lunar dust simulants in a representative environment. High vacuum is reached with a turbomolecular pump that allows reaching vacuum down to 10^{-6} mbar. Radiation sources, such as a VUV lamp or electron gun, are used to conduct dedicated experiments, from the study of the electrostatic charging of lunar dust simulant, to the use of a centrifugation system for the adhesion forces of regolith on technical materials.

Figure 9. DROP dusty vacuum chamber. Credit: Onera

3.1 Electrostatic charging

Electrostatic charging can be performed by two different irradiation means. On one hand a VUV lamp from Hamamatsu with 2 peaks at about 120 nm and 160 nm is used to represent the side of the moon that is illuminated by the solar wind. It has a photo-emission flux of about 1 nA/cm^2. On the other hand, an electron gun is used to represent the dark side of the moon. Fluxes of a few 100 pA/cm^2 are used to irradiate the samples, with energies ranging from a few tens of eV, up to a few hundreds of eV.

The surface potential is measured with a contactless Kelvin probe, before and after irradiation. It is used to study the charging levels obtained after irradiation as well as the relaxation dynamics of the samples, which has already been used in previous studies [3][4]. This approach is nonetheless macroscopic and does not give any precise information on the charge level of individual dust grains at local scale.

To overcome this limitation, a Faraday cup is connected to a transimpedance amplifier and polarized with respect to the ground. This polarization creates an electric field that exerts forces on the dust layer positioned right below the Faraday cup. Dust particles are attracted and go into the Faraday cup, which generates a signal, amplified by the transimpedance amplifier.

This signal is transmitted by a low noise coaxial cable to a data acquisition system that converts the analog signal to a digital one. It is further transmitted by an optical fiber to a dedicated software for data analysis, which makes the whole setup floating with respect to the ground. The background noise level is about +/- 20 mV after the amplification stage. A

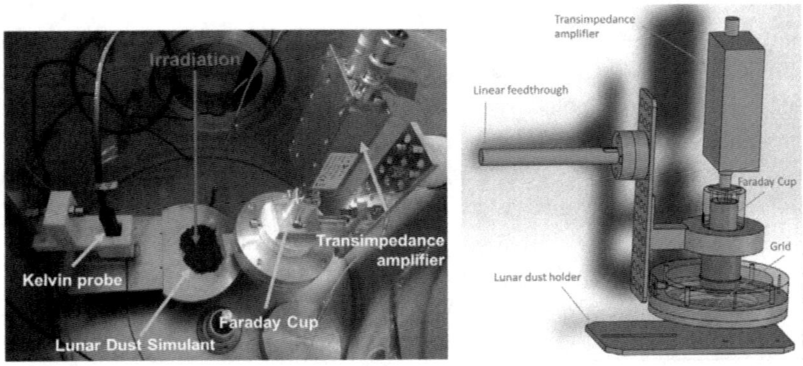

Figure 10. DROP vacuum chamber: the Kelvin probe is used to measure the surface potential of the dust layer. The dust layer is positioned at the centre, on a metallic plate connected to the ground. The Faraday cup and the transimpedance amplifier are connected to each other and can be moved by a linear feedthrough. A grid is used below the Faraday cup in order to homogenize the applied electric field. An electron gun is installed above the dust layer. Irradiation is performed perpendicularly, with respect to the dust layer. Credit: Onera

transimpedance amplifier with a gain of 10^{11} V.A^{-1} is used to amplify the signal and measure the charges of dust particles, down to 1 fC. The experimental setup is described in Figure 10 and the experimental protocol is illustrated in Figure 11.

A dust layer (here LHS-1 as an example) is positioned at the center of the DROP facility and lays on a metallic plate connected to the ground. The surface potential of the dust layer is measured for a duration of 1 minute at the center of the dust layer before irradiating the sample: it is about -2V. Irradiation is performed during 5 minutes with an electron gun with an energy of 30 eV and a flux of about 100 pA/cm^2. At the end of the irradiation, the surface potential is measured for a duration of 1 minute and is about -30V. The Faraday cup and transimpedance

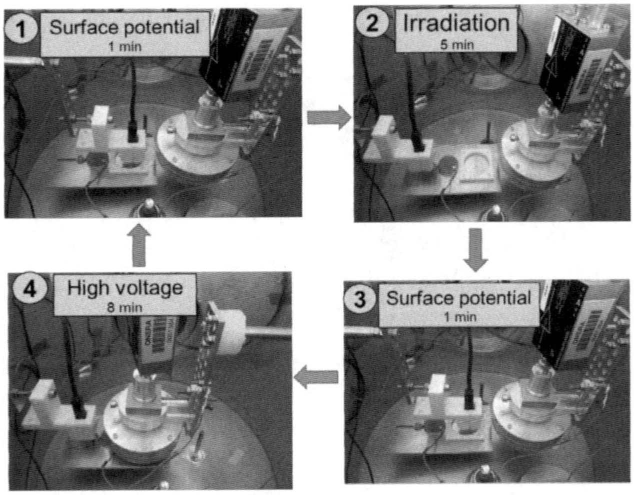

Figure 11. Illustration of the experimental protocol for the irradiation of lunar dust simulants for charge measurements. The surface potential of the sample is measured before and after irradiation. The Faraday cup and transimpedance amplifier are moved over the sample and polarized during a few minutes in order to attract dust particles. Credit: Onera

amplifier are then moved over the dust layer and polarized at +4000V, in order to apply an electric field between the Faraday cup and the dust layer.

An example of dust detection and charge measurement is shown below in Figure 12, done with LHS-1. It represents a dust particle entering the Faraday cup and generating a signal, of which the integral is charge dependent. In this example, the measured charge is -30 fC. Detection occured after 83 seconds, time during which the Faraday cup and transimpedance amplifier were polarized.

The Faraday cup, the transimpedance amplifier and the experimental protocole described in the aforementioned sections have been developed in a laboratory environment, wich does not yet fit the needs for future lunar missions. In the frame of an international collaboration (DUSTER project), the measurement technique as well as the control and data acquisition systems are further developed. It will be done with a dedicated data processing unit and a low and high voltage power supply. The long term objective is to analyse the charge of dust at the surface of the moon and understand its transport in a representative environment.

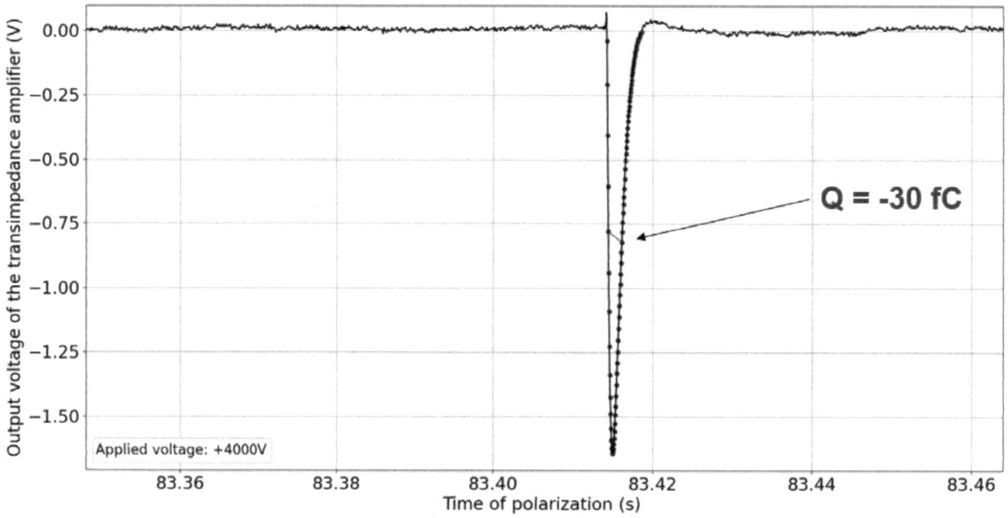

Figure 12. Output voltage of the transimpedance amplifier as a function of polarization time. A dust detection occurred after 83 seconds, time during which the detector is polarized at +4000 V. The signal corresponds to a dust grain with a charge of about -30 fC. Credit: Onera

3.2 Adhesion forces

A simple method for the calculation of adhesion forces is based on centrifugation [5]. A wheel is positioned in the DROP chamber and attached to a rotor. Lunar dust simulants are deposited with the DDC on technical materials (such as Kapton or Optical Solar Reflectors) and analyzed with the optical microscope in order to measure their size distribution before and after centrifugation.

Centrifugation is carried out in several steps, with different velocities for each step. The wheel can rotate up to 1500 rpm, corresponding to the maximum possible force exerted on the samples with this specific setup. At the end of each rotation, the samples are taken out and analyzed to measure the particule size distribution. When the test sequence is done, at the end of all rotation steps, the adhesion forces are calculated.

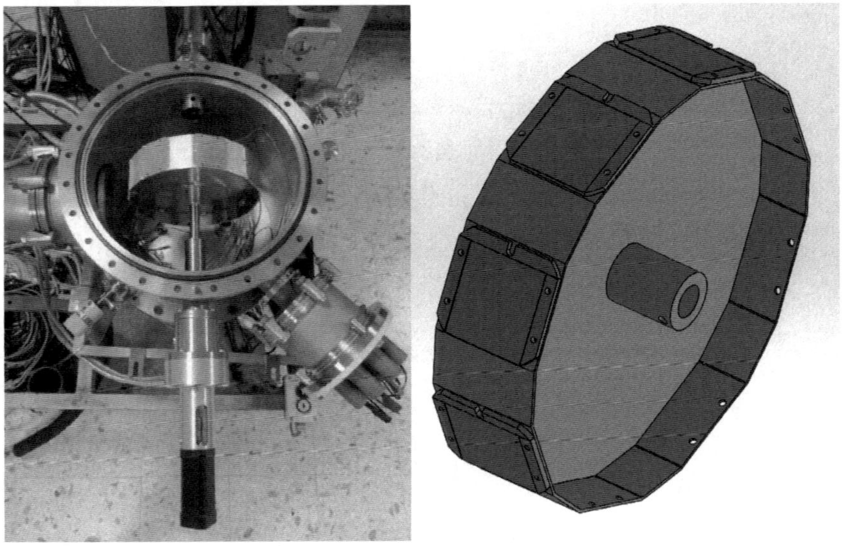

Figure 13. Centrifugation system. The wheel has a diameter of 11.6 cm. Up to 8 samples are studied at a time. Credit: Onera

The adhesion forces of detached particles are approximated to be equal to the force exerted by the centrifugation wheel. The latter depends on the centrifugation speed ω, the particle's radius r and mass m : $F_{adh} \approx F_C = mr\omega^2$. The parameters that influence the adhesion forces are the electrostatic charge of each particle as well as the forces of Van der Waals, which are dependent on the composition and surface roughness of the sample. Since the DROP vacuum chamber is equiped with an irradiation source, it is possible to study the impact of radiation (VUV or electron) on the electrostatic charging of dust simulants, and hence on their adhesion forces [6].

Acknowledgment

The developement of measurement capabilities has been possible with help of the Dust Study, Transport and Electrostatic Removal for Exploration Missions (DUSTER) project, financed by the European Union's HORIZON research and Innovation program under grant agreement n° 101082466. It has also been supported by the DPOWTHES study (Characterisation of Power and Thermal Functional Coatings and Surfaces Exposed to Lunar Dust for EL) funded by the European Space Agency in contract 4000139649/22/NL/AT.

The adhesion tests have recived a support from the European Space Agency Co-Sponsored Research Agreement No. 4000142578/23/NL/MGu/nh.

Special thanks also to Samuel Peillon and Cécile Bodiot from ASNR for their support to define and calibrate DDC, and to Emmanuel Porcheron for supervising PhD student activities in the field of dust particle adhesion.

References

[1] Ray C S, Reise S T, Sen S and O'Dell J S 2010 JSC-1A lunar soil simulant: Characterization, glass formation, and selected glass properties *J. Non Cryst. Solids.* **356** p 2369–2374
[2] Zeng X, He C, Oravec H, Wilkinson A, Agui J and Asnani V 2010 Geotechnical Properties of JSC-1A Lunar Soil Simulant, *J. Aerosp. Eng.* **23** p 111–116
[3] Champlain A, Matéo-Vélez J-C, Roussel J-F, Hess S, Sarrailh P, Murat G, Chardon J-P and Gajan 2016 A Lunar Dust Simulant Charging and Transport under UV Irradiation in Vacuum: Experiments and Numerical Modeling *J. Geophys. Res. Space Phys.* **121** p 103–116
[4] Oudayer P, Monnin L, Mateo-Velez J-C, Hess S L G, Sarrailh P, Murat G and Roussel J-F 2019 Multiscale Modeling of Dust Charging in Simulated Lunar Environment Conditions *IEEE Trans. Plasma Sci.* **47** p 3710–3716
[5] Barker D C, Olivas A, Farr B, Wang X, Buhler C R, Wilson J, Mai J 2022 Adhesion of lunar simulant dust to materials under simulated lunar environment conditions *Acta Astronaut.* **199** p 25–36
[6] Suarez Kahan A, Matéo-Vélez J-C, Porcheron E, Pacaud R, Moisset L, Delacourt B, Rampini R 2025 An investigation of lunar dust simulant adhesion using a centrifuge system under high vacuum and VUV irradiation conditions *IOP Conf. Ser.: Mater Sci. Eng.* accepted.

IOP Conf. Series: Materials Science and Engineering **1328** (2025) 012007 doi:10.1088/1757-899X/1328/1/012007

An investigation of Lunar dust simulant adhesion using a centrifuge system under high vacuum and VUV irradiation conditions

Alice Suarez Kahan*[1,2], Jean-Charles Matéo-Vélez[1], Emmanuel Porcheron[3], Rémi Pacaud[1], Lucas Moisset[1], Bruno Delacourt[2], and Riccardo Rampini[2]

[1] DPHY, ONERA – The French Aerospace Lab, Toulouse, France
[2] European Space Agency (ESA), European Space Research and Technology Centre (ESTEC), Keplerlaan 1, Postbus 299, 2200 AG Noordwijk, The Netherlands
[3] ASNR - Autorité de Radioprotection et Sureté Nucléaire et de Radioprotection, DRS/SCA, BP 68, 91192 Gif-sur-Yvette Cedex, France

*E-mail: alice.suarez@onera.fr

Abstract. Lunar dust contamination is of paramount concern for space agencies currently planning complex missions to the moon. The fine, sharp particles are known to damage mechanisms, cause wear, and deposit on surfaces thereby compromising optics, crippling power supplies, and causing thermal dysregulation. The adhesion of lunar dust is, however, not a well understood phenomenon. Lunar dust particles adhere through several mechanisms, including electrostatic forces due to their exposure to a radiative environment. This work was focused on centrifugation as a test method for determining adhesion. Lunar dust simulant (LDS) and glass bead particles were deposited in low concentrations on three types of substrates. Substrates were rotated under vacuum (10^{-6} mbar) in a centrifuge set-up, at incremental speed steps to a maximum acceleration of 277 g (1450 rpm). Tests were also carried out using a vacuum ultra-violet (VUV) irradiation step to cause the charging of the dust particles (as occurs on the Moon). Optical microscopy was used to count particles after each stage of rotation, allowing for the calculation of the adhesion forces present between detached dust particles and the substrate. The technique was validated for the characterisation of adhesion forces of larger lunar dust simulant particles (>30 μm) depending on the specific substrate. Forces on the order of 10-200 nN for particles between ~40-100 μm diameter were measured for LDSs once a particle volume correction was applied. The set-up is currently limited to this space of force and size.

1. Introduction

The aim of returning to the moon for longer missions and more sophisticated scientific objectives requires us to engage with the long-standing issue of lunar dust contamination which was identified already at the time of the Apollo missions.(1) This particle contamination can cause reduced performance and even failure of instrumentation and critical infrastructure such as solar panels for power generation. The lunar environment is favourable to the production of small highly charged dust particles due to the absence of atmosphere, which results in a heavy bombardment of the surface by UV radiation and solar wind electrons.(2) As well as charging from irradiation, particles can become significantly charged through tribocharging processes.(3)(4) Not only is there a net charge present on the particles, but local charge variation may be present on the surface of the grains. This can significantly impact electrostatic adhesion as high-charge-density areas will interact more strongly with other surfaces at small distances. The presence of ultra-high lunar vacuum has been found to affect the chemistry of the surfaces present as particles are clean from contaminants. This increases the adhesion present, in particular the electrostatic adhesion.(5) Van der Waals forces are always present and their contribution may become important especially at smaller particle size.(3) These forces are dependent on the surface energy of the materials as well as their geometric features. Not only is Van der Waals adhesion dependent on particle shape, but also on even nanoscale surface roughness.(6) The understanding of the adhesion behaviour of the non-spherical and dielectric particles with a large range of sizes and compositions is highly non-trivial and there is an urgent need for experimental data to contribute to models. Different forces of adhesion may dominate for different size regimes, depending on the interactions of many factors in these conditions.(3)

Centrifuge testing has been carried out in previous works researching lunar dust adhesion. A 2022 paper by Barker *et al.*(7) reports experiments carried using the centrifuge method under high vacuum using JSC-1A[1] simulant sieved to the <45 μm size fraction on various substrates (aluminium, anodised aluminium, Z93P painted aluminium and Ortho fabric) in presence and absence of UV radiation. In-situ data collection was carried out by using optical measurements (video), and calculating percentage coverage using image brightness. Several speed plateaus were measured, after which final particle size distributions of the remaining particles were measured. A similar study was carried out previously by Dove *et al.*(8) using a monolayer of JSC-1A particles below 25 μm on virgin and modified substrates, also including a UV comparison test. Ilse *et al.*(9) carried out centrifugal rotation experiments on a variety of terrestrial dusts under ambient pressure. Their analysis was used as a starting point for the calculations from the centrifuge tests used here.

Centrifuge testing has the benefit of providing information on a large population of particles which provides more averaged data. This however comes at the cost of having less information on each of the particles. Other methods of researching lunar dust adhesion include individual grain-level measurements by atomic force microscopy (AFM) such as those of Jardine *et al.*(10) and Xie *et al.*(11), which are usually carried out under less representative conditions (i.e. ambient pressure) and work such as that of Gaier and Berkebile(5) which was carried out using a higher fidelity environment (ultra-high vacuum and sputtering conditions) in order to research the effect of environment on the phenomena involved, but not using representative powder particles. Ideally a combination of higher fidelity conditions with high reliability adhesion data would be collected, but this is difficult to achieve. By testing a monolayer deposition under vacuum and VUV radiation, and measuring particle distribution at each rotation step (ex-situ) this work aims to combine elements of environmental fidelity with better statistical data collection to provide more insight into the Lunar adhesion problem.

[1] A NASA-produced simulant from the Johnson Space Centre, JSC-1A was made to replace JSC-1 which had depleted stocks

The first objective of this work was the validation of the centrifuge measurement techniques for adhesion of dust simulant particles. The second objective is to further the theoretical knowledge base on which parameters are most critical in determining the adhesion behaviour of dust particles, in particular the impact of the electrostatic charging of the dust particles. This paper presents the process used and preliminary results from the first campaign of centrifuge tests, with a view to reducing the major errors in future testing.

2. Materials and methods

The ONERA DROP (Dust regolith or particles) laboratory test set-up was used for testing. This is a dedicated dusty vacuum chamber capable of hosting a centrifuge. Centrifuges can be used to measure adhesion due to the effect of centrifugal force which detaches particles from previously prepared and characterised dusty samples. The change in particle size distribution on the substrate with rotation speed is measured using optical microscopy techniques and thus average adhesion force can be calculated.

The force of adhesion of the detached particle, i.e. the force experienced by that specific particle on a surface rotating fast enough for it to detach, will be the following equation:

$$F_{adh} \approx F_C = mr\omega^2$$ ***Equation** 1.*

where m is the particle mass, r the radius of rotation, ω the angular velocity. Assuming that the particle is a sphere, the equation can be expanded to use other inputs, d the particle diameter and ρ the particle mass density. Angular velocity can be re-expressed using f, the rotation frequency. This gives equation 2:

$$F_{adh} \approx F_C = \frac{2}{3}\pi^3 d^3\, \rho_{particle} f^2 r$$ ***Equation** 2.*

The force at which 50% of particles are lost (or retained) is a measure of the adhesion interaction between particle and surface. (12)

2.1 Materials

Four different lunar dust simulants were used in the tests: JSC-1A (Orbitec), and the others from Space Resource Technologies (previously Exolith); LMS-1 (Lunar Mare Simulant), LHS-1 (Lunar Highland Simulant), LHS-1-25A (LHS-1 modified with 25% anorthosite agglutinates), as well as soda-lime glass beads (Cospheric, ~10-100 µm; uncoated P2050SL-2.5 35-45 µm and P2075SL-2.5 65-75 µm mixed 1:1 by weight). Space Resource Technology simulants were purchased in early 2023. JSC-1A and LMS-1 were sieved to <50 µm and LHS-1 and LHS-1-25A were sieved to <100 µm.

Particles were deposited in sub-monolayer thickness on three types of sample: optical solar reflectors (OSRs) with CMX glass (150 µm) and mirror backing (120 nm Ag layer with overcoat of 50 nm Nichrome), standard uncoated or coated with ~12 nm Indium Tin Oxide (ITO), from Qioptics, and Kapton HN film (Dupont). OSR samples were of 4 cm x 4.2 cm, and Kapton film was cut to a similar size. The average surface roughness of the samples, Ra, is the following: Standard OSR: 0.5 nm, ITO OSR: 1.9 nm, Kapton: 3 nm (literature value(13)). See *supplemental information* for how the OSR roughness data was obtained by AFM.

2.2 Testing procedure

The centrifuge test wheel has a radius of 11.6 cm which is nearly the maximum that can fit inside the DROP chamber. The rotor has a maximum rotation speed of 1500 rpm. The pressure in the

IOP Conf. Series: Materials Science and Engineering **1328** (2025) 012007 doi:10.1088/1757-899X/1328/1/012007

chamber reaches $\sim 1\times10^{-6}$ mbar overnight. Samples were loaded onto the wheel and placed under vacuum ready for centrifugation the following working day. Rotation times were of 10 minutes. Charging conditions were not controlled as the use of an ionising pressure gauge was discovered *a posteriori* to induce positive charging on dielectric substrate samples when directly exposed in certain wheel orientations. For centrifuge tests carried out under VUV radiation a photoemission fluence of 1-2 nA.cm^{-2} was measured by Langmuir probe at the position of samples when passing under the VUV lamp (a Hamamatsu deuterium lamp emitting two intense spectral lines at 120 and 160 nm and a low continuum flux between 112 and 370 nm); in these cases substrate charging was measured to be very close to zero. The chamber was electrically grounded, as was the wheel.

Samples were glued to aluminium plaques (5.5 cm × 5.5 cm) which are made to fit to the aluminium test wheel. The wheel fits 6 samples which were split into 3 of each substrate used in that test. A silver nano-particle paste was applied to the edges of ITO samples in order to ensure their electrical grounding. Powder samples were baked out at 110 °C under low vacuum for >12 hours before use. A controlled deposition of particles was carried out in low vacuum, resulting in a mostly monolayer coverage, with most bigger particles not touching. In order to increase the likelihood of obtaining a pure monolayer, as well resolvable particles, a low coverage must be used (here between 4-8%). The dust deposit chamber mechanism is outlined in the supplementary information. Samples were inverted once slowly after deposit before first imaging. Substrates were either used pristine (soft isopropyl wipe on visible contamination) or in two cases recycled from previous deposits using a cleaning procedure: acetone wipe, ethanol ultrasound, demineralised water ultrasound, isopropyl alcohol wipe. Where cleaning was carried out this is specified. Figure 1 shows photographs of the set-up.

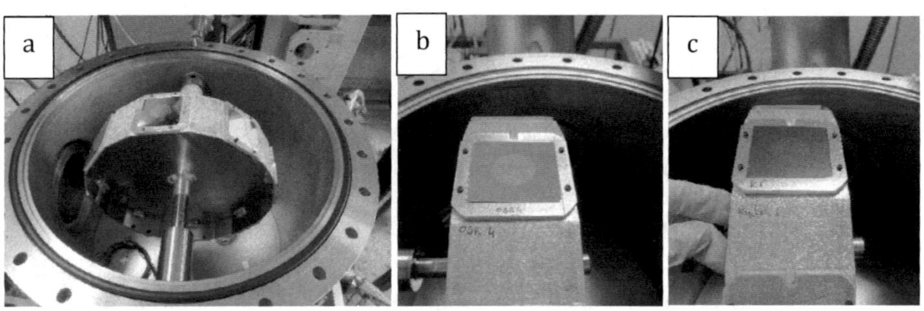

Figure 1. Vacuum centrifuge set-up a) overview from the top of the open chamber, b) closeup of OSR sample with deposit, c) closeup of Kapton sample with deposit. Credit: Onera

The central area of each sample in the test was imaged using optical microscopy (Nikon ECLIPSE LV100D-U/LV100DA-U, used with a magnification of ×5 under bright-field mode giving a resolution of 0.97 µm per pixel) before and after every rotation step. Particle measurement was carried out using the proprietary software of the microscope vendor Microvision; Ellix. A separation algorithm was used to try to separate particles that were in contact (this is not always successful due to the irregular shapes of the particles but does have some success rate). Closing and hole-filling macros were used and worked to also measure particles that have varying transparency, with few errors. The positions, sizes, and shape parameters of each particle that is detected can be extracted from each automated mapping. Their areas are summed to give percent

IOP Conf. Series: Materials Science and Engineering **1328** (2025) 012007 doi:10.1088/1757-899X/1328/1/012007

area coverage for each sample. The particle size distributions (PSD) of the particles on each sample after each rotation were treated using Python in order to calculate the force of adhesion. Mass was found using average grain density values, and centrifugal forces for detached particles could be calculated using equation 2. A mathematical correction was applied to volume which is detailed in the supplementary information. Data treatment was carried out following the example of Ilse *et al.*(9)

An initial test campaign was carried out using a small fraction of lunar dust simulants (<50 μm) on Kapton film and uncoated standard OSRs. Upon analysis of these first results a second test campaign was planned, with the optimisation of some parameters. For the second test campaign (using different substrates) a higher cut-off of particle size was used, and larger area of the sample was imaged (up from 1 cm^2 to 4.5 cm^2.), in order to increase available population data. Slightly different rotation speeds were used and imaging parameters were slightly tuned. Both sets of results are reported here as the materials used are different and the information may be of interest, but it is emphasised that the first tests (referred to as "pre-tests") have lower reliability data. Table 1 summarises which tests were carried out as a part of each campaign.

Table 1. Table summarising the materials and conditions of the tests carried out.

		Particle type				
Substrate type	Radiation conditions	LMS-1 (<50 μm)	JSC-1A (<50 μm)	LHS-1 (<100 μm)	LHS-1-25A (<100 μm)	Glass beads (<100 μm)
Kapton	No VUV	Pre-test 1	Pre-test 2			
	VUV					
Standard OSR	No VUV	Pre-test 1	Pre-test 2	Test 1	Test 2	Test 3
	VUV			Test 4		Test 5
ITO-coated OSR	No VUV			Test 1	Test 2	Test 3
	VUV			Test 4		Test 5

2.3 Methods

The data processing is described here using example figures from one of the tests carried out (test 1 on standard OSR). See the results section directly for the final data obtained for all tests described above.

An example image obtained by the microscope is shown in figure 2.

IOP Conf. Series: Materials Science and Engineering **1328** (2025) 012007 doi:10.1088/1757-899X/1328/1/012007

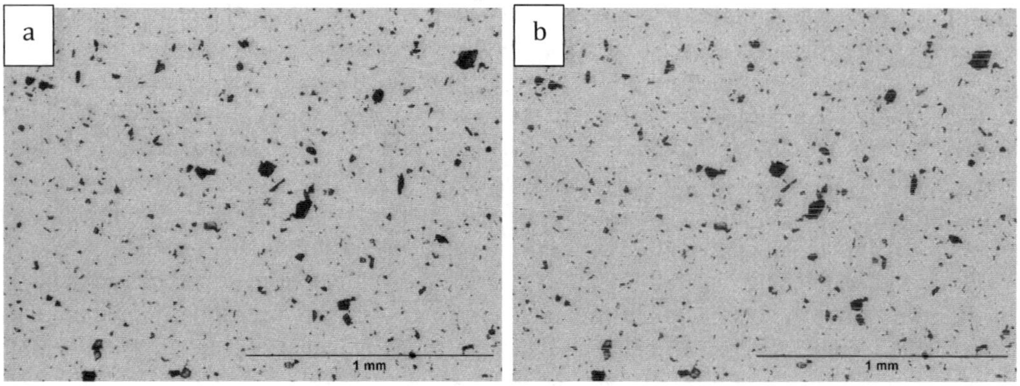

Figure 2. a) Microscope image of 2 x 1.5 mm from LHS-1 <100 µm deposit on Standard uncoated OSR, b) the same microscope image showing the particles counted by the software. They are outlined in green, and have horizontal hatching. It can be seen that some particles are separated, whilst it seems others are not. Credit: Onera

Microscope analysis produces a table of data points for each imaged particle. The measured area is used to produce an equivalent diameter value (the corresponding diameter of a circle that would have the same area). Python was used to plot PSDs for each initial deposit in order to check their nature. It is likely that some of the particles identified are in fact aggregates (especially towards the higher diameters). An example of a PSD graph is shown in Figure 3.

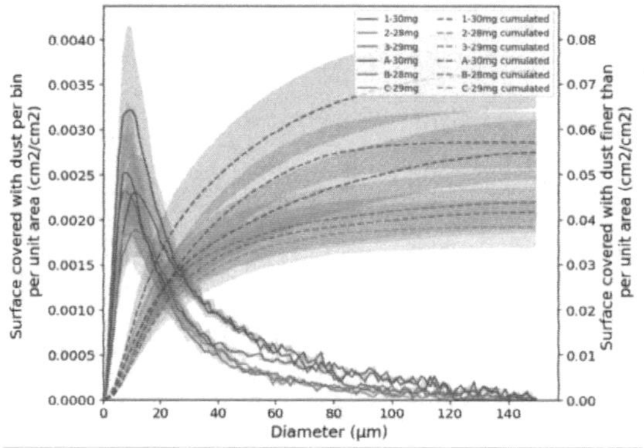

Figure 3. Graphs of particle size distributions of the initial depositions on the pristine surface for test 1 with LHS-1 on all 6 samples (labelled 1, 2, 3, for standard, A, B, C, for ITO). The envelopes around the lines show the accumulated error in surface coverage assuming an equivalent diameter of ±1 µm due to the resolution limit of the imaging used. The overall error from the imaging measurement is not known. Credit: Onera

These graphs can be plotted for all image series obtained (i.e. at each rotation step) but it is clearer and more relevant to the analysis to work with histograms such as the one in figure 4. For the main test campaign, the histograms from the initial deposition analysis were adjusted to produce a minimum of 100 particles per bin, leading to a large size of bin for the largest particles (which are rare, and may largely be aggregates). This was not always possible for the pre-tests: fewer than 100 particles per bin may be present for bin sizes above 42 μm in those samples.

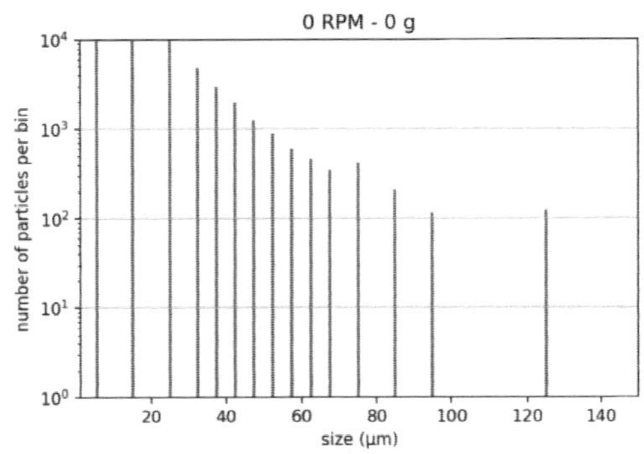

Figure 4. Histogram of the number of particles per bin for the summed initial sample PSDs using the bin sizes used to calculate adhesion forces in test 1 using LHS-1 simulant on uncoated standard OSRs. Credit: Onera

For each bin, the change induced in its population by the different rotation steps can be tracked. For this analysis of particle detachment, the samples can be analysed separately (there are 3 samples of each substrate type per test) or concatenated (all particle data for one substrate type treated together, to increase the population data available.) Separate sample data is shown in the supplementary information. Figure 5 shows the change in population data with acceleration.

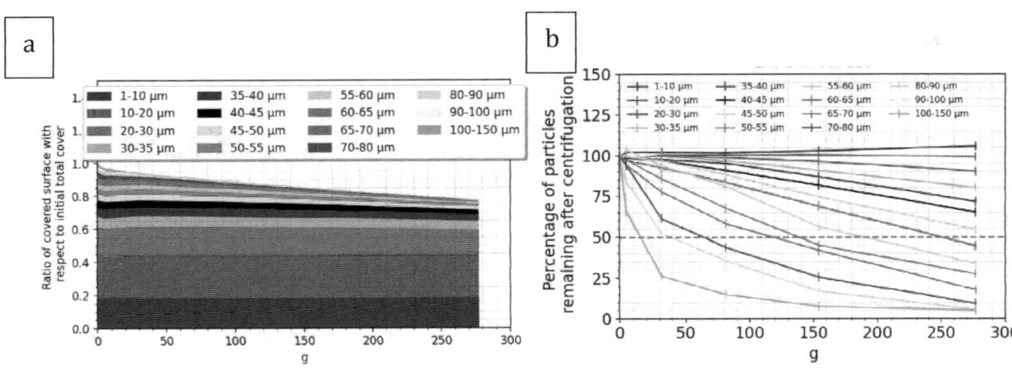

Figure 5. For LHS-1 test on Standard OSRs, a) the ratio of covered surface with respect to initial coverage, coloured by particle diameter bins, b) percentage of particles remaining per bin after each acceleration, connected by lines. Credit: Onera

The particle distribution present on the sample changes with each rotation step, as the acceleration causes preferential loss of larger particles. Using the intercept of 50% particle loss (for each bin) with acceleration, the acceleration needed to remove 50% of particles can be found, as long as enough particles in a bin are lost. Figure 6 shows the 50% removal intercepts with acceleration for each bin size plottable.

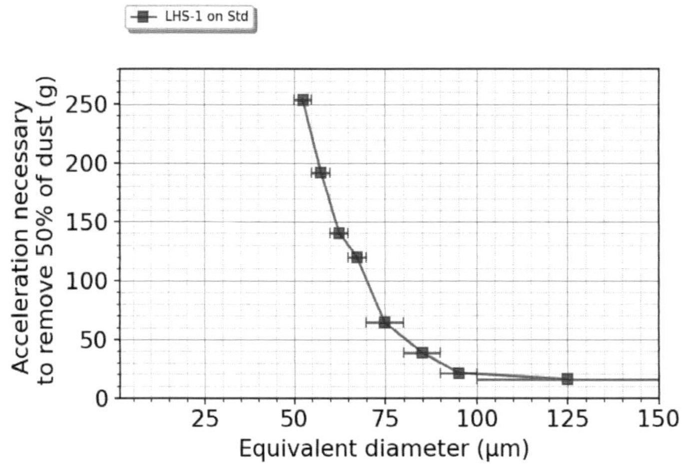

Figure 6. Acceleration necessary (g) to remove 50% of dust particles in LHS-1 simulant on standard (Std) OSRs. Credit: Onera

The errors in y (not plotted) are from errors in rotation speed and the size of the centrifuge, which are negligible. The horizontal lines plotted correspond to bin size. Errors in x (not plotted) are those from the particle size (each particle could have considered to have an approximate error in diameter of ±2 μm due to the 1 μm resolution limit) which determines which bin they are sorted in to.

The value of acceleration necessary is not as useful as the value of average force would be. In order to obtain a value for the force required to remove 50% of particles, acceleration must be multiplied by the mass of the particles (equation 1). Considering only 2D imaging data of the particles was obtained assumptions must be made to calculate force values. The simplest assumption is to treat each particle as a sphere and calculate its volume thusly (equation 2). The average grain density for the size fraction of the simulants used in this study were found using He-pycnometry (see supplementary information, except for the glass spheres where manufacturer data was used). Volume and density were combined to calculate force values. The resulting force graph is as shown in figure 7.

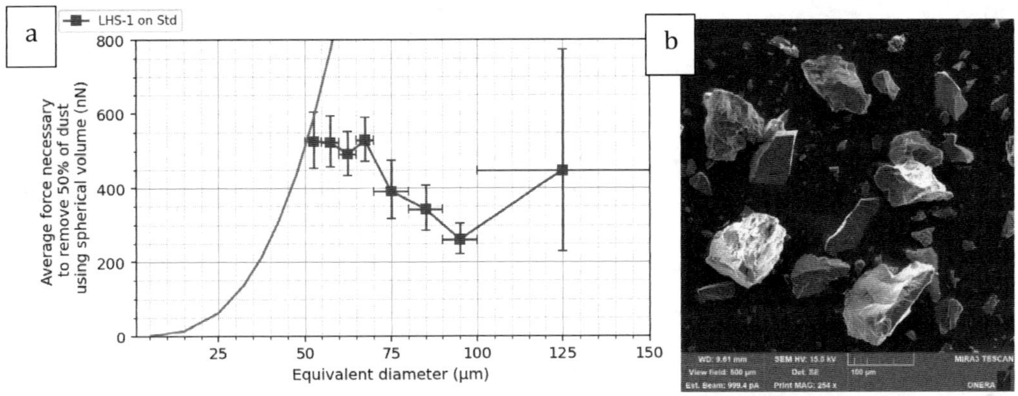

Figure 7. a) Average force necessary (nN) to remove 50% of dust particles in LHS-1 simulant test on standard uncoated OSR samples. Vertical "error bars" were calculated by assuming that the particle had the maximum or minimum size of the bin, as opposed to its centre. The light blue guideline represents the maximum force exerted on a sphere of that size based on the set-up, using the equation for centrifugal force. No data points above this line would be obtained. b) SEM image of the LHS-1 simulant. Credit: Onera

However, as illustrated in the above scanning electron microscopy (SEM) image, LDSs are not constituted of mainly spherical particles. This introduces a large error in the force calculation. Experiments were carried out to "calibrate" the mass information for each simulant whereby deposits were made on small witness samples, which were both weighed and imaged to provide a relationship between coverage and mass values. Using extrapolation, a value was found for the real expected mass of the samples used in the test (too large to be weighed) and all particle spherical volumes were adjusted by a factor which reduced the calculated total mass to match the real values. This factor is called the "volume factor" as it represents the smaller expected volume of each grain. Details of how the factor was calculated can be found in the supplementary information. Volume factors were not identical for each simulant, ranging between ~0.2-0.35 for the <50 µm fractions and 0.15-0.25 for the <100 µm fractions. This greatly reduces the calculated force. The equation for force calculation becomes the following:

$$F_{C\_Non-spherical\ particle} = F_{C\_Spherical\ particle} \times K \qquad \textbf{\textit{Equation 3}}$$

where K is the volume factor reducing particle volume to a fraction of that of a sphere.

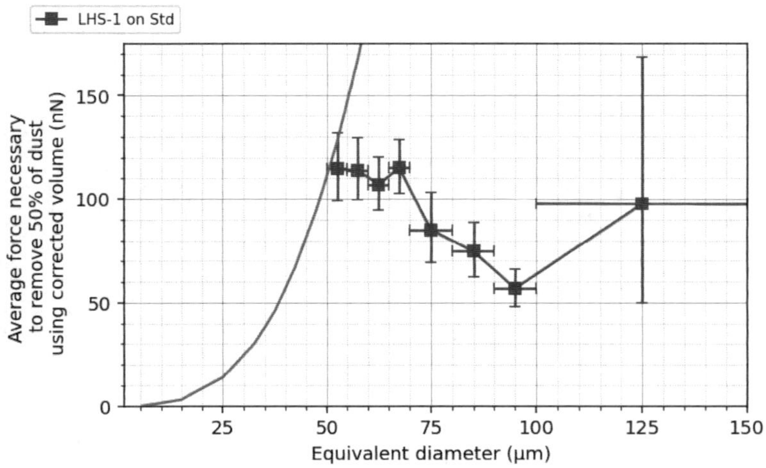

Figure 8. Average force necessary (nN) to remove 50% of dust particles in LHS-1 simulant test on Standard uncoated OSR samples using a volume correction factor of 0.22. Credit: Onera

3. Results

Glass bead detachment was not readily achieved with the forces reachable in the current set-up. Some data could be extracted by analysing the acceleration/force necessary to remove only 10% of the particles. It was qualitatively observed that adhesion seemed lower on ITO-coated substrates than on Standard uncoated OSR substrates, as higher percentages of dust removal were achieved. These results are not provided here. The results for the other tests are shown in the graphs below in figure 9 to figure 11.

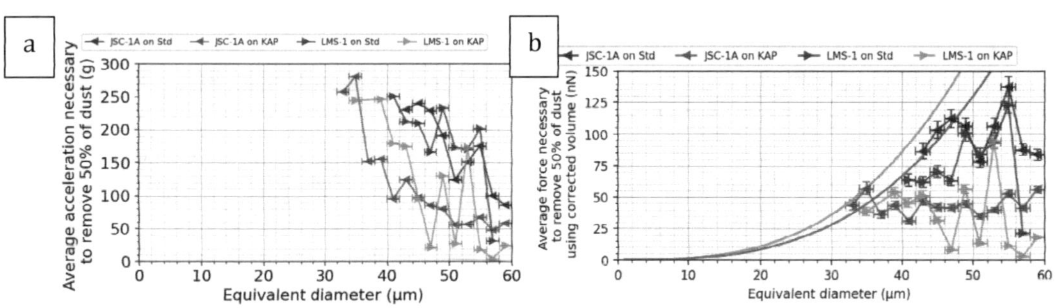

Figure 9. for LMS-1 and JSC-1A simulant on Standard OSRs and Kapton film in absence of VUV radiation, a) acceleration necessary to remove 50% of dust in particle diameter bin, b) force necessary to remove 50% of dust in particle diameter bin. LMS-1 has rightward triangle markers, JSC-1A leftward triangle markers, Kapton substrate: brown colour, Standard OSR: purple colour. The bars in x are the size of the bin and those in y represent the calculated values of force using the diameter from the edge of the bins. Credit: Onera

IOP Conf. Series: Materials Science and Engineering **1328** (2025) 012007 doi:10.1088/1757-899X/1328/1/012007

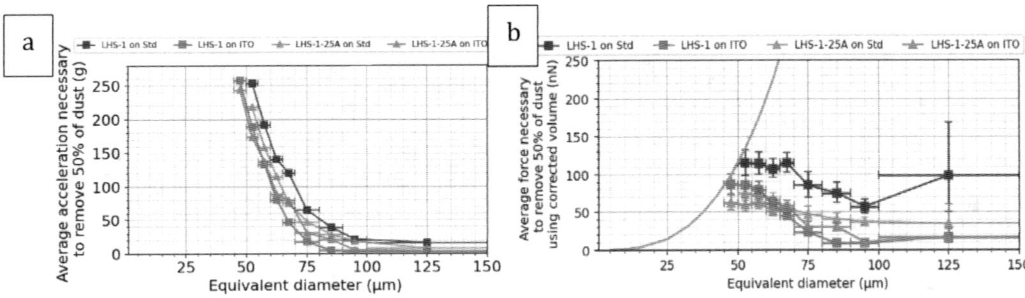

Figure 10. for LHS-1 and LHS-1-25A simulant on ITO and Standard OSRs in absence of VUV radiation, a) acceleration necessary to remove 50% of dust in particle diameter bin, b) force necessary to remove 50% of dust in particle diameter bin. LHS-1 has square markers, LHS-1-25A triangle markers, ITO substrate: red/pink colour, Standard OSR: blue colour. The bars in x are the size of the bin and those in y represent the calculated values of force using the diameter from the edge of the bins. Credit: Onera

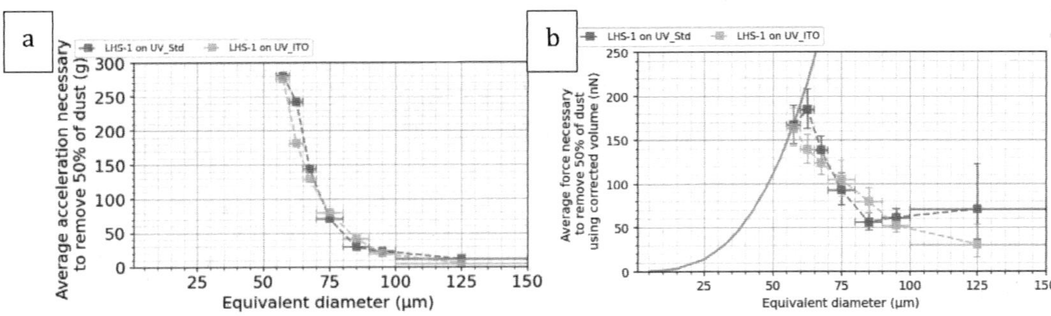

Figure 11. for LHS-1 simulant on ITO and Standard OSRs in presence of VUV radiation, a) acceleration necessary to remove 50% of dust in particle diameter bin, b) force necessary to remove 50% of dust in particle diameter bin. LHS-1 has square markers ITO substrate: yellow colour, Standard substrate: green colour. The bars in x are the size of the bin and those in y represent the calculated values of force using the diameter from the edge of the bins. Two out of the three ITO samples had to be cleaned and reused. Credit: Onera

There is a lower level of confidence in the 100-150 μm particle size bin of the main test campaign, so it is more appropriate to exclude them from analysis. Figure 12 shows an overview of all results plotted on the same graph, from 25 μm up to 100 μm.

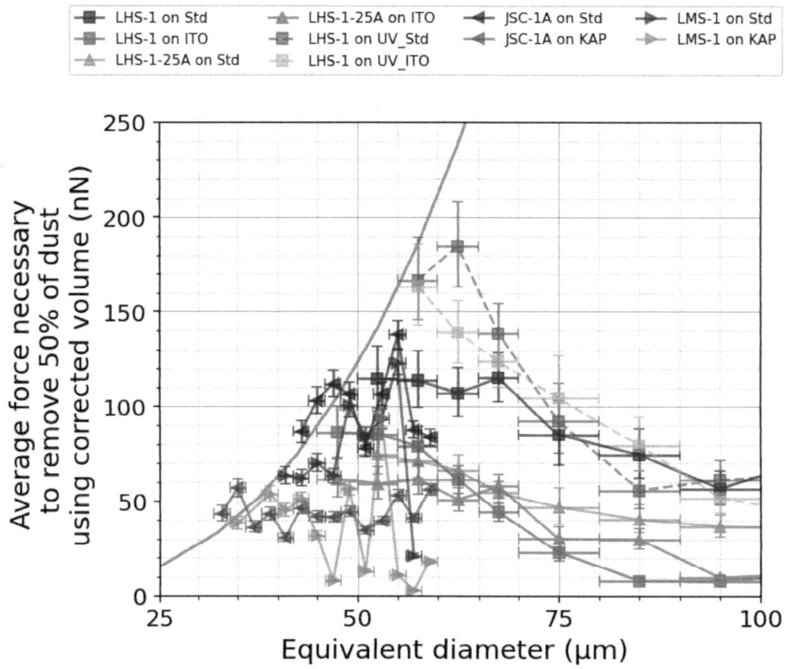

Figure 12. Average force necessary to remove 50% of dust (nN) for all centrifuge tests using the same legends as above particle data. Plotted from 25 - 100 μm equivalent diameter. After volume correction applied. The orange guideline of maximum force experienced was calculated using an average density value and average volume correction for all simulants. Credit: Onera

Adhesion forces were lowest on Kapton substrates. The other substrates are harder to differentiate. The least reliable data is from the pre-tests where low numbers of particles are present in each bin. We will leave them aside in our analysis except to point out that in the tests directly comparing Kapton and Standard OSRs, Kapton gave qualitatively lower adhesion (figure 9). There is also overlap between Standard OSRs and ITO OSRs in the summary figure above. However, in the separate tests using LHS-1 and LHS-1-25A, (both shown in figure 10) the ITO adhesion was lower each time.

With VUV in figure 11: the adhesion on ITO for LHS-1 increased to reach roughly the same as that of the Std OSR sample, which did not show as big of a change compared to the non-VUV measurement.

4. Discussion

4.1 Substrate charge

After centrifuge tests were completed, Kelvin probe measurements were made in the chamber on representative samples in order to measure the macroscopic charge present on the substrates in presence and absence of VUV. The Kelvin probe was set up to face one of the sample loading positions of the wheel, which was rotated to measure the different samples including an empty aluminium sample holder which was used to set the 0 V baseline. When no radiation was applied the dielectric samples (Kapton, Standard OSRs) were found to accumulate varying levels of positive charge (~5-15 and ~1-10 V respectively), with ITO samples (grounded) maintaining a level close to 0 V. An increase in the potential measured on dielectric samples was found to be due to samples' exposure to the ion flow from the ionising pressure gauge at every rotation cycle. In the presence of VUV radiation, the potentials of dielectric samples were measured much closer to zero volts. Slightly positive voltage (≤1 V) was measured in the dielectrics recently exposed to the ionising gauge, and slightly negative voltage (≥-1 V) was measured where the samples were rotated in the opposite direction and approached the Kelvin probe without passing in front of the pressure gauge (likely due to a reabsorption of emitted electrons from the chamber walls). The average ITO sample charge was still close to zero.

4.2 Impact of materials on adhesion

The difference of adhesion between JSC-1A and LMS-1 simulants on the same substrates cannot be compared using this data due to the high standard deviation (figure 9). For both simulants however, Kapton substrates had lower adhesion values. Kapton has higher surface roughness compared to the OSR samples, which could be responsible. Charge levels on Kapton were measured as being similar to standard OSR samples, possibly more positive depending on the specific samples. No value was found in the literature for the Hamaker constant of Kapton. Any combination of these factors could contribute to the lower adhesion measured.

The decision to use LHS-1 and LHS-1-25A under non-VUV conditions was for the purpose of comparing the adhesion of those two simulants on the same substrates. A difference was more noticeable in one case; deposit on standard substrate, and less clearly seen for the ITO substrate (figure 10). They had similar PSDs when characterised as powders, but slightly different levels of deposition were achieved which could be a source of error; influencing imaging and hence particle loss measurements. A difference in adhesion might be possible considering variation in size and shape distribution of the simulants, as well as the difference in their mineral composition but this hypothesis could not be supported without more data points.

4.3 Impact of VUV on adhesion

In order to better compare the difference in adhesion of LHS-1 on ITO and Standard samples, the impact of VUV radiation can be used to help separate the contributions of electrostatic adhesion and van der Waals adhesion (dependence on surface roughness and Hamaker constant). The graph of only LHS-1 adhesion is provided for clarity in figure 13 below. We have again ignored the largest bin; 100-150 μm.

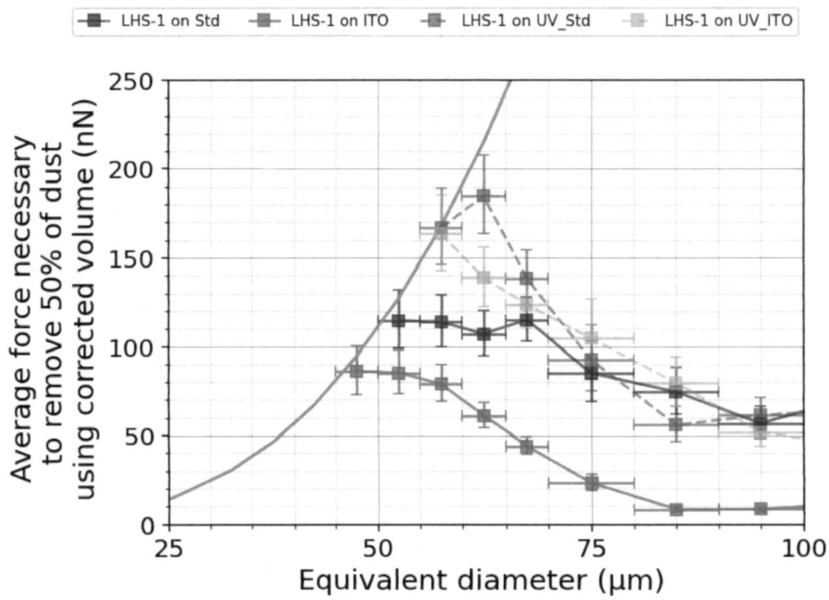

Figure 13. Average force necessary to remove 50% of dust (nN) for all LHS-1 centrifuge tests using same legends as above using aggregated particle data. Plotted from 25 - 100 μm equivalent diameter. After volume correction applied. The orange guideline of maximum force experienced was calculated using an average density value for LHS-1 and the volume correction factor. Credit: Onera

In the absence of VUV the standard substrates were positively charged, whereas the ITO substrates were not. (There is no information available on particle charge.) Adhesion was higher on the standard samples. This could be primarily due to the substrate charge accumulation, the lower surface roughness, or the compositional differences. The relationship of composition with adhesion is understood using the Hamaker constant of a material. No values could be found for the Hamaker constant of the CMX coverglass. A value which might be similar is that of amorphous SiO_2 (14): 6.4 x 10^{-20} J, which is lower than the value cited for ITO (15): 15.1 x 10^{-20} J. Considering the short range of action of van der Waals adhesion forces, the Hamaker constant of the 12 nm ITO layer would be expected to dominate over the glass characteristics for the ITO sample(16), and in theory would favour higher adhesion.

In the presence of VUV charge was cancelled out on the standard substrate surfaces but it is unclear whether the adhesion on the standard samples shifts or not with VUV (there is an increase in adhesion for lower particle size but it seems to be a feature only in one of the three samples, see *supplemental information*). Adhesion increased on the ITO samples to become more similar to that of the standard samples, especially from 65 μm on. As there is no change in surface charge for the ITO samples (as they are grounded) then the implication is that the charge of the particles under VUV has changed. This is a possibility, as they have low electrical conductivity and can accumulate charge through photoemission depending on their initial charge state.(4) A difference in charge between the substrate and the particles could lead to electrostatic adhesion forces being present. This could also be the case in the standard substrates as differential charging of the dust might occur versus that on the substrate.

As seen in figure 13, the difference between the substrates was much reduced under VUV, implying that their differing surface composition and roughnesses had a smaller impact than their charge state (or that a combination of charge and roughness factors combined to give similar adhesion). As seen in the literature, understanding the dependencies of the effect of VUV on substrate type is non-trivial due mainly to unknowns for the initial charge states of particles.(7)(8)

Several limitations of the experiment were brought to light throughout the test campaign. Some of these were limitations from our set-up:

- The maximum force imposed did not detach small LDS particles, nor most glass bead particles. The speed of centrifugation (or size of wheel) needs to be increased to be able to reach the important smaller size range.
- Limitations were present in that environment not perfectly simulated (cleanliness, vacuum level) which can impact the relative contribution of different adhesion forces. Any improvements that can be made to the fidelity of the environment would improve representativeness of the results.
- The ionizing pressure gauge used caused variation on dielectric substrate charge (now resolved by simply moving the gauge to behind a bend). For each test there were variations at each speed of the charge on each Kapton or standard OSR sample which must have contributed to error. The supplementary information shows the variation between adhesion data for ITO samples is much lower. This contributes to our hypothesis of the importance of the charge state on adhesion.
- There was no information gathered on particle charging during the tests, and tribocharging from deposition was not controlled. It may be dissipated in the ambient environment but this is unconfirmed. This means that the identification and contribution of electrostatic forces is made more difficult as hypotheses have to be made using only the substrate information and the adhesion behaviour in the VUV vs no VUV case.

Some limitations were inherent to the experiment design:

- 2D Microscopy limits; having no 3D data creates huge uncertainty in force value. There are general limitations to the microscopy including the difficulty in separating touching particles of irregular shape.
- The time-intensive procedure meant that few test runs / samples are available for each test. Outlier results were difficult to properly explain or exclude.
- The comparisons between substrates were only consistent for specific test runs with one simulant under certain conditions. This limits the usefulness of the data.

5. Conclusions

The main conclusions were the following:

- Forces were on the order of 10-200 nN between ~40-100 μm once a volume factor correction was brought into play. The set-up is currently limited to this space of force and size.
- Large errors in force calculation for our set-up come most importantly from volume/mass approximation. The lack of 3D information or better correction is quite limiting.

- There is a need to better control test conditions (including particle charging) in order to interpret our adhesion results with respect to lunar conditions and expected real adhesion values.

6. Outlook

Several improvements have been identified for the centrifuge system. These will be investigated in future tests at ONERA. The main modifications to be put in place are:

- Better control of charging environment – already achievable by modification of the position of the pressure gauge.
- Better measurement of particle charge state using electrostatic dust lofting techniques(17) and better understanding of these phenomena through the adaptation of numerical models developed for dust beds (18) to the present situation of dust monolayers.
- Reach of a greater range of particle sizes e.g. by using a modified wheel geometry that allows for larger diameter of rotation and hence higher centrifugal force
- Better characterisation of deposits could be achieved. Ideally in situ microscopy could be used to also limit risks from sample handling (although likely loss in resolution). Alternatively, the obtention of 3D particle information would also improve force values obtained. In the absence of a change in imaging set-up, volume correction could at least be achieved through obtaining real shape information at different sizes, or repeating the calibration mass measurements for sieved dust of narrow size ranges.

7. Acknowledgements

The authors would like to thank several people for their contributions. The team at ONERA Châtillon: Aurelie Jankowiak, Katia Dennis and Quentin Barres for the SEM images of the particles and the particle average grain density measurements and Yuriy Butenko at ESA for AFM support. From the ASNR (previously IRSN) thanks to Samuel Peillon for advice on how to deposit particles and characterise adhesion and Cécile Bodiot for calibration of the dust deposit chamber, and from ONERA Toulouse Gaël Murat for the set-up of the centrifuge system. The authors acknowledge support from the European Space Agency in the form of Co-Sponsored Research Agreement No. 4000142578/23/NL/MGu/nh.

References

[1] Gaier J R, 2021 The impact of dust on lunar surface equipment during Apollo *The Impact Of Lunar Dust on Human Exploration* ed JS Levine chapter 4 (Newcastle upon Tyne: Cambridge Scholars Publishing)

[2] Heiken G, Vaniman D, French B M, 1991 *Lunar sourcebook: A user's guide to the Moon.* (Cambridge: Cambridge University Press) p 47-56

[3] Walton O R, 2007 *Adhesion of Lunar Dust* NASA report NASA/CR-2007-214685

[4] Sickafoose A A, Colwell J, Horányi M, Robertson S, Photoelectric charging of dust particles in vacuum. *Phys Rev Lett.* 2000 **84** 6034–7

[5] Gaier J, and Berkebile S 2012 March. Implication of adhesion studies for dust mitigation on thermal control surfaces In *50th AIAA Aerospace Sciences Meeting including the New Horizons Forum and Aerospace Exposition* p 875
[6] Rabinovich Y I, Adler J J, Ata A, Singh R K and Moudgil B M, 2000 Adhesion between nanoscale rough surfaces: II. Measurement and comparison with theory. *Journal of colloid and interface science* **232** p17-24
[7] Barker D C, Olivas A, Farr B, Wang X, Buhler C R, Wilson J and Mai J, 2022 Adhesion of lunar simulant dust to materials under simulated lunar environment conditions *Acta Astronautica* **199** p 25-36
[8] Dove A, Devaud G, Wang X, Crowder M, Lawitzke A and Haley C, 2011 Mitigation of lunar dust adhesion by surface modification *Planetary and space science* **59** p 1784-1790.
[9] Khan I K, Lange M Z, Gurumoorthy K, Naumann H N, Hagendorf C V and Bagdahn J 2020 Rotational force test method for determination of particle adhesion—from a simplified model to realistic dusts *Journal of renewable and sustainable energy*, **12** 043503
[10] Jardine K, Dove A and Tetard L 2022 AFM force measurements to explore grain contacts with relevance for planetary materials *The Planetary Science Journal*, **3** p 273.
[11] Xie H, Wu Q, Liu Y, Xie Y, Gao M and Li C 2023 Direct measurement and theoretical prediction model of interparticle adhesion force between irregular planetary regolith particles *International Journal of Mining Science and Technology*, **33** p 1425-1436.
[12] Zimon A D 1982 *Adhesion of dust and powder* 2nd edn (New York: Consultants Bureau) p 14
[13] Arana G, González-Díaz M O, Castillo-Atoche A and Avilés F 2023 Improving adhesion between polyimide surface and carbon nanotube arrays for strain-sensing devices *Materials Today Communications* **34** 105008.
[14] Bergström L 1997 Hamaker constants of inorganic materials *Advances in colloid and interface science* **70** p 125-169
[15] Pan Y, Khan N, Lu M and Jeon J 2015. Organic microelectromechanical relays for ultralow-power flexible transparent large-area electronics *IEEE Transactions on Electron Devices* **63** p 832-840
[16] Walton O R 2008. Review of adhesion fundamentals for micron-scale particles *KONA Powder and Particle Journal* **26** p 129-141
[17] Pacaud R, Duzellier S, Matéo-Vélez JC, Murat G, Jankowiak A, Suarez Kahan A, et al. 2025 Testing with Dust at the DROP lab of ONERA: facilities and approach *IOP Conf. Ser.: Mater Sci. Eng. accepted.*
[18] Champlain A, Matéo-Vélez, J C, Roussel, J F, Hess S, Sarrailh P, Murat G, Chardon J P, and Gajan A 2016. Lunar dust simulant charging and transport under UV irradiation in vacuum: Experiments and numerical modeling *Journal of Geophysical Research: Space Physics* **121** pp.103-116.

IOP Conf. Series: Materials Science and Engineering **1328** (2025) 012008 doi:10.1088/1757-899X/1328/1/012008

Overview of the Lunar Gateway External Contamination Environment

Crystal A. Quiroz[a]*, Courtney A. Steagall[b], Brian Tulaba[c], Frederick Lutfy[c], John T. Yim[e], John M. Alred[f], William A. Hoey[f], Maxwell G. Martin[f], Carlos E. Soares[f]

a Oceaneering, JETS Contract, Houston, TX, USA
b NASA, Johnson Space Center, Houston, TX, USA
c Jacobs, JETS Contract, Houston, TX, USA
e NASA, Glenn Research Center, Cleveland, OH, USA
f Jet Propulsion Laboratory, California Institute of Technology, Pasadena, CA, USA

* Email: crystal.a.quiroz@nasa.gov

Abstract. As a part of the Artemis mission, Gateway will be a long duration space station in a near rectilinear halo orbit around the Moon. Gateway will be exposed to a variety of external contamination sources, which can degrade external hardware / surfaces and impact science utilization objectives. Requirements and methodologies addressing material outgassing, chemical and electric thruster plumes, vacuum venting, and visiting vehicle interactions have been developed by the Gateway Induced Environments team to ensure vehicle performance and mission success. The external contamination requirements and integration process are described along with required data deliverables. Integrated external contamination analyses are conducted by the Gateway Induced Environments Team to assess compliance with external contamination requirements. This paper provides a current status of integration activities and analysis results, as well as future plans to improve external contamination characterization.

1. Introduction

International and domestic partners are collaborating with NASA to establish Gateway as a long duration space station in a near rectilinear halo orbit (NRHO) around the Moon. The Power and Propulsion Element (PPE) utilizes chemical and electric propulsion thrusters to transport Gateway to the lunar orbit as well as to perform orbital maintenance maneuvers and provide attitude control. Solar arrays on PPE provide power for the space station. Early habitable utilization is provided by the Habitation and Logistics Outpost (HALO). Attached externally to HALO, the Lunar Link serves as the communication system of Gateway [1]. The Lunar I-Hab provides a secondary habitable volume for astronaut living quarters along with additional docking ports to allow Gateway to accommodate visiting vehicles or multiple configurations [2]. The Lunar View is a habitable volume that will perform refueling operations for PPE and will allow astronauts to view space from one of its many windows [3]. Extravehicular activities (EVA) will be possible with the Crew and Science Airlock module (ALM). The ALM will accommodate multiple aspects of EVA abilities including, but not limited to, EVA tool/suit storage and pre- and post-EVA activities. The Gateway External Robotic System (GERS) will assist in Gateway maintenance, crew during EVA missions, capture visiting vehicles, and other science operations. The Deep Space Logistics module (DSL) provides additional space for transporting cargo and payloads to Gateway. Similar to other visiting vehicles, it operates independently from Gateway with its own power, thermal, propulsion, communication, and navigation systems.

Gateway will have a 15-year mission lifetime and will be constructed over the course of several Artemis missions. Two modules, PPE and HALO, will be launched co-manifested together in a configuration referred to as the Co-Manifested Vehicle (CMV) [4]. Subsequent Artemis missions will launch additional Gateway modules until it reaches an assembly complete configuration. Orion will provide the propulsion system, power, and communication to the co-manifested payload (CPL) during transit to Gateway.

Over the assembly and lifetime of Gateway, the station will be exposed to a variety of external contamination sources, which can degrade external hardware/surfaces and impact science utilization objectives. The Gateway Induced Environments Team has developed requirements and methodologies addressing material outgassing, chemical and electric thruster plumes, vacuum venting, and visiting vehicle interactions to ensure vehicle performance and mission success. A brief introduction of all the Gateway induced environments is provided in Section 2, the current analyses used in the system studies are presented in Section 3, and current external contamination activities is presented in Section 4. Particulate contamination in the form of lunar dust is anticipated on Gateway external surfaces, however, molecular contamination topics are the focus of this work.

2. Gateway Induced Environments

The Gateway Induced Environments include external lunar dust transport, chemical and electric propulsion thruster plumes, materials outgassing, and vacuum venting. These environments are defined in GP 10057 'Gateway Space Induced Environments Requirements' [5] which species:

- Lunar dust contamination due to transfer from of a returning Lunar Surface excursion vehicle,
- Electric propulsion plume induced environments, including sputter erosion and plasma, and
- A molecular contamination limit of 250 Angstroms per year on contamination sensitive surfaces from all sources.

Exposure to these induced environments can impact vehicle performance, utilization, and mission success. For instance, absorptivity degradation of a material is directly related to sun exposure and contaminant deposition [6].

Gateway system-level induced environments are verified by The Gateway Induced Environments Team at NASA Johnson Space Center (JSC). The Gateway Induced Environments Team works closely with hardware owners to gather detailed contamination source and receiver characterization information to support integrated system analyses to confirm that the induced environments for the as-designed vehicle are within the limits in GP 10057. Similarly, Gateway hardware owners must verify hardware compatibility with induced environments defined in GP 10057 and are briefly described here.

2.1 Lunar Dust

Lunar dust will be introduced to the Gateway station as particulate contamination by the Human Landing System (HLS) upon return to Gateway following a lunar surface mission. Spacecraft plume surface interactions eject the lunar dust from the topsoil of the regolith and create plumes of dust that contact external surfaces [7]. Lunar dust adheres to some surfaces due to the electrostatic properties created by the environment. While lunar dust transport on the external surfaces of Gateway is estimated using the Gateway On-Orbit Lunar Dust Map Analysis Program (GOLDMAP) [8], there is some amount of ejecta anticipated by plume surface interactions [9].

On the Gateway station lunar dust is a unique problem, causing both internal and external contamination. Lunar dust external to the spacecraft can impact performance of external components such as radiators or solar arrays. Internal lunar dust can impact crew health and internal systems. GOLDMAP estimates the transport of lunar dust from HLS to Gateway for various scenarios including HLS docking to Gateway. Mitigation efforts, dust transport modeling, and testing is underway to characterize risks to hardware and evaluate candidate solutions [10].

2.2 Thrusters

Gateway thruster operations include orbital maneuvers, attitude control operations, and visiting vehicle thruster firings during rendezvous, docking, and undocking with Gateway. Thruster plume induced contamination and other effects depend on the thruster type, usage, and position/orientation with respect to Gateway surfaces of concern.

PPE has electric propulsion (EP) thrusters that serves as the primary mode of orbit raising during the CMV transit to NRHO. Once it arrives to the lunar orbit, the EP thrusters have the capability to be used for orbit maintenance maneuvers. Use of the PPE electric thrusters results in sputtering/erosion of nearby surfaces, as well as redeposit of sputtered material onto other surfaces. EP thrusters induce a plasma environment which can affect surface charging; however, this is not expected to be significant compared to natural environment charging.

PPE also incorporates chemical (bipropellant) thrusters which will support nominal attitude control operations while Gateway is in NRHO. These bipropellant thrusters impinge directly on the solar arrays. An example hemispherical view of a PPE chemical thruster is presented in Figure 1. Bipropellant thruster plume induced contamination and erosion/pitting due to droplet impacts has been observed on space shuttle flight experiments [11].

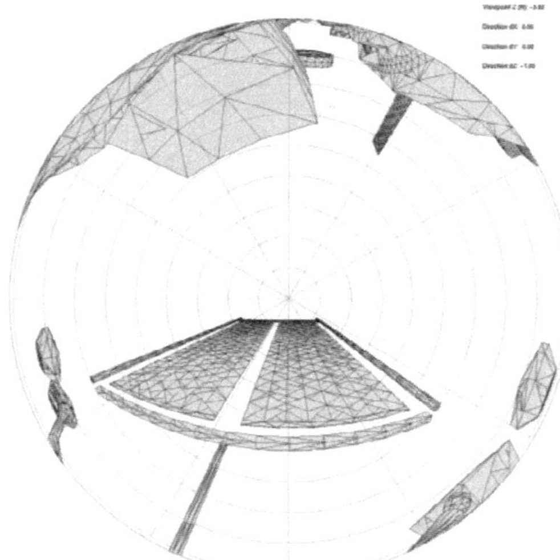

Figure 1 Example of a hemispherical view of a chemical thruster on the PPE module (own work)

2.3 Vacuum venting

Gateway utilizes multiple vacuum vents to release Environmental Control and Life Support Systems (ECLSS), utilization, and other unwanted products overboard. Released products may contact external surfaces and need to be evaluated for contamination and material compatibility. Multiple studies assessed gas and multi-phase vacuum vents for the International Space Station [10].

Most of the vacuum venting on Gateway is expected to involve release of inert gases or water vapor, which will not condense on or react chemically with exterior surfaces. However, venting of liquid/solid phase constituents is a potential concern for contacting and degrading external surfaces. For instance, propellent purge vents which will be performed following fuel transfers could impact or degrade solar array surfaces, as presented in the example hemispherical view in Figure 2. Review and characterization of vent products is crucial in determining the risk of contamination and degradation to exterior surfaces.

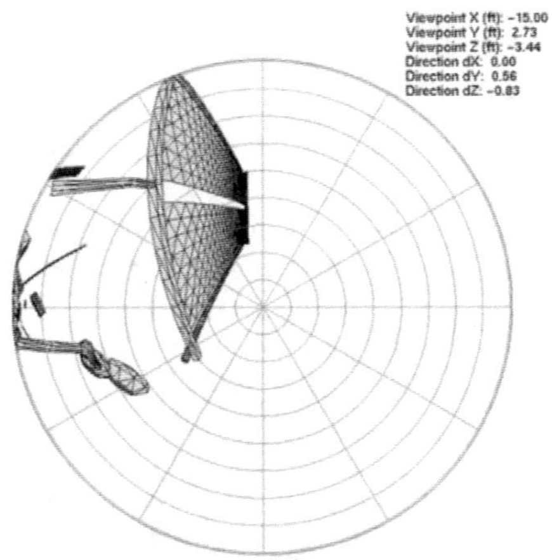

Figure 2 Example of a hemispherical view of a propellent purge vent on the PPE module (own work)

2.4 Materials Outgassing

Non-metallic materials in a vacuum environment release molecules from the microstructure which can be transported to and deposit on other spacecraft surfaces. High view factors from source to receiver surfaces increase the likelihood of contamination from outgassed material. Materials outgassing is the dominant source of external contamination on the International Space Station (ISS) and is modeled as a diffusion-driven process using ASTM E1559 test data [12].

Similar to the ISS, materials outgassing is expected to be the dominant molecular contamination source for Gateway, due to its large size and high view factors. For example, Gateway has body mounted payloads with large view factors to adjacent surfaces. Figure 3 shows an example hemispherical view. Early characterization of Gateway external contamination sources and receivers is essential for successful integration.

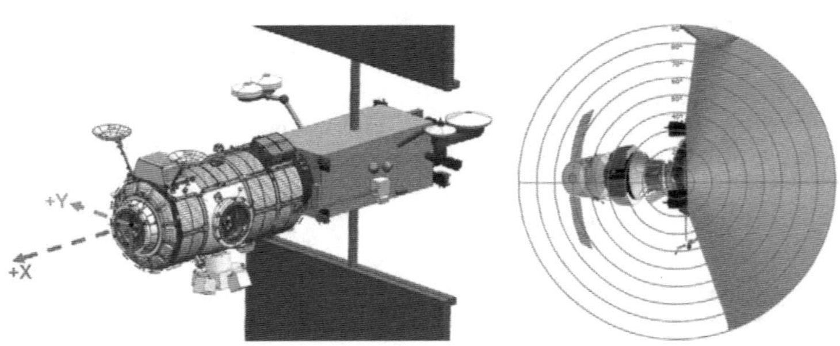

Figure 3 Example of hemispherical view of an adjacent surface of a body mounted payload (own work)

3. Gateway External Contamination System Studies

The Gateway Induced Environments Team, in collaboration with the Contamination Control Team at the NASA Jet Propulsion Laboratory (JPL) and the Gateway Ion Propulsion Team at the NASA Glenn Research Center, has performed an external contamination system assessment based on ISS heritage models and software. These early look Gateway integrated external contamination assessments are developed to inform requirement allocations for elements and interfaces. System studies help identify potential integration issues and risks as well as formulate long term strategy for addressing external contamination for verification purposes. An example assessment is presented here and includes a summation of all induced molecular contamination sources, including materials outgassing, chemical thruster plumes, and redeposited sputter material from electric propulsion thrusters. At this time, contamination due to vacuum venting is mitigated with requirements, so it will be neglected in this example.

Sensitive surfaces for the Gateway station include radiators, windows, solar arrays, and payload sites present on each module. These surfaces are assumed to be sensitive to molecular contamination for early look assessments, however hardware owners have the responsibility of identifying surfaces sensitive to contamination for final analysis and verification purposes. An assembly complete configuration of Gateway is presented in Figure 4.

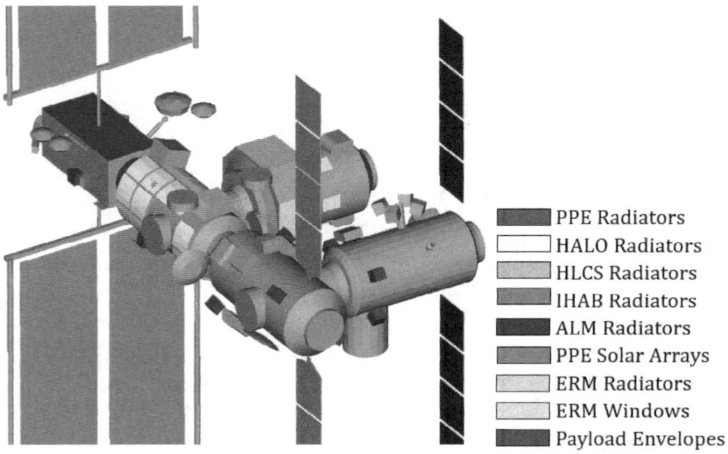

Figure 4 An example assembly complete configuration with sensitive surfaces identified as payload sites (purple), solar arrays (pink), radiators (orange, blue, yellow, and green) and windows (light blue) (own work)

3.1 Materials Outgassing

Gateway uses a predictive model developed by the Contamination Control Team at JPL [13] to estimate the contaminant deposition due to materials outgassing. The tool allows manipulation and scaling of analysis inputs for early look assessments. It incorporates a Gateway view factor matrix to compute molecular transport for outgassing from source to receiver surface combinations. An example view factor visualization is presented in Figure 5. The solar arrays, highlighted in green, represent an outgassing source with direct views to the surrounding receiver surfaces.

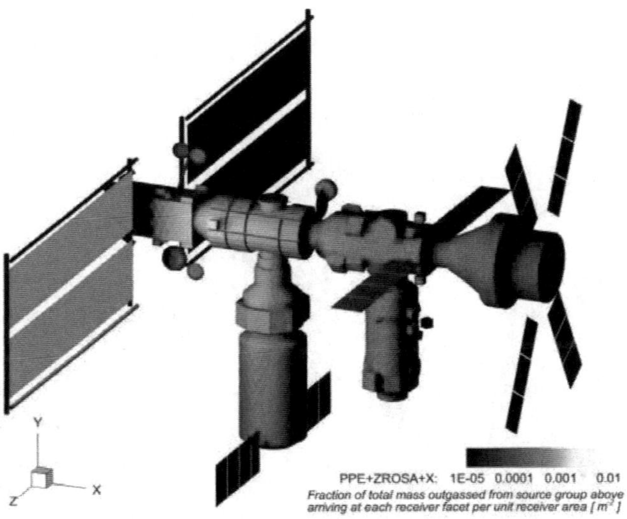

Figure 5 Visualization of Direct Flux from a Single Outgassing Source generated by the Contamination Control team at JPL. Figure by W A Hoey et al 2023 IOP Conf. Ser.: Mater. Sci. Eng. **1287** 012002 and is licensed under Creative Commons Attribution 3.0 licence. No changes were made to the figure.

The tool provides material outgassing deposition results based on current outgassing rate assumptions and mission durations. This allows a time phased arrival of elements and visiting vehicles to simulate partial configurations of Gateway as well as assembly complete configuration. Outgassing rate source terms are based on similarity to heritage materials and ASTM E1559 outgassing rate test data. Contamination due to reflected flux is neglected from this study and only direct flux is considered. A more thorough look into the outgassing model workflow and the methodology used by JPL for this assessment can be found in reference [13].

3.2 Chemical Thruster Plumes

Gateway collaborated with the JPL Contamination Control team on a simplified thruster plume contamination model for conducting parametric assessments for Gateway and Orion thrusters with line of sight to Gateway sensitive surfaces [13]. The thruster plume model is based on heritage ISS bipropellant plume contamination models. Analysis data is provided in tabular format which allows for manipulation of thruster on-time assumptions for various analysis scenarios. An example of a contamination analysis contour plot generated by JPL is presented in Figure 6. The figure shows a thruster that has a direct line of sight to the solar arrays. An example hemispherical view of a similar thruster can be seen in Figure 1. The PPE thrusters are directly impinging on the PPE solar arrays with some views of other surfaces. A more thorough look into the plume effects modeling used by JPL for this assessment can be found in reference [13].

IOP Conf. Series: Materials Science and Engineering **1328** (2025) 012008 doi:10.1088/1757-899X/1328/1/012008

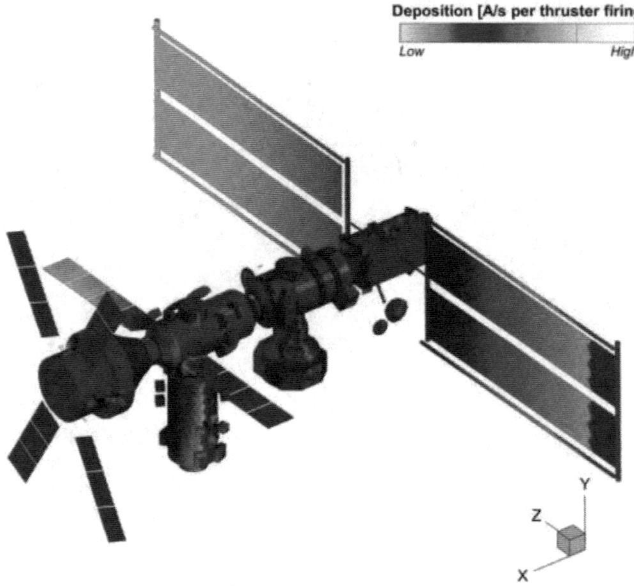

Figure 6 Example plume contamination analysis plot generated by the JPL Contamination Control team (Å/sec per thruster firing). Figure by W A Hoey et al 2023 IOP Conf. Ser.: Mater. Sci. Eng. **1287** 012002 and is licensed under Creative Commons Attribution 3.0 licence. No changes were made to the figure."

3.3 Redeposited Sputter Material

The Gateway ion propulsion system is primarily used for the lunar transit, NRHO orbit maintenance maneuvers, cislunar orbit transfers, and end of life disposal. Sputter erosion from EP plume impingement and redeposit of sputter material is modeled and analyzed by the Gateway Ion Propulsion team at NASA Glenn Research Center (GRC). The ion thrusters will be used heavily during the lunar transit phase, primarily affecting HALO and PPE. Most contamination due to redeposited sputter material will occur during this phase, while two orbit transfers will provide an additional amount. Chemical thrusters instead of ion thrusters will be used during most nominal operations. Figure 7 shows example analysis results performed by the Gateway Ion Propulsion team. An assembly complete configuration similar to Figure 4 is considered. The deposition values of select surfaces are presented for the 15-year mission lifetime.

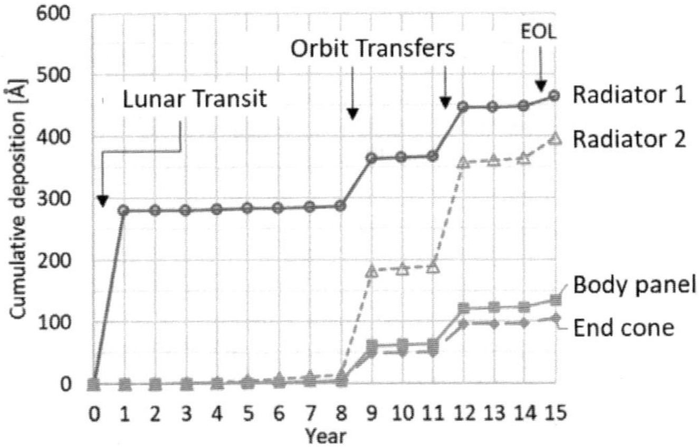

Figure 7 Example EP plume analysis generated by the Gateway Ion Propulsion team. The analysis shows the contamination (Å) over Gateway's 15-year life due to redeposited sputter material (own work).

3.4 System Level Contamination

System level contamination results are generated by accumulating the contamination estimates from sources such as materials outgassing, PPE chemical thruster plumes, and redeposited sputtered material analyses. Only nominal operations are considered for the system level contamination analysis, so contamination due to orbit transfers are neglected. The estimates are compared to the 250 Å/year system limit to assess possible contamination exceedances. A condensed example of the system level results is provided in Table 1.

Table 1 Example system level contamination (Å) results. Results include molecular sources from outgassing, PPE chemical thruster plumes, and redeposited sputter material from PPE ion thrusters

Receiver Surface	Outgassing	Chemical Thruster Plumes	Redeposited Sputter Material	Grand Total
PPE Radiator 1	118	0	0	118
PPE Radiator 2	117	0	0	117
HALO Radiator	254	0	1	255
HLCS Radiator	237	0	0	237
IHAB Radiator 1	191	0	1	191
IHAB Radiator 2	199	0	1	200
ERM Radiator 1	53	0	1	53
ALM Radiator	215	0	1	216
PPE ROSA 1	108	21	0	129
PPE ROSA 2	108	23	0	131
ERM Window 1	102	0	0	102
ERM Window 2	63	0	0	64
PPE Payload 1	350	0	5	355
PPE Payload 2	261	0	3	264
HALO Payload	117	0	2	119
IHAB Payload 1	306	0	0	307
IHAB Payload 2	197	0	0	197
ALM Payload 1	206	0	0	206

The example results show that materials outgassing is the dominant source for molecular contamination. There are localized exceedances, notably on the surfaces that might have the highest view factors such as body mounted payload envelopes. These are areas for further model refinement or detailed analysis. Despite the local exceedances, with the current assumptions incorporated in the analyses, the example results show that the 250 Å/year allocations stated in the requirements are appropriate and achievable.

4. Other External Contamination Activities

In addition to conducting integrated system studies, the Gateway Induced Environments Team is engaged in multiple efforts to improve understanding of the external contamination environment at Gateway and associated effects on hardware. For example, the Gateway Induced Environment Team collaborated with the Gateway Passive Thermal Control System Team and the JPL Contamination Control Team to develop a tool to estimate absorptivity changes due to molecular contamination on Gateway thermal control surfaces [14] [15]. The tool is based on an empirically derived optical property model that considers the equivalent sun hours, contaminate deposition on receiver surfaces, and initial solar absorption of the material [6]. This tool supports end-of-life heat rejection assessments.

In collaboration with JPL, ESA, and the Gateway Induced Environments Team, an effort is underway to evaluate the equivalency between isothermal testing and stepped temperature testing to determine outgassing rate data for materials. JPL developed a multi-species model [16] that could bridge the gap between these two processes, which could mean a significant process improvement as well as reduced testing costs to determine outgassing rate data. Testing efforts are underway at JPL to validate the multi-species model with outgassing rate test data.

Another joint effort between ESA, JPL, and the Gateway Induced Environments Team is a bipropellant thruster plume characterization test [17]. The test was performed in the High-Vacuum Plume Test Facility for Chemical Thrusters at Deutsches Zentrum für Luft-und Raumfahrt (DLR) in Göttingen, Germany. Thruster plume induced degradation of select spacecraft materials is considered, including MLI, Kapton, and solar cell coverglass [18]. Sample analysis is still underway. The goal of the test is to improve modeling of spatial distribution of plume effluents for structural, thermal, and induced environments disciplines to support modeling and analysis of contamination and erosion effects on solar arrays and other Gateway surfaces.

5. Conclusion

The Gateway Induced Environments team defines the induced space environment, conducts integrated induced environments analyses, and provides technical and integration support for the Gateway system. Requirements and methodologies for managing induced environments have been established for managing induced environments within acceptable system limits. Forward work remains to refine current understanding of the induced environment, develop, and validate models, and finalize integration and verification workflow with hardware providers and international partners.

Acknowledgements

A portion of this research was carried out at the Jet Propulsion Laboratory, California Institute of Technology, under a contract with the National Aeronautics and Space Administration (80NM0018D0004).

IOP Conf. Series: Materials Science and Engineering **1328** (2025) 012008 doi:10.1088/1757-899X/1328/1/012008

References

[1] European Space Agency (ESA). Gateway: Lunar Link 2025 [Online]. Available: https://www.esa.int/Science_Exploration/Human_and_Robotic_Exploration/Gateway_Lunar_Link.

[2] European Space Agency (ESA). Gateway: Lunar I-Hab 2025 [Online]. Available: https://www.esa.int/Science_Exploration/Human_and_Robotic_Exploration/Gateway_Lunar_I-Hab.

[3] European Space Agency (ESA). Gateway: Lunar View 2025 [Online]. Available: https://www.esa.int/Science_Exploration/Human_and_Robotic_Exploration/Gateway_Lunar_View.

[4] National Aeronautics and Space Administration (NASA). NASA's Lunar Exploration Program Overview 2020 [Online]. Available: https://www.nasa.gov/wp-content/uploads/2020/12/artemis_plan-20200921.pdf

[5] National Aeronautics and Space Administration (NASA) 2025 GP 10057 Space Induced Environments Requirements

[6] Alred J W and Soares C 2001 Solar absorptivity as a function of spacecraft external contamination *Int. SAMPE Tech. Conf.* **33** 1581-1584

[7] Heiken G,Vaniman D and French B 1991 *Lunar sourcebook: A user's guide to the Moon* Cup Archive

[8] Lee RG, Worthy ES, Willis EM, Brown GL, Cipriani F, Barker DC 2023 Development of a comprehensive physics-based model for study of NASA gateway lunar dust contamination *Acta Astronautica* **210** 616-626

[9] Batcheldor D, Mantovani J, and Wittal M 2021 Lunar regolith trajectories as a result of plume surface interactions *Lunar Surface Science Workshop*

[10] Lee R, et al 2024 Status of Gateway Lunar Dust Modeling and the Study of Dust During Docking *16th Int. Sym. on Mat. in the Space Env.*

[11] Soares C, Barsamian H and Rauer S 2002 Thruster Plume Induced Contamination Measurements From the PIC and SPIFEX Flight Experiments *Proc. SPIE* **4774** 200

[12] Huang A, et al 2016 Materials Outgassing Rate Decay in Vacuum at Isothermal Conditions. *Proc. of SPIE* **9952** 995206-1

[13] W A Hoey et al 2023 *IOP Conf. Ser.: Mater. Sci. Eng.* **1287** 012002

[14] Zinecker A, et al 2023 *IOP Conf. Ser.: Mater. Sci. Eng.* **1287** 012009

[15] Hoffman B and Zinecker A 2024 Estimation of End of Life Solar Absorptivity for Complex, Multi-Phased Space Missions *Proc. of the 16th Int. Sym. on Mat. in the Space Env.*

[16] Alred J M, et al 2022 Development of Multispecies Model for Comprehensive Characterization of Outgassing Molecular Constituents *Proc. of the 15th Int. Sym. on Mat. in the Space Env.*

[17] Soares C 2024 The EPIC-2 Bipropellant Plume-Induced Contamination Test Program and Application to Europa Clipper, Lunar Gateway and the International Space Station Programs *Proc. of the 16th Int. Sym. on Mat. in the Space Env.*

[18] Steagall C, et al 2024 The Effect of Impingement Angle on Bipropellant Thruster Plume Degradation of Spacecraft Materials *Proc. of the 16th Int. Sym. on Mat. in the Space Env.*

IOP Conf. Series: Materials Science and Engineering **1328** (2025) 012009 doi:10.1088/1757-899X/1328/1/012009

Estimation of End-of-Life Solar Absorptivity for Complex, Multi-Phased Space Missions

B Hoffmann[1*] and A Z Howard[2]**

[1] Jacobs, 2224 Bay Area Blvd, Houston, TX 77058 (USA)
[2] NASA Johnson Space Center, 2101 E NASA Pkwy, Houston, TX 77058 (USA)

[*]E-mail: brandon.m.hoffmann@nasa.gov
[**]E-mail: abigail.a.zinecker@nasa.gov

Abstract. Solar absorptivity degradation over time is an important contributor to the thermal performance of a spacecraft at the end of its life. When environments are complex and include many different factors known to impact solar absorptivity degradation, such as UV, charged particles, contamination, or lunar dust coverage, combining data from several different studies can be done to estimate total degradation. While the most conservative way is to stack delta absorptivity from each factor to come up with a total delta, this may be overly conservative for long or complex missions with multiple mission phases. As absorptivity increases the rate of degradation decreases, and synergistic effects need to be taken into account. In this report an alternate method is proposed using Gateway materials as an example. This method determines which environmental factor will be dominant at what time (such as contamination or charged particles) and uses that degradation rate for that mission phase. For the next mission phase, the new absorptivity can be used instead of the original BOL absorptivity to determine degradation rate. This approach prevents overestimating optical degradation when considering degradation from interconnected sources.

1. Introduction

As humans push the bounds of exploration, vehicles encounter new environments and longer durations of travel which our spacecraft materials must withstand. Thermal engineers require End-of-Life (EOL) solar absorptivity estimates to predict thermal performance of the vehicle, but testing to predict material performance degradation cannot always be done in the realistic environment and over the needed lifetimes due to cost, schedule, and the availability of test facilities which can replicate multiple different environmental factors simultaneously. Summing up delta absorptivity from various factors can sometimes be used for End of Life (EOL) predictions, but for long duration missions with many different sources of degradation this will lead to overly conservative estimates at best or estimates greater than 1, which is physically impossible, at worst. This results in overconservative design or inability to meet heat rejection requirements. Degradation of optical properties are extremely difficult to verify until the vehicle is already in flight. On projects that have extremely complex environments and tight heat rejection margins, better estimations of coating optical property degradation are needed. A new method has been developed for Gateway, which has a unique and challenging environment, to attempt to synthesize material degradation from multiple different tests and analyses without being overly conservative.

2. Background

Gateway is a deep space station that will orbit the Moon for 15 years and provide support for a sustainable, long-term presence on the lunar surface. Similar to the International Space Station (ISS), Gateway will be assembled piecewise in orbit, and most modules are launching at different times with different transit orbits to reach the destination. Figure 1 shows how the station will be assembled, starting from the first launch of the Co-Manifested Vehicle (CMV), which includes the first two modules, the Power and Propulsion Element (PPE) and the Habitation and Logistics Outpost (HALO). International Habitat (I-HAB), ESPRIT Refuelling Module (ERM), and the airlock will launch separately on Artemis IV, V, and VI. Orion, Logistics Modules, and Human Lander System (HLS) will visit Gateway multiple times over the lifetime in Near Rectilinear Halo Orbit (NRHO). Thus far, only the 55-pound CubeSat, Cislunar Autonomous Positioning System Technology Operation and Navigation Experiment (CAPSTONE), has flown in NRHO before, so this environment is new for large spacecrafts like Gateway.

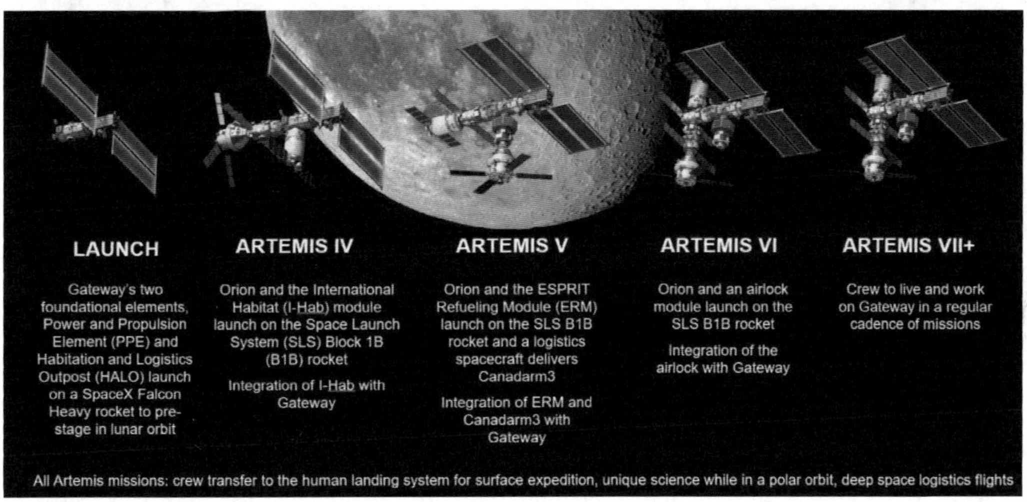

Figure 1. Gateway will be assembled in NRHO across different Artemis missions [1] credits: NASA, reproduced with permission.

Both transit and NRHO environments will degrade Gateway surfaces over exposure time. Degradation of surfaces is expected to be caused by exposure to natural environments such as ultraviolet (UV) radiation, charged particles from solar wind and inside the Van Allen Belts, vacuum, and micrometeroids as well as induced environments such as molecular outgassing contamination, Electric Propulsion (EP) plume sputtering from the PPE thrusters, crew interaction and ground handling, and even from lunar dust contamination carried back from the lander which will dock to Gateway.

Since no data on optical property degration from CAPSTONE is available, there is currently no flight data for degradation of optical properties in NRHO for comparison. The transit environments complicate matters further. The first two modules, PPE and Habitation and HALO will be launched together and will be propelled to NRHO in an electric orbit raising spiral shown in Figure 2, taking over a year. On this journey, the vehicle will pass through the Van Allen Belts multiple times, being bombarded with high and low energy particles. Over 95.5% of the charged particle dosage received by CMV surfaces will be due to this transit. Because of these new and complicated environments, much of Gateway's estimation for degradation of optical properties must then come from analysis and ground testing of materials under conditions from both transit and NRHO environments. The Gateway team performed several different analyses and tests to evaluate all of these different environmental factors which are described in the Methodology Section. The Gateway Passive Thermal team integrated results from these analyses and tests to

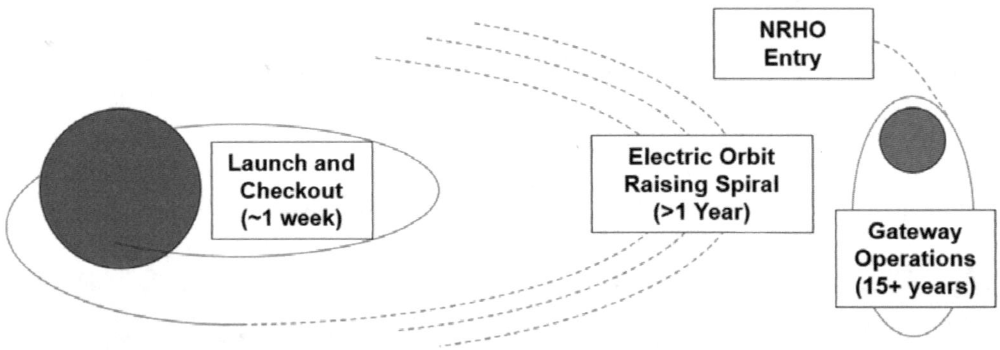

Figure 2. The CMV will spend over a year in transit to NRHO due to the low thrust engines employed by PPE, spiralling out from Earth and passing in and out of the Van Allen Belts multiple times [2]. credits: NASA, reproduced with permission.

develop predictions of EOL properties which can be used in thermal models.

3. Methodology

Many different analyses and tests were carried about by the larger Gateway team to enable estimation of EOL optical properties. This includes an Equivalent Solar Hour (ESH) Analysis, UV/Charged Particle Testing, Molecular Contamination Assessment, EP Plume Sputtering Testing and Analysis, the Gateway On-orbit Lunar Dust Modelling and Analysis Program (GOLDMAP), and finally an integrated optical property estimation model.

The first step in estimating EOL properties is to determine solar exposure, measured in ESH. One ESH is defined as the radiation experienced by a surface in 1 hour at 1 AU away from the sun. For Gateway, this was done with the Equivalent Solar Hour (ESH) Analysis performed in Thermal Desktop, which is described in detail in Spivey, Zinecker, and Smith [3]. The ESH estimates include both direct incident solar and indirect solar as shown in Equation 1, meaning surfaces can reflect sunlight onto other Gateway surfaces. This analysis took the phased arrival of modules and visiting vehicles into account and assumed that the attitude was kept in Solar Pressure Equilibrium Attitude (SPEA) based on the configuration being analysed. Eclipse durations were neglected. Figure 3 shows the total ESH for Gateway surfaces including transit and NRHO.

$$ESH = \frac{Duration}{Solar\ Flux}\left[Q''_{Sol\ Direct} + Q''_{Sol\ Indirect}\right] \tag{1}$$

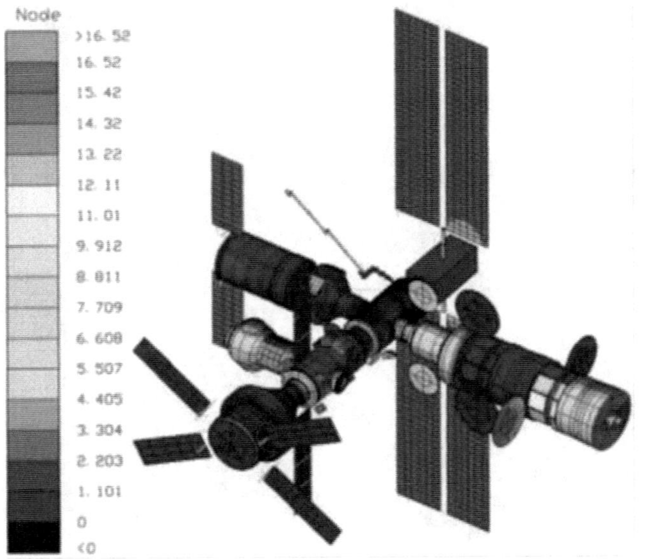

Figure 3. Gateway ESH estimates are shown after the 15-year mission (scale is in years) [3]. credits: NASA, reproduced with permission.

After ESH was calculated for both transit and NRHO portions of the mission, degradation from UV could be estimated. UV/Charged Particle Testing was performed on over a dozen proposed Gateway materials at Goddard Space Flight Center as described in Zinecker, Spivey, Jayne, & Hasegawa [4]. It was decided that test surfaces would be exposed to UV and expected Gateway charged particles simultaneously in order to capture synergistic effects and to avoid overconservatism when combining degradation from multiple sources. It was unnecessary to test for the full expected ESH duration as some surfaces expect to see multiple years worth of solar exposure and degradation from UV decreases as exposure time increases. At some point UV degradation begins to level off and it becomes easier to extrapolate to higher ESH values. UV exposure in ground testing can be achieved at a higher rate than 1 ESH per hour due to high intensity lamps, but this test was limited to 2.6 ESH per hour, or 2.6 suns. 5000 ESH was targeted in these tests in order to ensure greater extrapolation capabilities. As discussed in the Introduction, the vast majority of charged particle exposure occurs during the transit period for the first two Gateway modules. The charged particle exposure was based on Total Ionizing Dose (TID) from the CMV environmental bounding mission through the Van Allen Belts. Charged particles were applied in two short burtsts over the testing duration due to ion and electron beam capabilities. The charged particles were applied on the order of several hours which means that while UV was still being applied, it was a relatively low amount. This allows for separate calculation of degradation caused by UV and charged particles without compromising on synergystic effecs, which is helpful for other Gateway modules which will have lower TID exposure levels. Four total rounds of testing were completed, each with 12 samples, 8 of which were able to be measured in-vacuum. Round 1 samples arranged in the sample plate are shown before and after testing in Figure 4 below.

Next, NASA Gateway Environments worked with Jet Propulsion Laboratory (JPL) Contamination Control Team to build a Gateway Contamination Model, described in Hoey et al. [5]. The molecular contamination model predicted contaminant thickness on critical Gateway

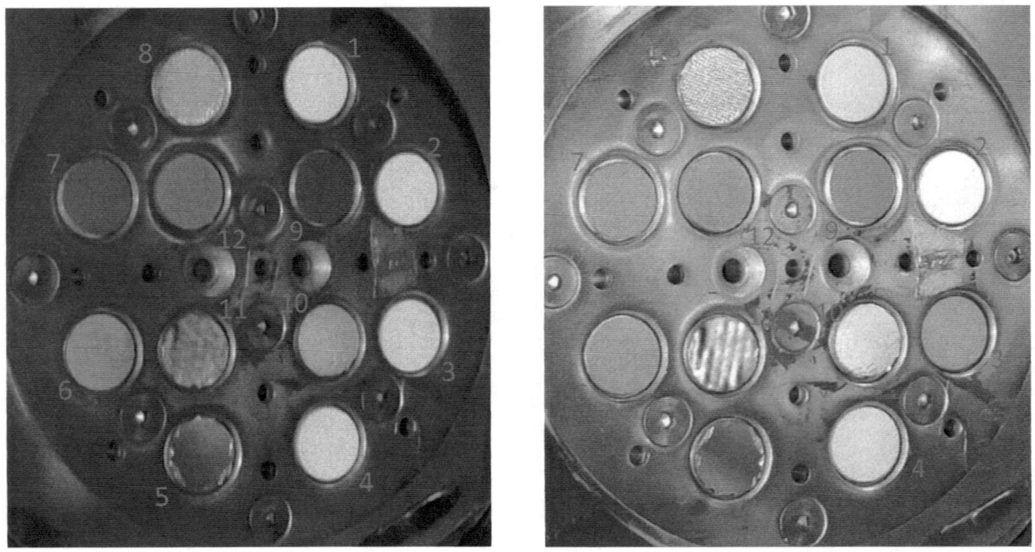

Figure 4. Images of the sample test plate from Round 1 of UV/Charged Particle Testing at GSFC Solar Wind Test Facility before and after environmental exposure [4]. credits: NASA, reproduced with permission.

surfaces based on Gateway geometry and materials using known outgassing rates, expected surface temperatures, and ESH. Self-contamination and contamination from other modules was included in this analysis. View factors from each surface to every other surface were calculated and outgassing rates were applied to those values to determied flux from one surface to another. Figure 5 shows an example of mass of contaminant outgassed onto Gateway from a single source, in this case, one of the PPE solar arrays. Total contamination is then the sumation of contaminants from each of the sources. After contaminant deposition thicknesses were calculated, optical property degradation could be assessed using an ISS heritage model shown in Equation 2 [6]. The model estimates solar absorptivity as a function of ESH and contaminant thickness where $\alpha(t)$ is the solar absorption at time t, α_0 is the BOL solar absorption, $K_{mixture}$ is a semi-emperical parameter specific to contaminant type and the substrate material, ESH(t) is ESH at time t, and d(t) is the contaminant thickness at time t. The model parameters assumed similarity to ISS materials with a contaminant mixture of 50% silicone and 50% hydrocarbon and a contaminant density of 1 g/cm^3. Separate assessments were completed for the CMV and the total Gateway configuration to account for the long duration transit of CMV.

$$\alpha(t) = \alpha_\infty \left(1 - e^{-2K_{mixture}\sqrt{ESH(t)d(t)}}\right) + \alpha_0 e^{-2K_{mixture}\sqrt{ESH(t)d(t)}} \qquad (2)$$

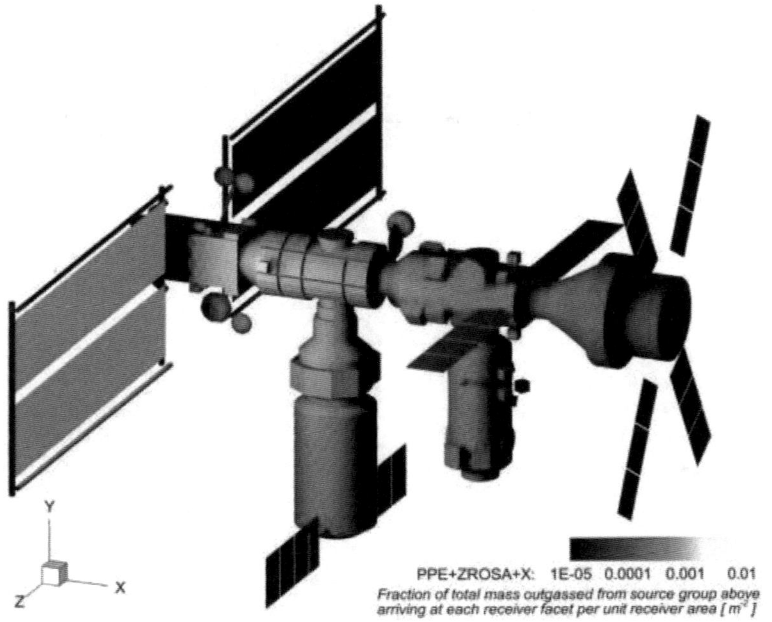

Figure 5. Visualization of Direct Flux from a Single Outgassing Source. In this example, a source group representing the PPE ROSA +Z +X face (highlighted in green) outgasses and deposits mass in the direct sense onto all Gateway elements in its line-of-sight [5]. credits: NASA, reproduced with permission.

Another critical contamination source on Gateway is from PPE's EP plumes, which can impinge on spacecraft deployed geometry and cause erosion which then redeposits onto other surfaces. Redeposition can happen on any spacecraft surface with a view to the impinged surface, as shown in Figure 6 below. To estimate degradation from this source, several steps were taken. First, testing was conducted in a vacuum chamber at Glenn Research Center (GRC) where Xenon ions were directed at several different surfaces which could be impinged on Gateway. Materials were selected based on structural and solar array materials which could be exposed to EP plume and testing was conducted in an Ion Tech Dual Ion Beam System [7]. In the test, eroded materials were sputtered onto certain surfaces and erosion yields were calculated. Next, a Gateway plume model was built and the calculated erosion rates were input to determine how much mass would be deposited on nearby surfaces. Lastly, further testing on certain Gateway receiving surfaces measured change in solar absorptivity relative to the contaminant type and thickness. With the estimated thickness from all three sputtered contaminants generated from the model, a change in solar absorptance of Gateway surfaces could be calculated.

Lunar Dust represents the biggest unknown in the Gateway induced environment. The Human Landing System (HLS) will dock with Gateway after ascending from the lunar surface, potentially brining back lunar dust, which could then transfer to Gateway sensitive surfaces. The amount of dust that lands on HLS, the amount that stays on HLS during ascent, and the amount that then transfers to Gateway is a knowledge gap that must be filled by analysis with support from ground testing. The Gateway On-Orbit Lunar Dust Modeling and Analysis Program (GOLDMAP), developed by the Gateway Systems Engineering and Integration (SE&I) team, seeks to close some of these gaps. The GOLDMAP is a comprehensive, physics-based model which

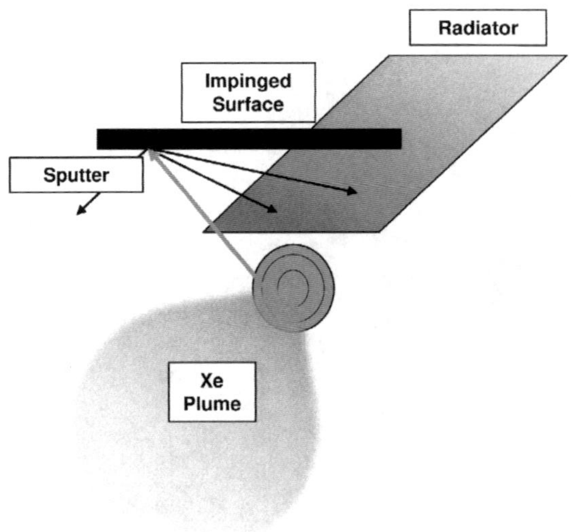

Figure 6. EP plume sputter deposition can occur on Gateway surfaces dependent on proximity and field of view [2]. credits: NASA, reproduced with permission.

considers station geometry and configuration, materials and adhesive properties, lunar orbital environment, and dust particle and surface charging to predict dust transfer and settling onto external Gateway surfaces from the HLS. A detailed description of the model and limitations can be found in Lee et al, 2023 and subsequent papers [8]. Outputs of the model were dust concentrations on Gateway surfaces in kg/m^2 and Percent Area Coverage (PAC); Figure 7 shows an example of how dust mass is represented. The thermal impact of dust itself is also a topic of research and is difficult to evaluate [9]. For this optics assessment, the rule of mixtures was used to determine delta absorptivity assuming a dust absorptivity of 0.8 for conservatism. The rule of mixtures is shown in Equation 3, where α_{new} is the dusted surface absorptivity, PAC_{dust} is the percentage of surface area covered by dust provided by GOLDMAP, α_{dust} is 0.8, and $\alpha_{pristine}$ is the absorptivity of the underlying surface, which may be already degraded by other factors depending on when in the mission the dusting is taking place.

$$\alpha_{new} = PAC_{dust}\alpha_{dust} + (1 - PAC_{dust})\alpha_{pristine} \tag{3}$$

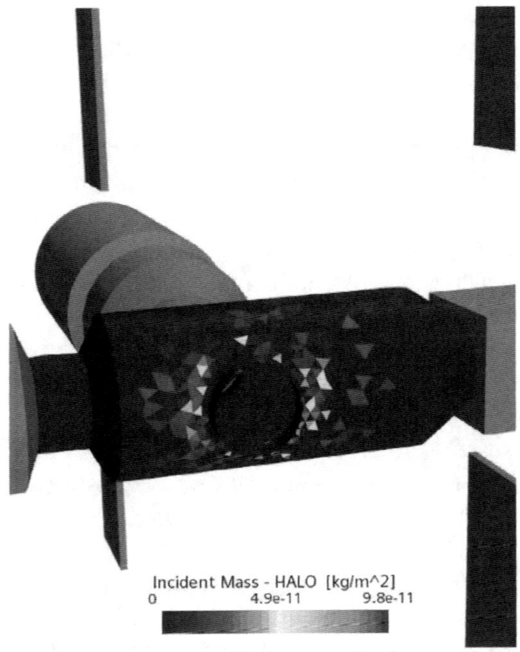

Figure 7. GOLDMAP incident dust mass map [8]. credits: NASA, reproduced with permission.

It was stated earlier that summing delta α from different sources, especially with a mission lifetime of 15 years like Gateway, can lead to overestimation of EOL properties, so other methods were needed. The integration of all testing and analysis results to estimate EOL absorptivity was done using the flowchart shown in Figure 8. The process was developed based on two different ideas: multiple mission phases have different sources of degradation, and during different phases some degradation source may be more significant or more dominant than another. Solar

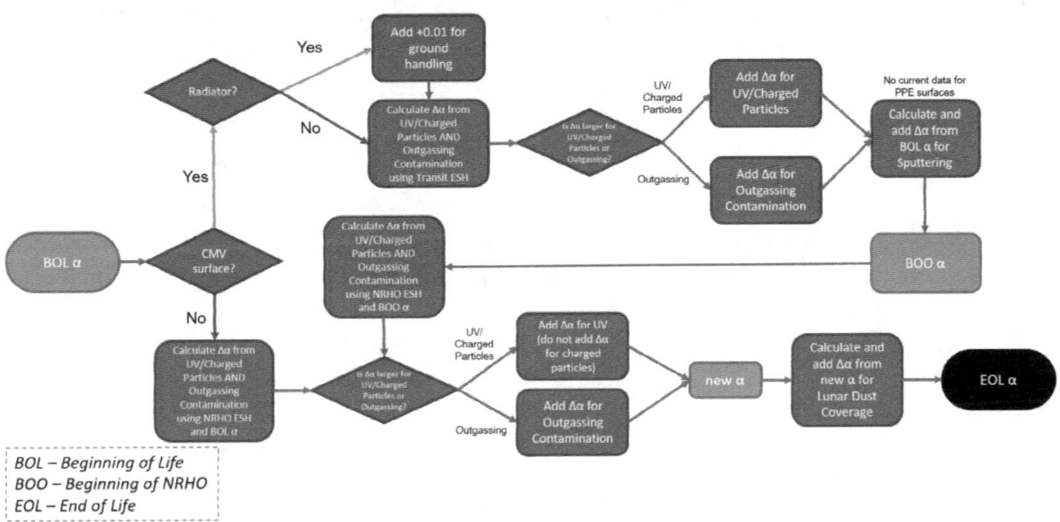

Figure 8. This decision tree can be used to calculated EOL α for both CMV and non-CMV materials.

IOP Conf. Series: Materials Science and Engineering **1328** (2025) 012009 doi:10.1088/1757-899X/1328/1/012009

absorptance will tend to increase more over its lifetime compared to the delta when BOL solar absorptivity is higher, or when some degradation has already occurred on the surface. Further degradation from a new source will not be as high as it might have been on a pristine surface. In this case, it makes sense to act as if the starting α for a later mission phase is actually the degraded α from the previous mission phase. For Gateway, the two main mission phases are transit and NRHO, though visiting vehicle interactions could also be considered new mission phases. The Gateway Thermal team defines the solar absorption after the transit period and at the beginning of the NRHO period to be Beginning of Orbit (BOO). Degradation can also occur from multiple sources at the same time, such as UV, molecular contamination, and charged particles. For a single mission phase, the analyst has the option to sum Δα from each factor or to choose the most dominant factor, or that which results in the largest Δα. For the transit phase of CMV, there were four factors degrading surface optics: UV, charged particles, molecular contamination, and EP plume sputtering. The Gateway UV/Charged Particles Testing included both factors while the Gateway Contamination assessment included degradation from UV and molecular contamination. If the Δα were summed from these assessments, UV degradation would have been counted twice. It was therefore determined that the larger of the two would be used and the other neglected. Then Δα from EP plume sputtering was then added on top of this degraded value. Degradation of solar absorptivity for non-CMV Gateway surfaces may be calculated similarly, but without EP plume sputtering. This step was not included in Figure 8 because no information was available on transit times or attitudes for non-CMV modules at the time, though they were expected to be short.

For the NRHO portion of the mission, again the dominant Δα was used between the UV/Charged Particle Testing and the Contamination Assessment based on higher ESH values (over 15 years) determined from the ESH Analysis. The final EOL α was determined by adding degradation caused by lunar dust coverage as estimated from the GOLDMAP and using the rule of mixtures where αpristine was the degraded value calculated in the last step. This step could be repeated for multiple dust transfers caused by multiple HLS dockings to Gateway.

There may be synergistic effects from these multiple sources which may be difficult or impossible to simulate with ground testing where surface properties are altered differently than they would have been from different combinations of those sources, and this methodology cannot account for that. It is expected that this method would be conservative because it still uses plain addition to include testing that evaluates total Δα from a single source starting from BOL optics (such as EP plume sputtering testing) which generally results in higher degradation than it would if it started from a pre-degraded state.

4. Results

Table 1 below shows an example of what the BOO calculation would look like for a radiator surface on a CMV. Note, these numbers are examples only and are not necessarily representative of real Gateway surface optics or materials. The beginning of orbit value generated in the above table can then be used to start the EOL calculation seen below in Table 2. While the two previous examples are for HALO which is part of CMV, the final example in Table 3 is for the I-HAB. For this vehicle, the transit is unknown but expected to be significantly shorter than that seen by a CMV. Due to the short transit, the BOO calculation is unnecessary and the calculation of EOL α follows the lower pathway shown in Figure 8.

Table 1. BOO absorptivity calculation is shown for a CMV radiator.

Surface	Material	BOL α	ESH Transit (hr)	Contam. Deposition (Angstrom)	Sputtering Deposition (Angstrom)	Ground Handling	Δα UV+ TID	Δα Sputter Redep.	BOO α
HALO Radiator	White Paint	**0.13**	500	100	50	0.01	0.13	0.011	**0.28**

Table 2. EOL absorptivity calculation is shown for a CMV white paint radiator.

Surface	Material	BOO α	ESH (hr)	Contam. Deposition (Angstrom)	Lunar Dust PAC	Δα UV/Contam.	Δα Lunar Dust	EOL α
HALO Radiator	White Paint	**0.28**	1500	2000	3.00 %	0.023	0.014	**0.32**

Table 3. EOL calculation is shown for a non-CMV radiator material.

Surface	Material	BOL α	ESH (hr)	Contam. Deposition (Angstroms)	Lunar Dust PAC	Δα UV/Contam.	Δα Lunar Dust	EOL α
I-HAB Radiator	Mirror	**0.11**	2500	4000	0.00 %	0.055	0.000	**0.16**

5. Conclusions and Future Work

The estimation of Gateway external surface solar absorptivity degradation is complicated by the various environmental factors that degrade materials, the complex phasing of the mission, and Gateway's long lifetime once in NRHO. Solar absorptivity for all major surfaces were predicted by an integrated model which included inputs from an ESH analysis, UV/charged particle testing results, a molecular contamination assessment, EP plume sputtering test results, and dust coverage predictions from the GOLDMAP using the rule of mixtures to evaluate optical degradation from dust. Some caveats to the Gateway analysis presented here are as follows:

- EP plume sputtering test data is only available for two materials, which are not necessarily representative of all surfaces which may be impacted
- Not all materials were able to be included in UV/charged particle testing
- Due to the complexity and unknowns regarding lunar dust, and unknowns in how much dust will adhere to HLS in the first place, there is still a high level of uncertainty in GOLDMAP lunar dust coverage estimates
- Molecular contamination and UV/Charged Particle degradation will both take place at the same time for CMV surfaces, but this assessment chooses to use only the more significant factor since degradation is not additive

Despite the caveats, the Gateway Passive Thermal team believes this methodology and these results to be the best available at this time and will continue to improve this analysis in future design cycles. Future work will involve updates to all analyses as the Gateway design develops and as modeling methods improve. Since the time of this integrated analysis, further testing has been completed by Gateway regarding lunar dust adhesion and transfer, and investigating thermo-optical property impact with dust coverage. Data from these tests will be used to update the GOLDMAP and the estimates in change of solar absorptivity due to dust coverage. Ideally, these results will be validated when Gateway is in operation using correlated thermal models and on-orbit temperature monitoring to calculate solar absorptivity.

This methodology has applications beyond Gateway and could be applied to almost any long complex, multi-phased mission, given some modifications. The key to this methodology is separating the mission into phases where environmental factors causing degradation of solar absorptivity are different and estimating degradation for each phase where the starting absorptivity is that of the degraded surface from the previous phase. Degradation at each phase can be estimated by either choosing a dominant degradation factor or summing multiple factors if choosing a dominant one is impractical.

References

[1] National Aeronautics and Space Administration Gateway Communications, Public Affairs, and Outreach (CPAO) 2024 101 Roadshow
[2] Berg J 2024 Thermal Impacts of Lunar Dust on Rovers *Thermal Fluids Analysis Workshop 2024*
[3] Spivey B, Zinecker A, and Smith J 2021 Using Thermal Desktop to Determine Equivalent Solar Hours on Spacecraft Surfaces *Thermal Fulids Analysis Workshop 2021*
[4] Zinecker A *et al* 2023 *IOP Conf. Ser.: Mater. Sci. Eng.* **1287** 012009
[5] Hoey W A *et al* 2023 *IOP Conf. Ser.: Mater. Sci. Eng.* **1287** 012002
[6] Soares C and Alred J 2001 *33rd International SAMPE Technical Conference*
[7] Racine E and Molnar A 2019 Thermal Design Challenges for In-Flight Exposure to and Electric Propulsion Plasma Plume Environment *Thermal Fulids Analysis Workshop 2019*
[8] Lee R G et al 2023 Development of a comprehensive physics-based model for study of NASA gateway lunar dust contamination *Acta Astronautica* **210** 616-626
[9] Howard A and Stewart S 2024 Thermal Impacts of Lunar Dust on Rovers *Thermal Fluids Analysis Workshop 2024*

IOP Conf. Series: Materials Science and Engineering **1328** (2025) 012010 doi:10.1088/1757-899X/1328/1/012010

Continuation of Solar Absorptivity Degradation of Spacecraft Materials Due to UV and Charged Particles in the Gateway Environment

B Hoffmann*[1], A Z Howard[2], B Spivey[1]**

[1] *Jacobs, 2224 Bay Area Blvd, Houston, TX 77058 (USA)*
[2] *NASA Johnson Space Center, 2101 E NASA Pkwy, Houston, TX 77058 (USA)*
* brandon.m.hoffmann@nasa.gov
** abigail.a.zinecker@nasa.gov

Abstract. Gateway is a lunar orbiting platform that will be assembled in space similar to the International Space Station. Modules will arrive following different transit orbits, some of which have an extended duration in the Van Allen Belts. This means the spacecraft is exposed to significant charged particle radiation, after which it must operate in cislunar space for 15 years. Charged particle and UV degradation of potential Gateway materials was identified as a knowledge gap by the Passive Thermal Control System (PTCS) Group as end-of-life solar absorptivity greatly impacts thermal performance. Multiple rounds of ground testing have been completed at NASA Goddard Space Flight Center in which potential Gateway materials were subjected to expected worst-case transit charged particle fluence (9.86x1015 protons/cm2 at 2.5 keV and 3.1x1016 electrons/cm2 at 10 keV) and up to 5000 Equivalent Solar Hours (ESH) of UV exposure. This report builds on previous data from Zinecker et al. 2022 which presented the first round of ground testing. Additional materials have been tested and solar absorptivity degradation along with lessons learned are shared. Comparisons to previous results are also presented.

1. Introduction

Charged particles and ultraviolet (UV) radiation have been shown to significantly increase solar absorptivity (α) for some thermal control materials and coatings [1,2,3,4]. Most spacecrafts operate in a Low Earth Orbit (LEO) or Geostationary Orbit (GEO), therefore most of the data accumulated for material optical property degradation in space has been in these environments [1,3,4,5]. No spacecraft, barring the 55 pound CubeSat called Cislunar Autonomous Positioning System Technology Operation and Navigation Experiment (CAPSTONE), has flown in Near Rectilinear Halo Orbit (NRHO), and no data on optical property degradation from this system is available at this time [6]. The Gateway space station will be assembled and flown in NRHO for 15 years. Gateway materials need to be better understood in this environment before flight so thermal engineers can take degradation of optical properties into account for End Of Life (EOL) thermal performance modeling. In addition to the NRHO environment, some Gateway modules will also have a slow spiral transit from Earth to NRHO, taking a year or more to pass through the Van Allen Belts, which contain damaging charged particles [7]. Due to the long duration transit, significant degradation to optical properties is expected occur on external materials before the NRHO portion of the mission begins. Over 95% of the charged particle exposure for the 15 year mission occurs during the transit portion of the environmental bounding mission for these

IOP Conf. Series: Materials Science and Engineering **1328** (2025) 012010 doi:10.1088/1757-899X/1328/1/012010

modules. This test emphasises materials for these modules. Since α will impact thermal performance over the mission, the Gateway Passive Thermal Control System (PTCS) Group identified charged particle and UV degradation of Gateway materials as a knowledge gap. To close this gap, expected Gateway materials were subjected to bounding mission charged particle environments and simultaneous UV exposure at NASA Goddard Space Flight Center (GSFC). Change in reflectivity was measured at specified intervals during testing such that change in α could be plotted as a function of time of solar exposure.The first round of testing was shared by Zinecker, et al. in 2022 [8]. Round 2 of testing, which consisted of 3 separate tests with slightly different conditions, are shared in this paper.

2. Methodology

The Solar Wind Facility at GSFC was used for this testing due to their expertise in materials and equipment availability. Details on the test chamber, sample size and shape, UV application, and proton and electron beams can be found in other papers [3,4,9]. During the entire test series, 16 potential Gateway materials and coatings were tested, including thermal control coatings, radiator materials, multilayer insulation (MLI) outer materials, and structural surfaces. The materials tested and what rounds they were tested in are shown in Table 1. Three separate rounds of the same materials were tested in this series using fresh samples each time. The test team attempted to replicate the UV and charged particle environment to be experienced by Gateway in each of the three rounds, however, there were slight differences which are noted in Table 2.

Table 1. List of Material Samples with slot numbers and round

Chamber Slot	Material Name	Description	Round Tested
1, 9*	Beta Cloth	Fine woven silica fiber	2a-c
2	Ag FEP 10 mil	Silver Teflon flexible radiator coating	2a-c
3, 7,10*	Alion Z93C55 with MLOP300AZ Primer	Primer for adhesion with silicate based white paint on top, electrically dissipative	2a-c
4	Alion Z93C55	Silicate based white paint radiator coating, electrically dissipative	1a, 2a-c
5	CCAg	Composite coating of silver and oxide layers	2a-c
6	Sulfuric Acid Anodize (SAA) with hot water seal Indium Tin Oxide (ITO) coating	Corrosion resistant coating, common for exposed structural surface with electrically dissipative ITO coating	2a-c
8	SAA 6061-T6 with hot water seal	Corrosion resistant coating common for exposed structural surfaces	2a-c
11*	AgFEP 5 mil	Silver Teflon flexible radiator coating	2a-c

Table 2. Environments experienced by each round

ENVIRONMENT	2A	2B	2C
UV ESH (hours)	1600	2011	5000
Vacuum pressure (torr)	$2x10^{-5}$	$2x10^{-5}$	$1x10^{-6}$
Suns	2.5	2.5	2.5
TID Exposure 1	30% TID at 810 ESH	100% e- only at 2011 ESH	40% TID at 890 ESH
TID Exposure 2	90% TID at 1600 ESH	-	99% TID at 990 ESH

The samples were 19 mm diameter discs which were mounted inside the chamber and exposed to simultaneous proton, electron, and UV up to 5000 ESH while maintaining a constant vaccum at $2x10^{-5}$ Torr. The Energetiq EQ400 applied nUV light and the Hamamatsu 100W Deuterium lamp was the source of vUV light. The Deuterium lamp had a minimal infrared (IR) component to minimize heating of samples during the test. Perliminary testing showed that the samples did not reach above 40°C, but due to chamber limitations it was not possible to monitor sample temperatures during the test. The small IR component of the UV sources should not significantly affect the temperature of the samples, therefore we assume temperature is not a factor in material degradation. The solar lamps applied UV at ~2.5 equivalent suns, which was averaged over a wavelength range of 200 to 400 nm.

Total Ionizing Dose (TID) was applied with a 2.5 keV proton beam and a 10 keV electron beam. The total dose was calculated using a radiation model in software called NOVICE. The dose depth profile was approximated according to the worst-case Gateway transit environment particle fluences. The TID was then converetd to a particle fluence based on the material surface and energy levels of the particle sources. Based on these calculations, the fluence was determined to be $9.86x10^{15}$ protons at 2.5 keV and $3.10x10^{16}$ electrons at 10 keV. See Table 2 for the dose reached in each of the tests in this series.

A passive contamination monitor was placed in one of the sample slots. Changes in the contamination monitor were found to be within the uncertainty of the instrument, indicating the contamination environment in the chamber was not a factor in test results. Twelve samples can fit in the chamber at a time. Eight of those samples were measured in-situ; those shown in Table 1 with a * were not measured in-situ due to chamber restrictions.

Solar reflectivity of each sample was measured at ambient pressure, vacuum after an overnight pump down, multiple intervals of ESH, and after return to ambient pressure. TID application was sometimes split into two intervals to provide approximately half of the dose at one time and the other half at another time. Little to no UV was applied at the same time as the charged particles since they are applied more quickly than UV. Sensors were placed next to each sample in the chamber to monitor the approximate proton and electron dose received.

At 1600 ESH in round 2a, an internal component failed which prevented any future exposure to charged particles. This failure resulted in only 33% of electron exposure occurring. Another component failed at 2011 ESH in round 2b. This prevented the measurement of particle flux, and therefore protons could not be applied during this round. Round 2c reached 5000 ESH. Over the course of testing, two repressurizations occurred due to power issues. This resulted in temporary vacuum loss and minor bleaching of some materials, however this is not believed to impact solar absorptivity measurements or conclusion.

Reflectivity measurements were taken at wavelengths ranging from 250 to 2500 nm. Total α was calculated from these measurements by integrating 1-%R with respect to the solar spectral irradiance as defined by ASTM E490, assuming a transmittance of zero [10]. Percent R is defined as the reflectance measured at a given wavelength.

IOP Conf. Series: Materials Science and Engineering **1328** (2025) 012010 doi:10.1088/1757-899X/1328/1/012010

Modeling curves were generated for all tested materials in order to predict Δα for each of the materials listed in Table 1. This was done using Microsoft Excel. Only UV degradation data was used for these predictions; charged particle radiation damage was filtered out and so must be added at the end to make predictions for degradation on Gateway modules. For this calculation, a natural logarithmic curve was assumed. Because ln(0) is undefined, the initial ESH was set to 1 instead of 0. A starting ESH of 1 was shown to provide a curve that best matched the test data. The form of the modeling curves is shown in Eqn. 1 below where m and b are modeling parameters dependent on material.

$$y = m \ln(ESH) + b \qquad \text{Eqn. 1}$$

3. Results

Figure 1 and Figure 2 show the pre- and post-test images of the sample plate from Round 2c. The numbers next to each sample correspond to the those shown in Table 1. The sample from slot 4 is missing in the Figure 2 because it was accidentally dropped toward the end of the test. Looking closely, degradation occurs visibly in samples 7 and 9. Table 3 shows the model parameters for each material tested in Round 2c according to Eqn. 1.

Figure 1. Sample plate pre-test. *Credit: Own work*

Figure 2. Sample plate post-test. *Credit: Own work*

Table 3. UV degradation model parameters and TID degradation from Round 2c

Material	*m*	*b*	Δα due to TID
Beta Cloth	0.0046	0.3432	0.08
AgFEP 10mil	0.0039	0.1018	0.07
Primed Z93C55	0.0117	0.1661	0.06

Standard Z93C55	0.0108	0.1814	0.07
CCAg	0.0172	0.1198	-0.03
SAA ITO	0.0041	0.4105	-0.01
SAA no ITO	0.0097	0.3518	-0.01

Figure 3 plots the solar absorptivity of all Z93C55 samples that were tested in all rounds. Degradation of the material due to charged particles is removed in post processing and kept as a separate value, which is then added at the end and represented by the *b* term in Eqn. 1. This plot is the basis for which we develop UV degradation curves.

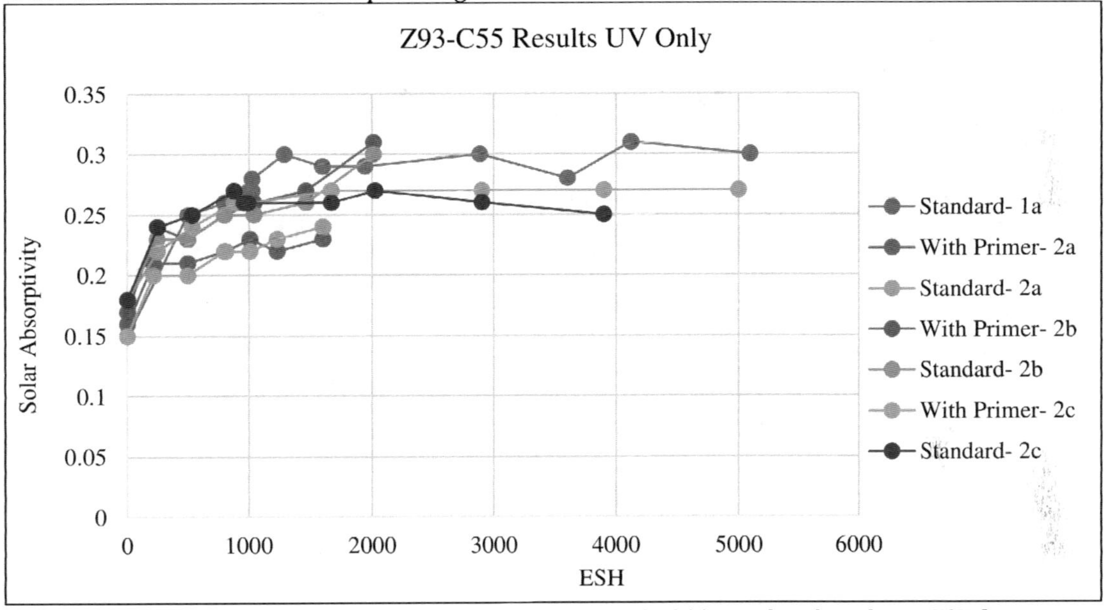

Figure 3. Solar Absorptivity degradation of 7 in-situ Z93C55 samples plotted over ESH from 4 total rounds. Degradation from charged particles have been removed since exposures were different in each round.

Figure 4 plots the solar absorptivity degradation for all SAA aluminum samples from all rounds. Round 1 SAA was sealed with nickle acetate while samples from Round 2 were sealed with hot water. Additionally, several rounds included SAA that had an ITO coating.

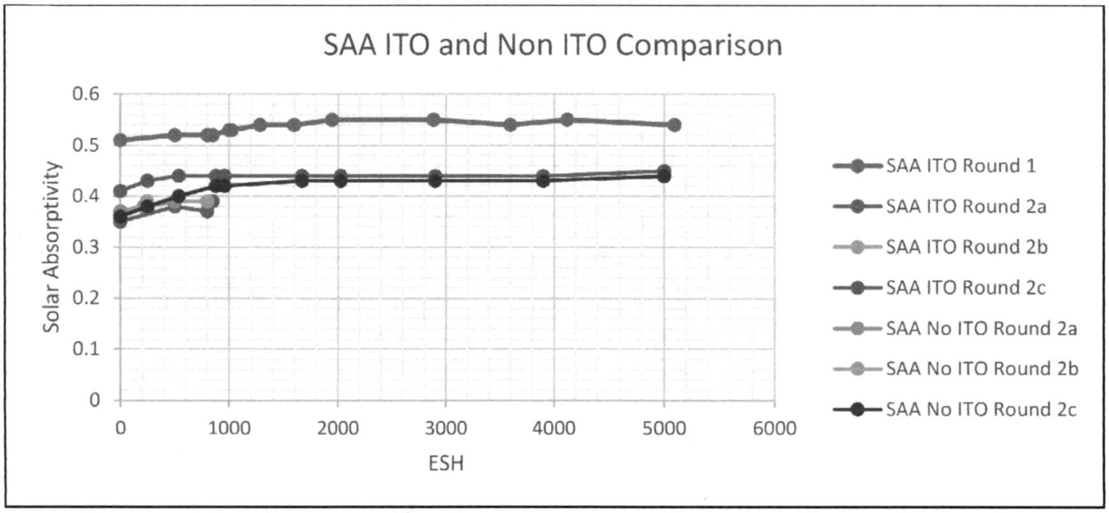

Figure 4. Solar Absorptivity degradation comparison of 7 SAA samples, 4 with ITO and 3 without. This chart contains UV data only.

The full data from each round of testing can be found below in tables 4 through 7. Note the TID degradation was removed in post processing and put into its own category. This is done so UV degradation fit curves can be generated without being influenced by TID degradation. During testing, the TID is applied separate from the UV, so removing the charged particles degradation from the UV is trivial in post processing. The initial absorptivity reading is listed at 1 ESH to create a more realistic fit curve.

Table 4. UV and TID degradation results for all samples from round 1.

	Round 1 UV Degradation Only													TID
Sample	1 ESH	500 ESH	800 ESH	850 ESH	1000 ESH	1025 ESH	1286 ESH	1595 ESH	1941 ESH	2885 ESH	3600 ESH	4120 ESH	5093 ESH	
Aptek 2719	0.22	0.34	0.36	0.36	0.36	0.36	0.37	0.37	0.38	0.39	0.39	0.4	0.41	0.09
Aptek 2711-1	0.18	0.18	0.17	0.17	0.17	0.17	0.15	0.16	0.17	0.14	0.13	0.18	0.18	0.11
DS13	0.14	0.37	0.38	0.38	0.38	0.38	0.38	0.38	0.39	0.39	0.4	0.41	0.42	0.13
Z93C55	0.15	0.25	0.26	0.26	0.27	0.27	0.3	0.29	0.29	0.3	0.28	0.31	0.3	0.04
OSR	0.09	0.12	0.12	0.12	0.11	0.11	0.14	0.12	0.1	0.1	0.13	0.18	0.18	0.01
SAA ITO	0.51	0.52	0.52	0.52	0.53	0.53	0.54	0.54	0.55	0.55	0.54	0.55	0.54	0.04
GBK	0.59	0.59	0.59	0.59	0.59	0.59	0.6	0.6	0.6	--	--	--	--	0.01
Astroquartz	0.35	0.44	0.46	0.46	0.46	0.46	0.48	0.49	0.5	0.52	0.52	0.53	0.54	0.01

Table 5. UV and TID degradation results for all samples from round 2a

	Round 2a UV Degradation Only								TID
Sample	1 ESH	215 ESH	500 ESH	800 ESH	810 ESH	1008 ESH	1230 ESH	1600 ESH	
Beta Cloth	0.28	0.32	0.32	0.33	--	--	--	--	--
AgFEP 10mil	0.1	0.14	0.08	0.09	--	--	--	--	--
Mod Z93C55	0.16	0.21	0.21	0.22	0.22	0.23	0.22	0.23	0.09
Std Z93C55	0.15	0.2	0.2	0.22	0.22	0.22	0.23	0.24	0.07
CCAg MAXAR	0.12	0.17	0.2	0.22	--	--	--	--	--
SAA ITO	0.35	0.38	0.37	0.39	--	--	--	--	--
SAA no ITO	0.37	0.38	0.39	0.41	--	--	--	--	--
CREO FF OSR	0.12	0.11	0.13	0.12	--	--	--	--	--

Table 6. UV and TID degradation results for all samples from round 2b

	Round 2b UV Degradation Only							
Sample	1 ESH	245 ESH	500 ESH	800 ESH	1040 ESH	1460 ESH	2011 ESH	TID
Beta Cloth	0.34	0.34	0.35	0.36	0.36	0.37	0.37	0.01
AgFEP 10mil	0.09	0.1	0.1	0.09	0.09	0.11	0.1	0.02
Mod Z93C55	0.17	0.24	0.23	0.25	0.26	0.27	0.28	0.03
Std Z93C55	0.18	0.23	0.23	0.25	0.25	0.26	0.29	0.01
CCAg MAXAR	0.14	0.19	0.2	0.21	0.24	0.24	0.25	-0.03
SAA ITO	0.36	0.38	0.39	0.39	0.41	0.42	0.43	0
SAA no ITO	0.36	0.39	Stuck	Stuck	Stuck	Stuck	Stuck	--
CREO FF OSR	0.13	0.14	0.14	0.15	0.15	0.16	0.16	-0.01

Table 7. UV and TID degradation results for all samples from round 2c

	Round 2c UV Degradation Only											TID
Sample	1 ESH	250 ESH	535 ESH	880 ESH	890 ESH	960 ESH	1667 ESH	2024 ESH	2900 ESH	3900 ESH	5000 ESH	
Beta Cloth	0.34	0.36	0.37	0.38	0.38	0.38	0.38	0.38	0.38	0.37	0.37	0.08
AgFEP 10mil	0.11	0.1	0.12	0.13	0.13	0.13	0.13	0.13	0.14	0.14	0.14	0.07
Mod Z93C55	0.17	0.21	0.23	0.25	0.25	0.25	0.26	0.26	0.26	0.26	0.26	0.06
Standard Z93C55	0.18	0.24	0.25	0.27	0.27	0.25	0.26	0.27	0.27	0.26	--	0.07
CCAg MAXAR	0.13	0.19	0.23	0.24	0.24	0.23	0.26	0.27	0.27	0.26	0.27	-0.03
SAA ITO	0.41	0.43	0.44	0.44	0.44	0.44	0.44	0.44	0.44	0.44	0.45	-0.01
SAA no ITO	0.36	0.38	0.4	0.42	0.42	0.42	0.43	0.43	0.43	0.43	0.44	-0.01

In Figures 5 through 18 below are the fit curves used to predict degradation due to UV exposure. With this curve, it is possible to input an ESH value and output the expected degradation amount. To prevent skewing the data, the TID degradation has been removed and is considered in its own category.

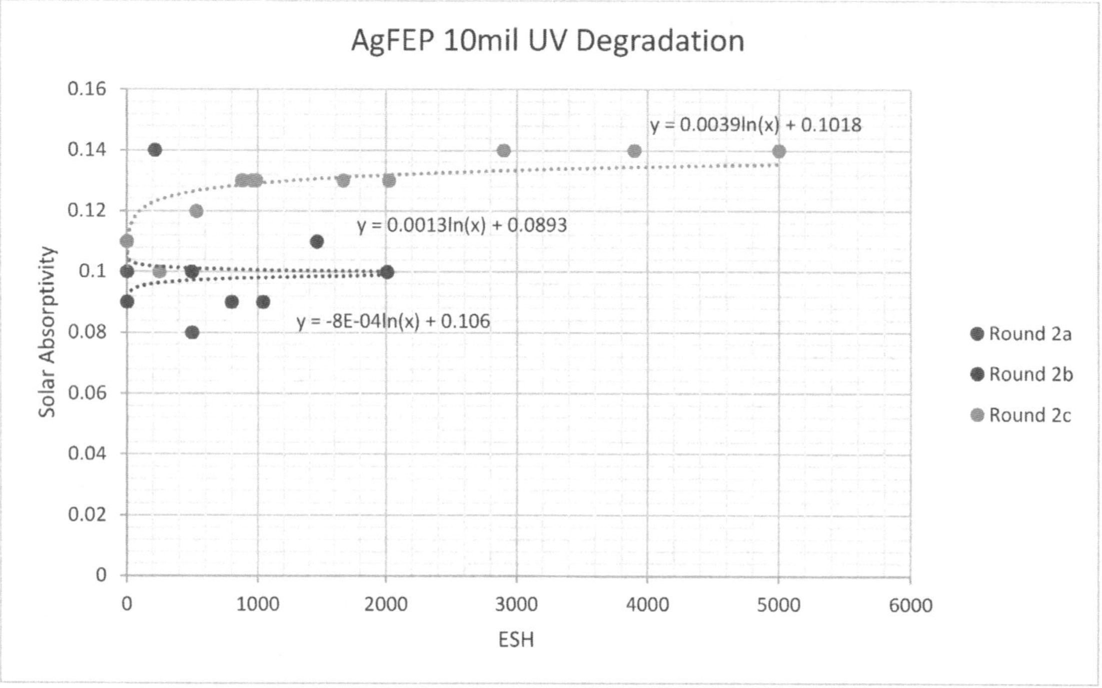

Figure 5. UV degradation and curve fit for AgFEP 10 mil

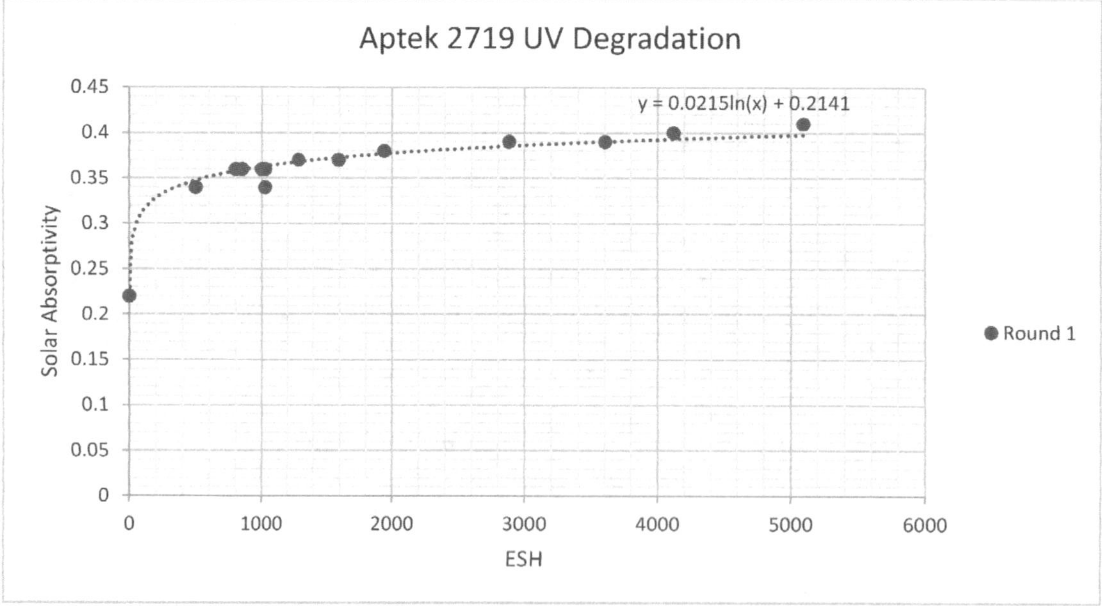

Figure 6. UV degradation and curve fit for Aptek 2719

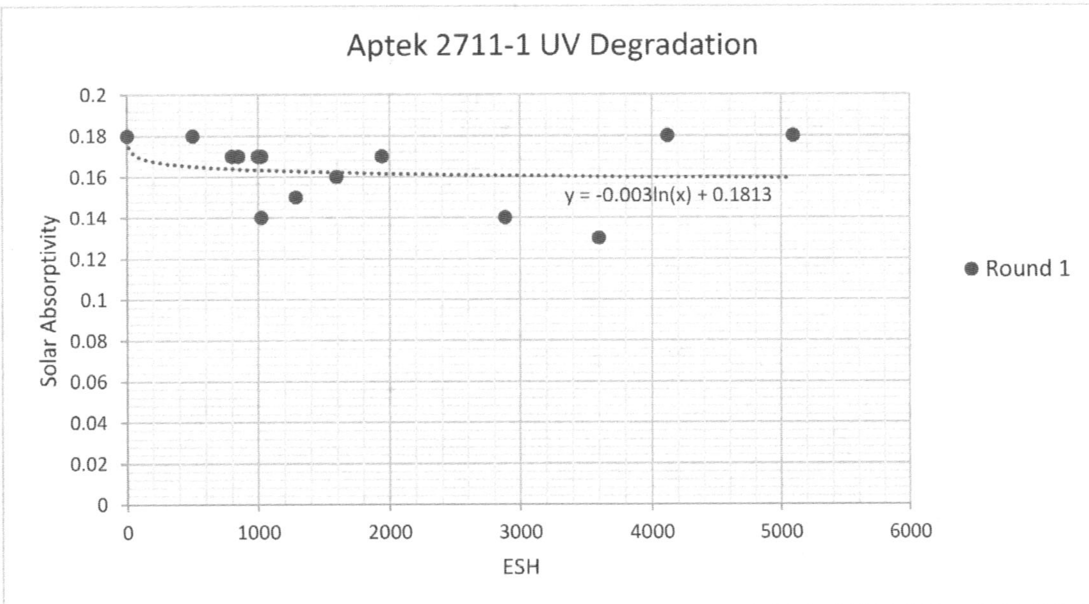

Figure 7. UV degradation and curve fit for Aptek 2711-11

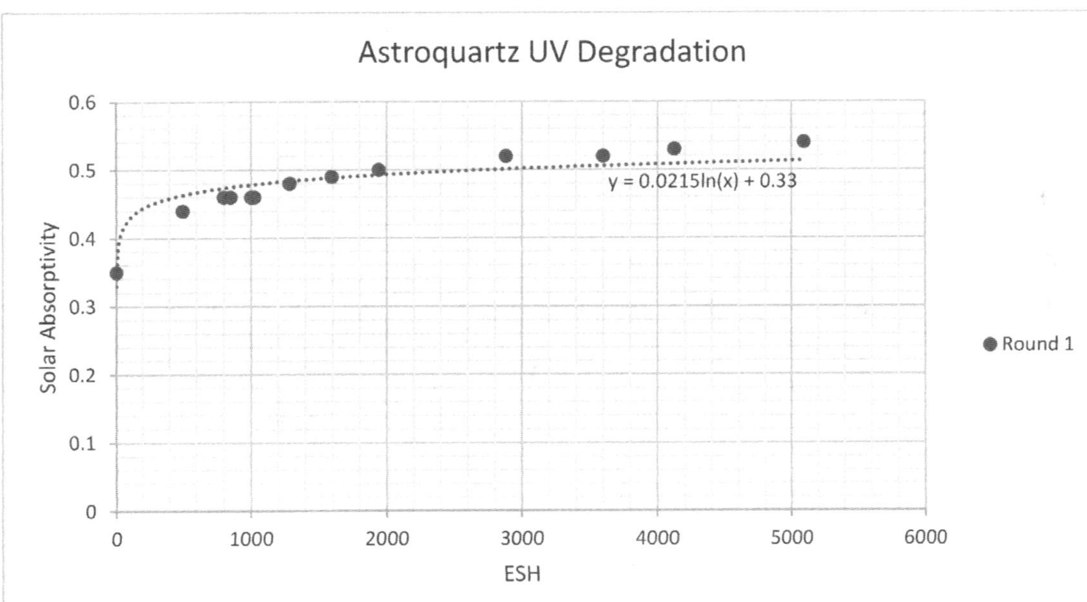

Figure 8. UV degradation and curve fit for Astroquartz

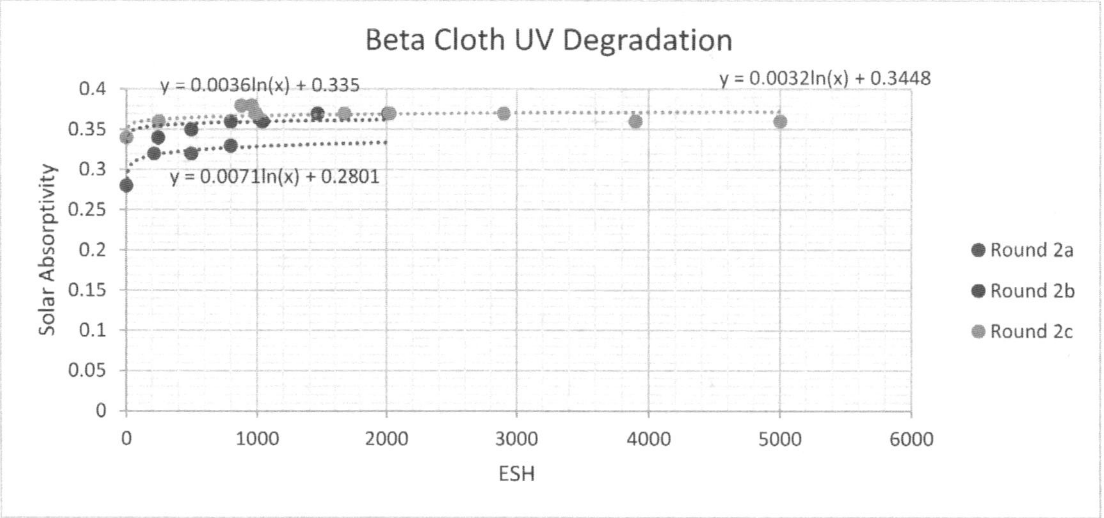

Figure 9. UV degradation and curve fit for Beta Cloth

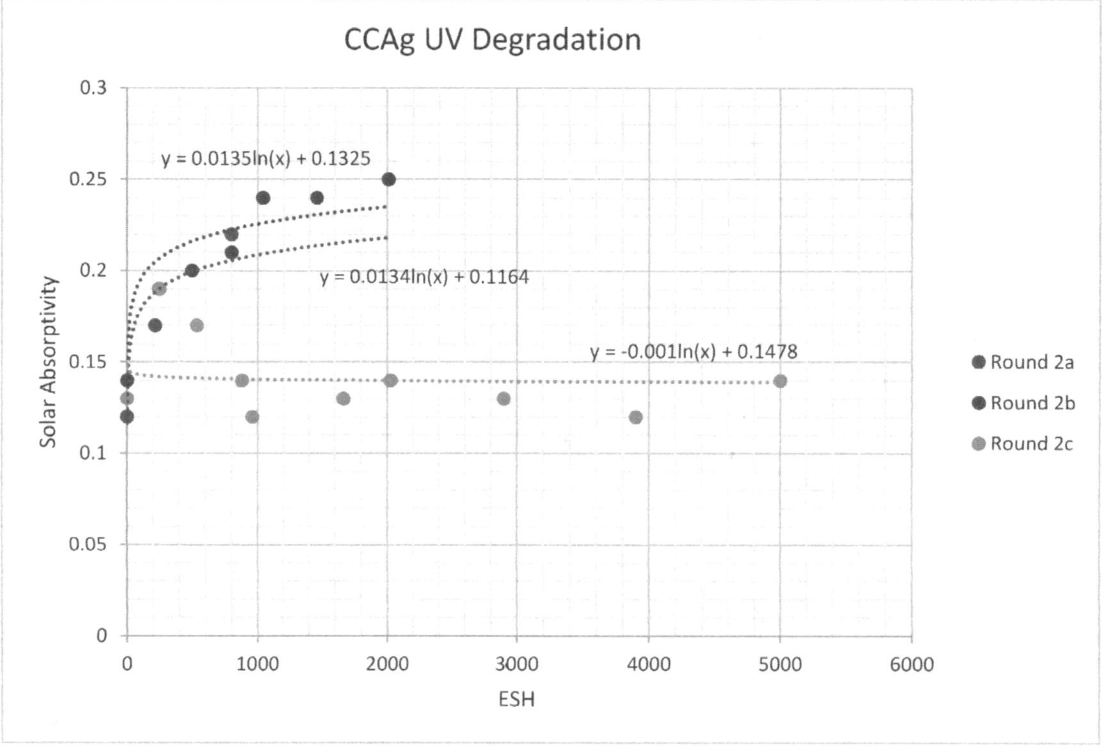

Figure 10. UV degradation and curve fit for CCAg

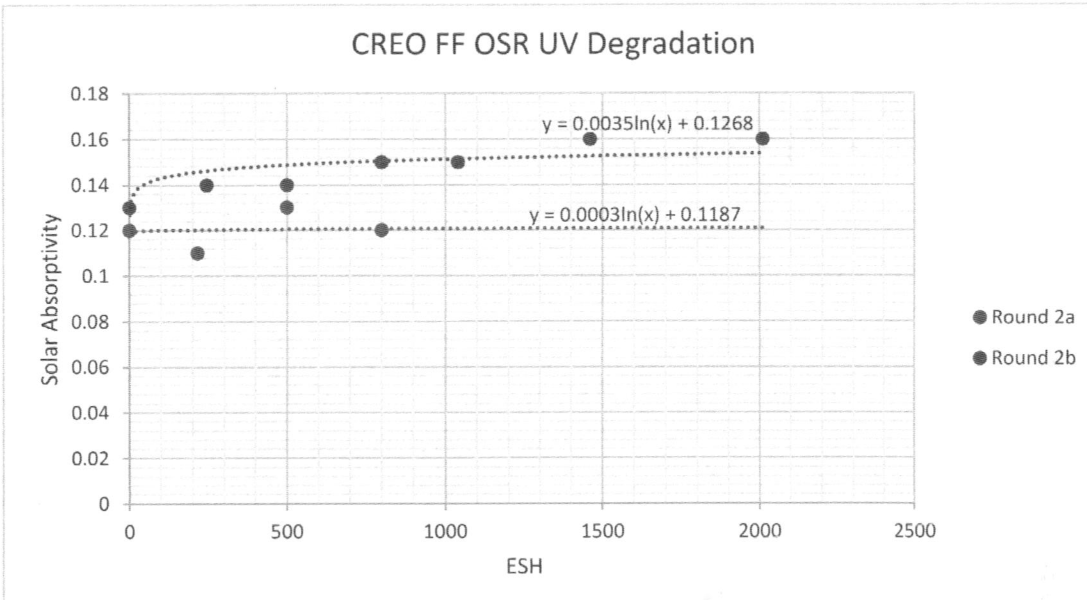

Figure 11. UV degradation and curve fit for CREO FF OSR

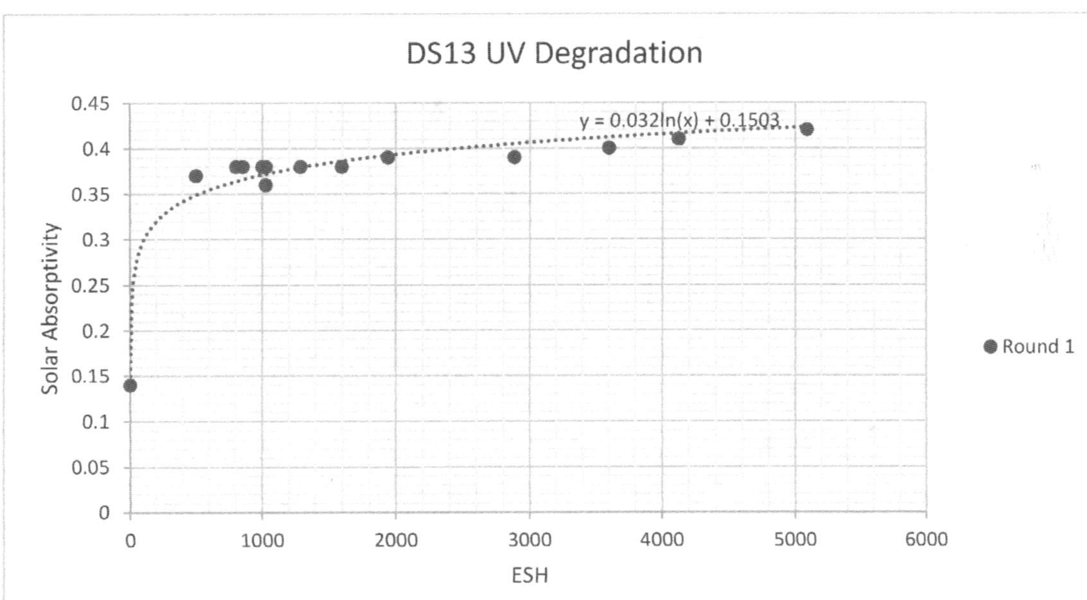

Figure 12. UV degradation and curve fit for DS13

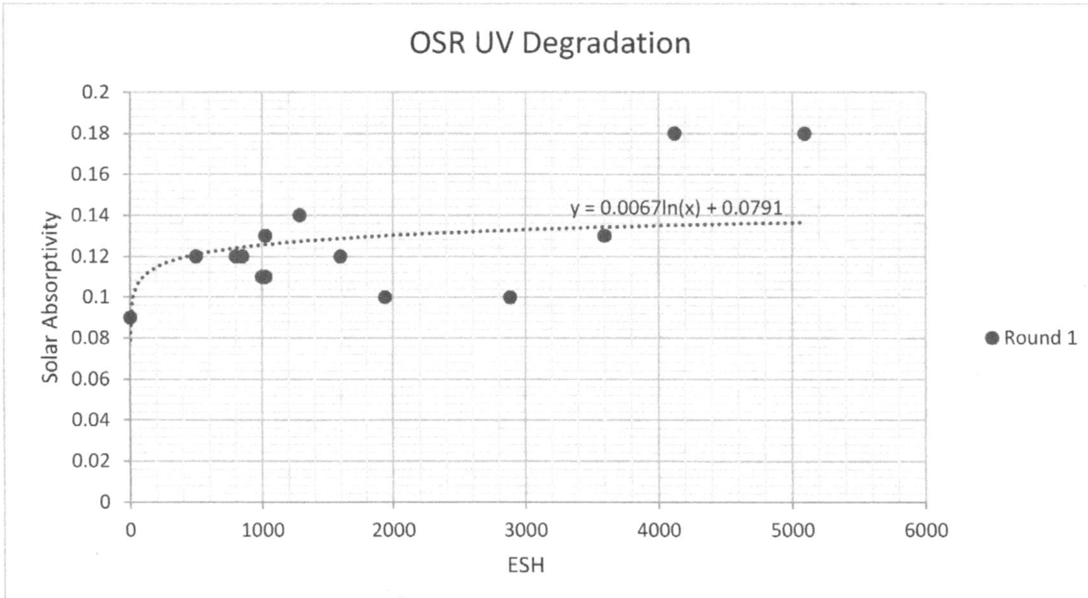

Figure 13. UV degradation and curve fit for OSR

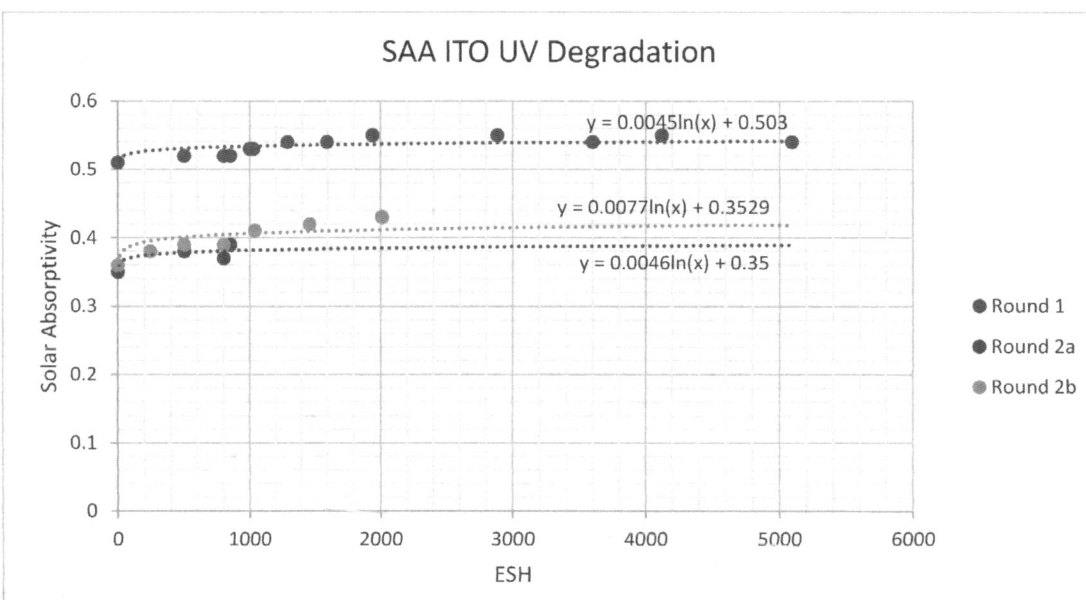

Figure 14. UV degradation and curve fit for SAA ITO

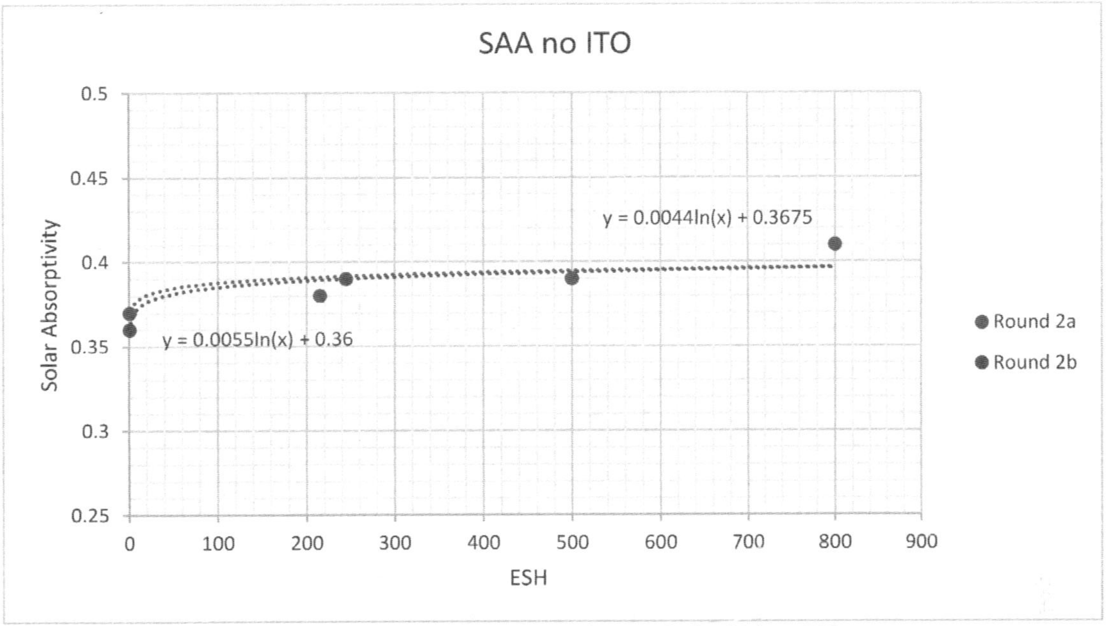

Figure 15. UV degradation and curve fit for SAA no ITO

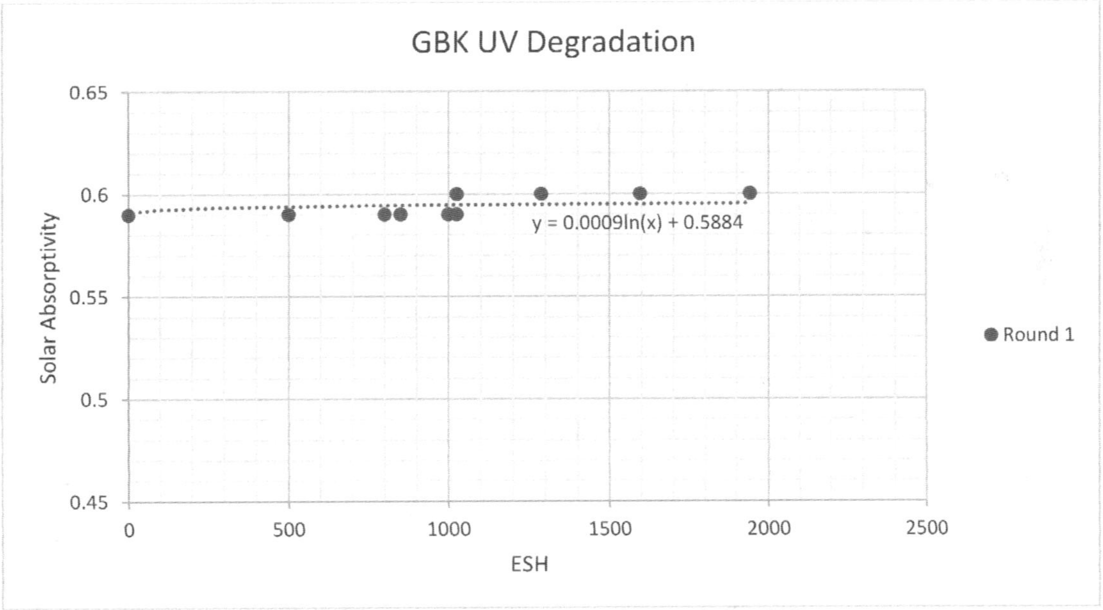

Figure 16. UV degradation and curve fit for GBK

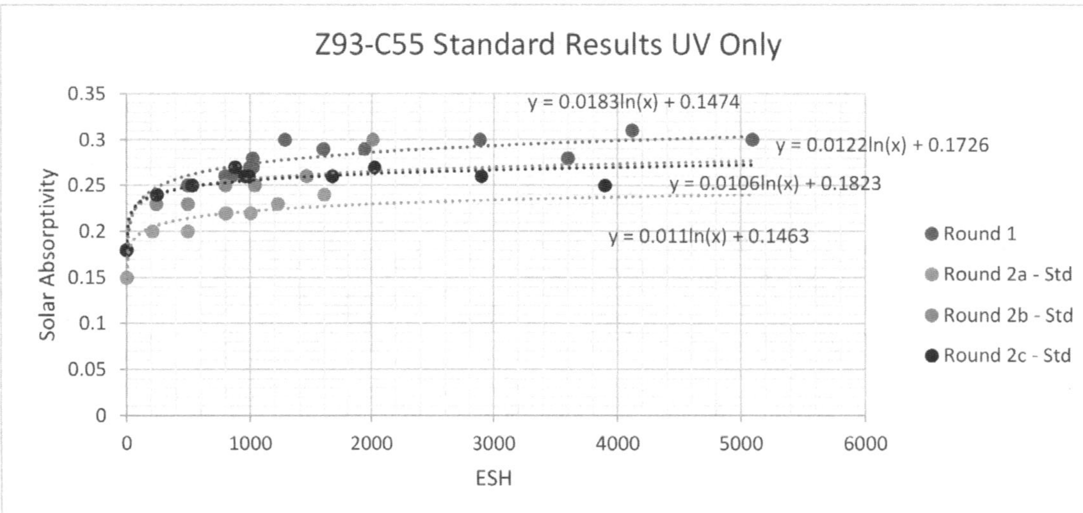

Figure 17. UV degradation and curve fit for Z93C55 Standard

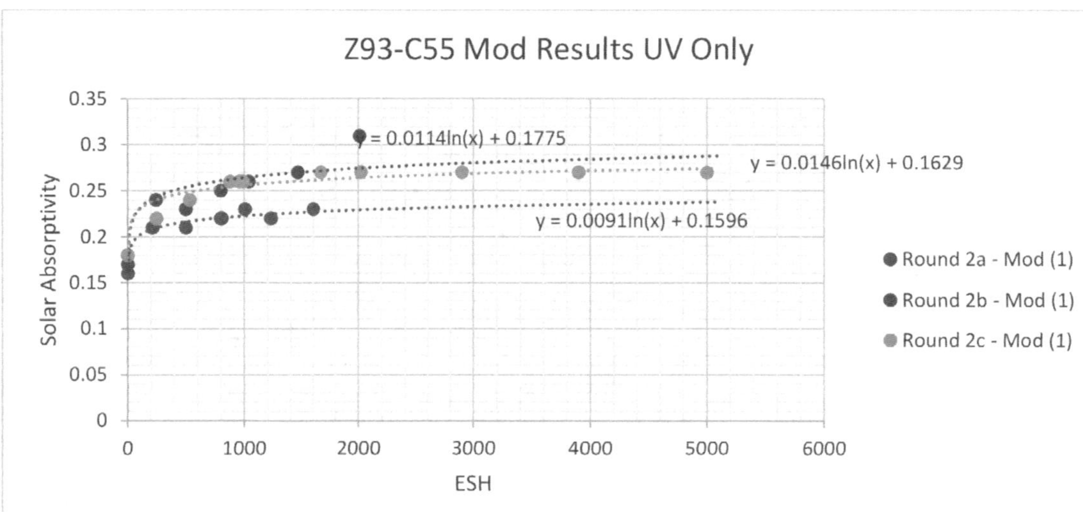

Figure 18. UV degradation and curve fit for Z93C55 Mod (Primer)

4. Conclusions and Future Work

This study continues previous work evaluating radiator coatings, MLI outer layers, and exposed structural materials for Gateway. Testing results have reduced uncertainty in material degradation in the Gateway environment by quantifying the charged particle and UV induced optical property degradation for critical thermal surfaces. Results from this test will be integrated with degradation resulting from other sources such as molecular contamination and lunar dust coverage to quantify expected EOL absorptivity for Gateway materials.

Combining all rounds of testing, Z93C55 EOL absorptivity degradation for UV was between 0.24 and 0.3. The samples reached their highest absorptivity between 1500 and 2000 ESH. When applying charged particles to Z93C55 the less degraded the sample was at time of exposure, the more the charged particles would degrade the surface. There did not appear to be a significant

IOP Conf. Series: Materials Science and Engineering **1328** (2025) 012010 doi:10.1088/1757-899X/1328/1/012010

difference in degradation of Z93C55 with and without the primer for degradation caused by both UV and charged particles. Solar absorptance measurements were within the error when comparing the standard and primed surfaces in the same round. Change in solar absorptance in Z93C55 was greater in Round 1 than in Round 2, even with similar environmental exposures. This phenomenon is under investigation, but it could be due to a slight variation in coating formulation from one batch to another.

Beta cloth and Silver Teflon saw significantly more degradation due to charged particles in round 2c than round 2b. Since round 2c included proton and electron exposure while round 2b only included electron exposure, these results may indicate that beta cloth and Silver Teflon charged particle degradation is primarily due to proton exposure. It could also be due to starting at different solar absorbtivities.

The SAA and CCAG samples decrease in absorptivity after charged particle exposure is likely due to sample rotation when measuring. As the sample pawn is moved to the measurment chamber the pawn can rotate, changing the location being measured. This locaton change can affect the optical values. Outside of rotation issues, these samples did not see significant degradation due to charged particles. The SAA with and without ITO saw similar degradation rates, the main difference between the two likely stemming from intial absorptivity.

Both Aptek materials and DS13 saw signifigant degradation due to charged particles. Aptek 2719 and DS13 both saw signifigant degradation from UV, while Aptek 2711-1 saw no degradation from UV.

As in Round 1, there is some uneven distribution of charged particles inside the chamber such that some samples received more or less dose than others. In round 2c, the samples received anywhere from 83.28% to 121.42% of the targeted proton exposure and 92.85% to 132.3% of the targeted electron exposure. Samples in the center of the plate receive higher charged particle flux than those at the edges. The energy spectrum from the Gateway environment was consolidated from a wide range of energetic protons and electrons to a single energy level for each. This is not truly flight-like, but it is a ground testing limitation and it is representative of the flight environment based on the step profile. It is not possible to measure the same location on each sample each time due to the arm manipulators' spatial accuracy. This leads to a small measurement uncertainty.

Future work for this testing would expand its scope to include materials planned for future Gateway modules and gathering more data on materials which have been studied already. As more accurate environmental data is acquired, future tests may have varied charged particle dosage amounts and energy levels.

References

[1] Hasegawa M and Babel H 1992 Estimation of the end-of-life optical properties of the Z-93 thermal control coating for the Space Station Freedom *Materials Specialist Conference – Coating Technology for Aerospace Systems* AIAA-92-2168-CP

[2] Hasegawa M, Freese S, Kauder L and Triolo J, 2011 End-of-life optical property predictions of white conductive thermal control coatings through analysis of on-orbit and ground based testing data *AIAA Atmospheric and Space Environments Conference*

[3] Jaworski D A and Kline S E 2008 Review of End-of-life thermal control coating performance

[4] Soares C E, Mikatarian R, Schmidl W, Boeder P and Alred J 2006 Optical characteristics of materials in a space environment *10th International Symposium on Materials in a Space Environment*

[5] Finckenor M M, Golden J L, O'Rourke M E and Kravchenko M 2012 Results of international space station vehicle materials exposed on MISSE-7B *National Space and Missile Materials Symposium 2012*

[6] NASA 2025 What is capstone?. *NASA*. https://www.nasa.gov/smallspacecraft/capstone/

[7] McGuire M L, McCarty S L, et al 2021 Overview of the lunar transfer trajectory of the co-manifested first elements of NASA's Gateway *AAS/AIAA Astrodynamics Specialist Conference 2021* AAS 21-697

[8] Zinecker A et al 2023 IOP Conf. Ser.: Mater. Sci. Eng. **1287** 012009

[9] Kauder L 2009 Space environmental testing at GSFC *Contamination, Coatings and Materials workshop*

[10] ASTM 2019 Standard extraterrestrial reference spectrum ASTM E490-00a

IOP Conf. Series: Materials Science and Engineering **1328** (2025) 012011 doi:10.1088/1757-899X/1328/1/012011

On-ground and in-orbit experiments: new insight to reveal Atomic Oxygen Induced Contamination

Riyo Yamanaka[1*], Delphine Faye[2]

[1] Japan Aerospace Exploration Agency, Tsukuba, Japan
[2] Centre National d'Etudes Spatiales, Toulouse, France

*E-mail: yamanaka.riyo@jaxa.jp and delphine.faye@cnes.fr

Abstract. Lots of satellites with equipment more and more sensitive to contamination operate in Low Earth Orbit and can be exposed to high Atomic Oxygen fluence. Therefore, it is paramount to highlight the risks of damage according to the chemical nature of materials and quantify the level of potential induced contaminants produced by erosion and oxidation that may alter the surface properties in close vicinity[1, 2]. AIC (Atomic oxygen Induced Contamination) topic is part of the research themes defined for collaborative studies between Japanese and French space agencies[3, 4]. First AIC test campaigns were performed in JAXA facility to identify the nature of the contaminants potentially generated by AO impingement on polymer materials and to characterize their effects. As a result, it was clarified that some materials commonly used on spacecraft could generate mainly particles. Next step is to compare the effect of AO preliminary observed during ground tests on various polymers such as those used on the International Space Station or on LEO satellites with the real degradation of the same kinds of polymers that will be exposed on EMA (Euro Material Ageing)/SESAME experiment[5]. This experiment was launched by SpaceX on the 5th of November 2024 and installed on ram face of ISS for more than one year. Thus, it will offer a unique opportunity to better understand the impact of space environment parameters and especially the effect of AO and AIC generation. As for ground tests where conductive carbon double-sided tapes were used as collector materials to capture AIC contaminants, the same double-sided tapes have been placed on EMA/SESAME to evaluate the possible redistributed particulate contamination coming from the eroded polymers. This paper presents the results on AIC from previous ground test campaigns and will focus on the analysis plan to characterize the double-sided tapes not only after ground tests but also after EMA/SESAME retrieval.

1. Introduction

Space is a hostile environment both complex and dynamic. Synergistic effects of plasma, charged particles, thermal cycling, photons... can cause severe damage on spacecraft. Atomic oxygen (AO) resulting from the photodissociation of oxygen molecules by UV radiation is of particular interest as it is the major constituent in Low Earth Orbit whose density greatly depends on solar activity. Due to its strong chemical reactivity and high-speed impact (8 km/s), it is well known that AO embrittles any organic material exposed on ram faces by oxidation and significant erosion[1,2].

However, are all polymers degraded in the same way? What happens to the potential microparticles or chemical byproducts that may be generated? Could such AO induced contaminants (AIC) affect the performance of equipment in close vicinity? Thus, since it is paramount to highlight the risks of AIC, JAXA and CNES decided to jointly carry on activities on this specific issue as one of the focused topics of their research cooperation through ground test campaigns and ISS materials exposure experiment.

2. On ground experiment

2.1 Concept of AIC test

The conceptual diagram of our atomic oxygen induced contamination (hereafter AIC) test is shown in Figure 1. Collector plates consisted of aluminum plate and carbon double-sided tape are set on the base jig and the material sample under evaluation is fixed on the jig with two strip-shaped aluminum masks and installed diagonally. In addition, direct exposure to AO is avoided by the AO shield jig which is set over the surface of collector plates as shown in Figure 1. If particulate AIC is induced by AO, AIC can be collected by collector plates. Figure 2 shows the image of AIC test jig.

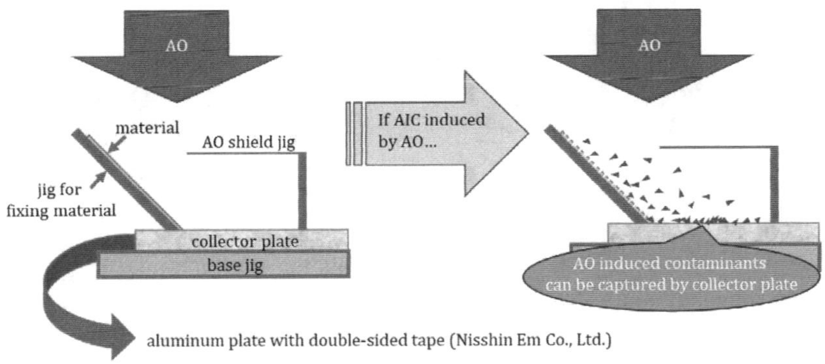

Figure 1. Conceptual diagrams of AIC test (© JAXA)

Figure 2. Image of AIC test jig (© JAXA)

2.2 Exposed materials

15 types of materials used as insulation, thermal control coatings, composites and adhesives on spacecrafts were evaluated in the viewpoint of AIC. The detailed information is as follows:

- Thermal insulation: Beta Cloth CF500 F (Polytetrafluoroethylene-impregnated glass cloth),
- Thermal control coatings: MAP paint SG121FD, MAP SCK5N (silicone),
- Thermal control films: Ag coated FEP, FEP(fluoropolymer), BSF-30 (polyimide), Upilex-S 125 μm (polyimide), PEEK (polyaryletherketone),
- Structural adhesive: Scotchweld EC2216(epoxy),
- Composites for high dimensional stability structure: CFRP G969/RTM6 (Carbon Fiber Reinforced Epoxy polymer), CFRP EX1515/M55J (Carbon Fiber Reinforced Cyanate-ester prepreg resin),
- Solar cells adhesives: Nusil CV 2566, RTV-S691, MAPSIL QS 1123 (silicone)
- Potential space material candidate with very high strength and excellent thermal stability: Zylon (poly(p-phenylene-2,6-benzobisoxazole) (PBO fiber))

2.3 Test conditions

AO irradiation tests were planned by using Combined Space Effects Test Facility at Tsukuba Space Center (see picture on Figure 3). The facility can irradiate materials with AO under high vacuum environment ($< 10^{-2}$ Pa) at approximately room temperature[4]. All of the 15 types of materials which are explained in section 2.2 were irradiated by AO in this facility.
Six AO irradiation tests were performed. For these tests, AO beam average speed was 8.11 km/s and AO fluence approximately 1.0×10^{21} atoms/cm^2.

Figure 3. Picture of Combined Space Effects Test Facility @ JAXA, Tsukuba Space Center (© JAXA)

2.4 Example of results for Beta cloth[3, 4]

Among the tested materials, the irradiated composites were significantly affected that was directly confirmed not only by visual change of the surface between an exposed and unexposed sample but also with mass measurement indicating a non-negligible mass loss of the irradiated sample with respect to a masked reference sample inside the chamber.
Particles and mainly fibers, some of several hundreds of µm long and even more, were captured by the adhesive collector plate located just below the irradiated samples.

As an example, some noteworthy results regarding especially Beta cloth material, a fluoropolymer coated woven fiberglass fabric, are presented hereafter that demonstrated the generation of AIC contaminants.
Irradiated samples and particles were inspected with optical means and analyzed by different combined techniques such as Scanning Electron Microscopy (SEM) and Energy Dispersive X-ray Spectroscopy (EDS) (see Figures 4 and 5). This showed that the PTFE coating of Beta cloth was indeed eroded, thus revealing the fibers more resistant to AO but more easily frangible.
As it can be seen on Figure 6, fluoride and carbon elements from PTFE coating disappeared on the exposed side whereas silicon, oxygen and other chemical elements (e.g. calcium) of the glass fibers appeared.

Figure 4. Visual change between exposed and unexposed sides of Beta cloth sample (© CNES)

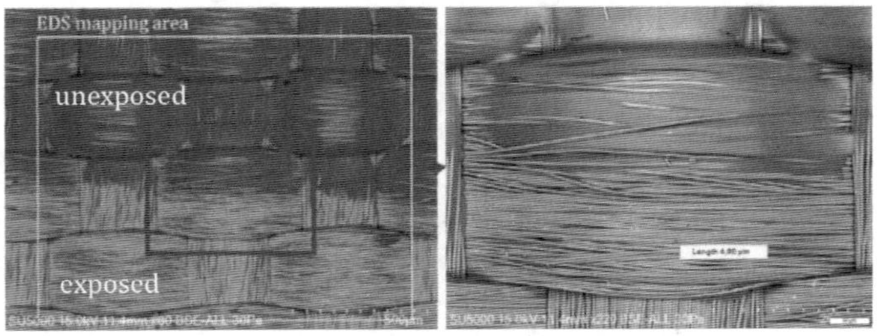

Figure 5. SEM images of exposed and unexposed sides of Beta cloth sample (© CNES)

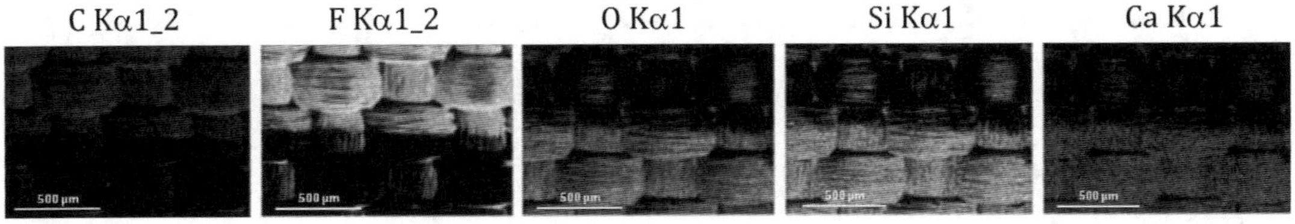

Figure 6. SEM/EDS analysis results on Beta cloth (Hitachi SEM and Oxford EDS spectrometer, 15 kV and working distance of about 10 mm) (© CNES)

The chemical identification of the captured fibers composition and their comparable diameter with the diameter of the eroded fibers remaining on the irradiated samples (see Figure 7) confirmed their origin as AIC particles coming from the embrittled Beta cloth material.

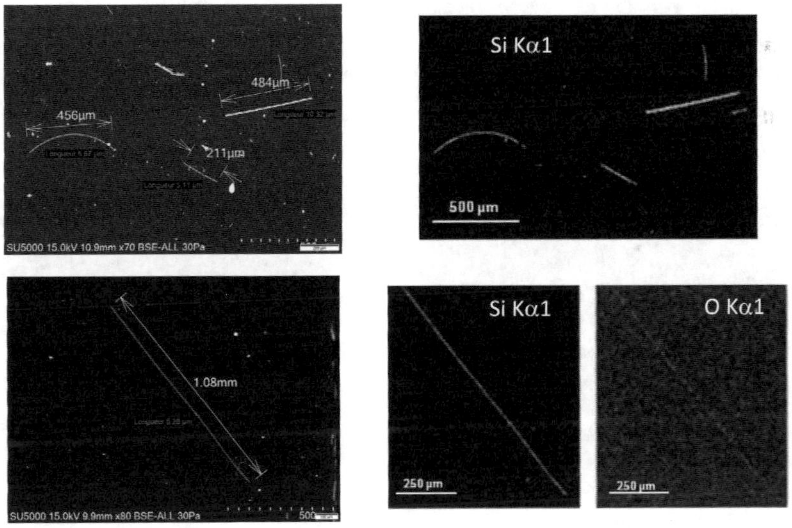

Figure 7. Curved and other straight fibers of different lengths, captured on collector plate (left) and corresponding chemical composition (Silicon and Oxygen) (right) (© CNES)

3. Are double-sided tapes compatible for flight?

As a new opportunity of materials exposure in Low Earth Orbit was offered with SESAME experiment, one of the two payloads aboard Euro Material Ageing (EMA), a platform developed by CNES to be installed on the Bartolomeo payload hosting facility on the International Space Station, it was suggested to position some similar double-sided tapes at different locations on SESAME structure as collector materials (material reference: carbon double-sided tape from Nisshin Em Co., Ltd., Japan, same batch as for EMA/SESAME) . Indeed, in case of degradation of the various materials exposed to atomic oxygen in ram face (more than 140, mainly polymers), AIC contaminants may be generated and opportunely captured on tapes in vicinity of eroded surfaces like it was specifically observed on ground with composites and other fabrics. After retrieval about one year after launch, the material samples and collector tapes will be characterized with identical analytical techniques to compare the degradation and AIC to on-ground results. However, the main issue before using the double-sided tapes in orbit was to confirm their compatibility with the harsh environment of vacuum and AO exposure in terms of adhesion properties and their ability to keep trapping particulate contamination.

3.1 Compatibility assessment of the double-sided tapes as AIC collectors

In order to assess their capability to hold particles when exposed on EMA to LEO environment, the double-sided tapes were tested under several vacuum conditions. A thermal cycling test was performed at CNES as a mandatory project qualification test to evaluate the risk of detachment of the adhesive samples. Other tests were also performed at JAXA with sequences combining exposure to vacuum with and without (w/o) AO at ambient temperature.

3.1.1 Results after thermal vacuum cycling test

A series of 100 thermal cycles were performed under vacuum (pressure < 10^{-4} Pa) in the temperature range between -50°C and +70°C on two representative substrates as those foreseen for the location of the tapes on EMA. Therefore, several samples were sticked on the mock-up, Figure 8, both on aluminum surface passivated with Surtec 650® treatment and on black anodized aluminium.

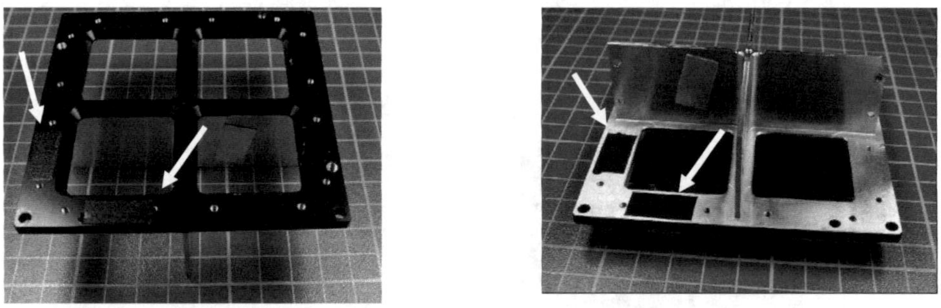

Figure 8. View of sticked samples on black anodized aluminium (left) and on passivated aluminium with low of particles Surtec 650® treatment (right) (© CNES)

Calibrated beads of two sizes (20 µm Borosilicate glass and 100 µm Polystyrene DVB microspheres from Thermo Scientific) were used to assess the holding capability of the double-sided tapes (see Figure 9). They were deposited on a clean microscope glass slide and sprinkled on the samples after the thermal vacuum cycling test. Then, pictures were compared before and after a manual tapping test on the back of the substrate. No change in particles position was observed, as shown on Figure 10.

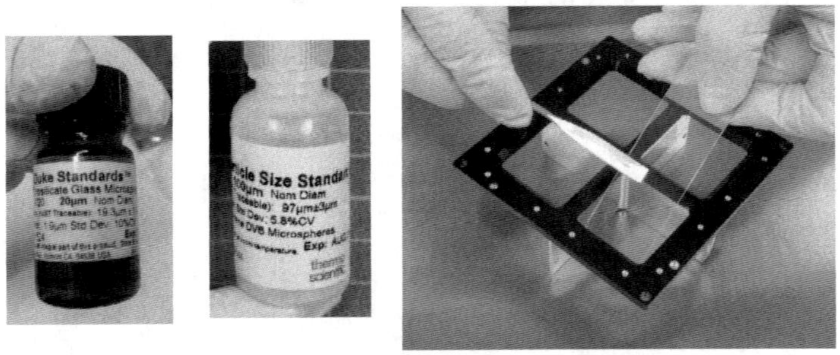

Figure 9. Calibrated microspheres and sprinkling method (© CNES)

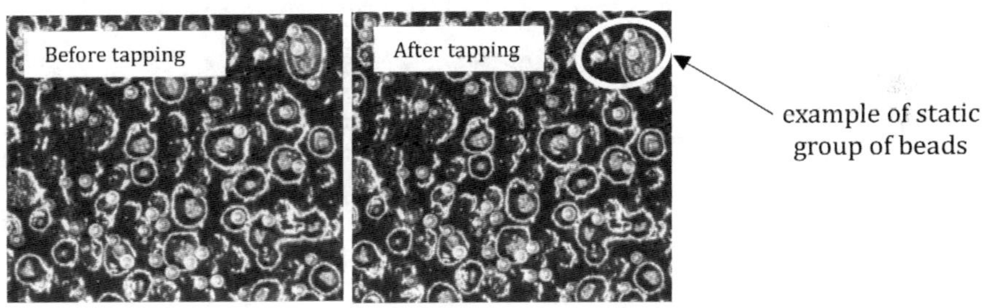

Figure 10. Microspheres under binocular magnifier before and after the tapping test (© CNES)

3.1.2 Results after vacuum and AO exposure tests

Figure 11 shows the flow of environmental tests for determining the particles holding capability of double-sided tapes after exposure to vacuum environment at a pressure lower than 10^{-2} Pa, under less than room temperature and with or w/o AO.

First, 4 samples with the same double-sided tape area were prepared using the method displayed in Step 1 using SUS (stainless steel material) plates and rings. After that, 2 samples were irradiated by AO (ID: 1 and 2) under vacuum and 2 others (ID: 3 and 4) were exposed to vacuum only at Step 2. At Step 3, particles (zirconia beads) were sprinkled and the particles were counted both before and after upside down the samples at Step 4. All samples (ID: 1,2,3 and 4) were irradiated by AO under vacuum at Step 5. 2 samples (ID: 1 and 3) were irradiated by AO under vacuum and 2 others (ID: 2 and 4) were exposed to vacuum only at Step 6. Finally, the particles retained on each of 4 samples were counted after once upside down them at Step 7.

Table 1 shows the results of Step 4 and Step 7. By comparing the number of particles on Step 4 (before and after upside down samples) between the 2 samples (ID: 1 and 2) and the 2 samples (ID: 3 and 4), it becomes obvious that vacuum environment has no effect on the particles holding capability of the double-sided tapes whereas AO reduced holding capability of double-sided tapes. However, it was clarified that once particles are held by the double-sided tapes, vacuum environment and AO have little effect on the particles holding capability of double-sided tapes by comparing the number of particles between the 4 samples (ID: 1,2,3 and 4) of Step 4 (after upside down samples) and the 4 samples (ID: 1,2,3 and 4) of Step 7.

IOP Conf. Series: Materials Science and Engineering **1328** (2025) 012011 doi:10.1088/1757-899X/1328/1/012011

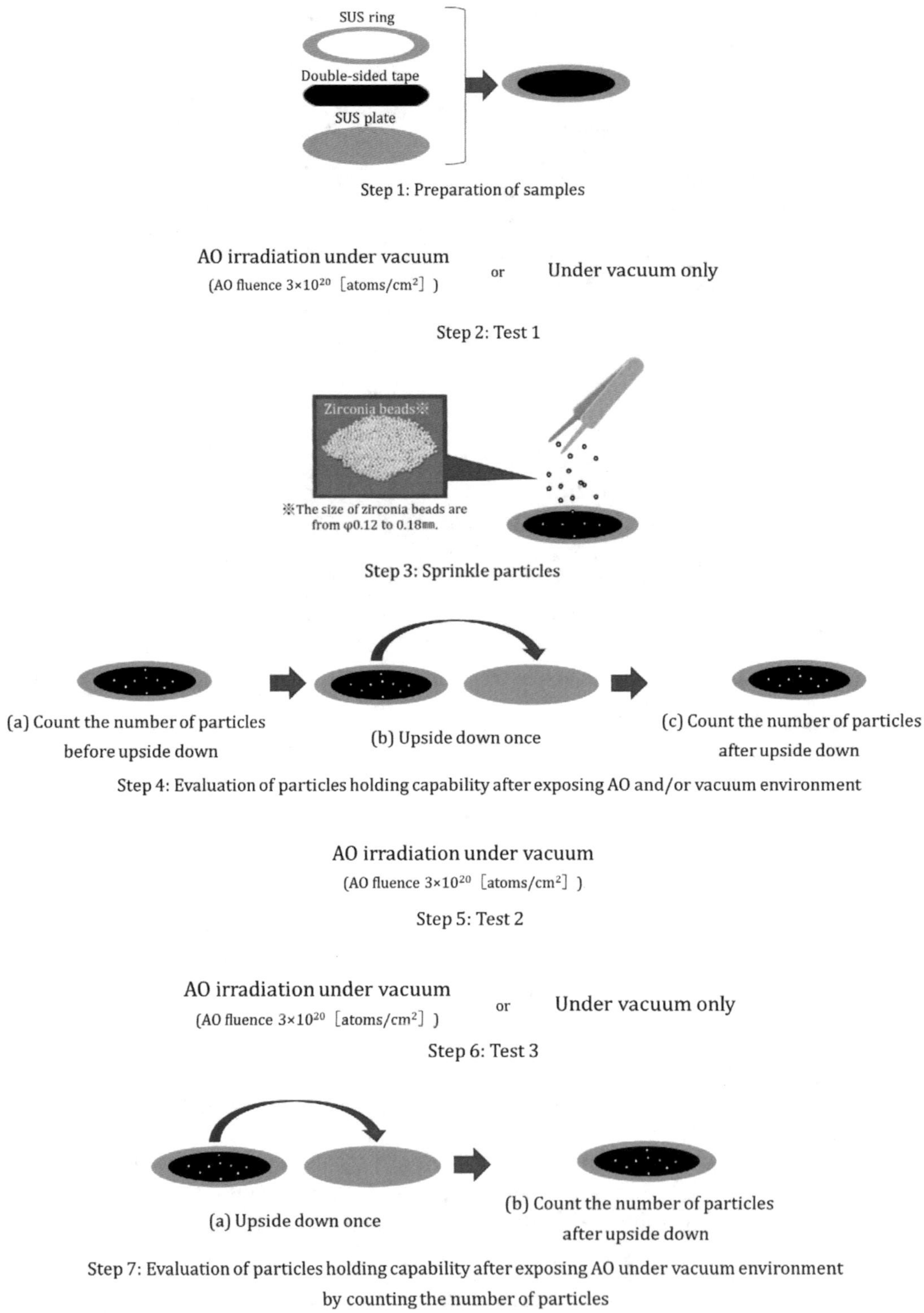

Figure 11. Flow of particles holding capability tests on double-sided tapes under vacuum and or w/o AO (© JAXA)

IOP Conf. Series: Materials Science and Engineering **1328** (2025) 012011 doi:10.1088/1757-899X/1328/1/012011

Table 1. Results of compatibility assessment of the double-sided tapes as AIC collectors after vacuum and AO exposure tests

Sample ID	Step 2 Test 1	Step 5 Test 2	Step 6 Test 3	Number of particles		
				Step 4 Before upside down	Step4 After upside down	Step 7 After all steps
1	AO	AO	AO	63	8	7
2	AO	AO	Vac.	180	15	13
3	Vac.	AO	AO	113	113	113
4	Vac.	AO	Vac.	97	97	97

Further to the qualification vacuum tests at different temperatures, with and w/o AO, the double-sided tapes were not damaged and their adhesive properties were confirmed. It was also demonstrated that particles of different chemical natures and sizes could be kept trapped without moving and thus they were accepted for flight.

3.2 Implementation of double-sided tapes on EMA/SESAME

Ten pieces of carbon double-sided tapes were allowed to be installed as AIC collectors at recommended locations not only on the structure in ram face of the flight experiment but also on other accessible surfaces on another faces (see Figure 12). Their original protective cover was kept in place for their integrity during launch until the last operations in ISS and for this specific purpose, they were numbered and covered by a kapton® tape designed to be easily removed by the astronauts.

zoom on a sample holder

on the edges of the sample holders (in red)

close to the electrical connector (back side)

Figure 12. Locations of double-sided tapes (© CNES)

IOP Conf. Series: Materials Science and Engineering **1328** (2025) 012011 doi:10.1088/1757-899X/1328/1/012011

4. Analysis plan

After retrieval, all double-sided tapes will be characterized in a clean environment to avoid any parasitic effect of ground contaminants and then obtain confidently the occurrence of AIC in the space environment on the ISS.
Therefore, the methodology for analyzing the retrieved double-sided tapes is very important. The main analysis methods and their objectives at this time are shown in Table 2. These analysis methods will continue to be reviewed and updated to select the most appropriate ones for the exposed tapes.

Table 2. Analysis plan for double-sided tapes after EMA/SESAME retrieval

Analysis methods	***Objectives***
Inspection (photography, microscopy)	Appearance, color change → Evaluation of space environmental effect
Inspection (binocular magnifier)	Number and shape of particles → Classification and distribution of particles, evaluation of AIC environment on ISS orbit
Fluorescence @ different wavelengths	Particle characteristics → Search for the origin of particles
Topography (profilometry)	Erosion depth on materials → Evaluation of space environmental effect
Topography (Atomic Force Microscopy)	Detailed shape measurement of particles → Classification of particles, evaluation of space environmental effect
Elemental analysis (SEM/EDS)	Particle elemental data → Search for the origin of particles, evaluation of space environmental effect
Elemental analysis (Electron Probe Microanalyzer)	Mapping elemental data → Search for the origin of particles, evaluation of AIC environment on ISS orbit

5. Conclusion

AO ground test campaigns confirmed the generation of AIC (particles and fibers) mainly coming from degraded composite materials. This kind of contamination should occur in orbit.
Then, the double-sided tapes on EMA/SESAME experiment should capture AIC particles. After retrieval, they will be characterized according to the proposed analysis plan and the results compared to those of another traps such as aerogels. That will help to confirm the origin and the redistribution of AIC particles. Additionally, adsorbents, optical witnesses, as well as a microbalance will also be exposed to collect potential molecular contaminants. The in-orbit data will be compared to the results obtained from ground tests with the same materials and shared to the space community to improve the understanding of the degradation of the materials and the induced effects such as AIC.

Acknowledgements

The authors would like to thank Mecano-ID, MAP Coatings and CNES for providing the different materials for the irradiation tests as well as the staff from the thermal department at CNES for the supervision of the thermal cycling test. They are grateful to ELEMCA, THALES SIX GTS FRANCE SAS for SEM/EDS analyses. The authors wish also to thank the staff of Advanced Engineering Services Co., Ltd., for their support in this study.
Finally, the authors appreciate EMA project team for giving them a great opportunity to evaluate AIC environment on the ISS orbit.

References

[1] Tagawa M and Yokota K 2008 *Acta Astronautica* 62 203–211
[2] De Groh K K, Banks B A, Miller S K R and Dever J A 2018 Degradation of Spacecraft Materials Handbook of Environmental Degradation of Materials *3rd ed Elsevier Science*
[3] Yamanaka R, Faye D, Kimoto Y and Maillard N 2023 *IOP Conf. Series: Mater. Sci. Eng.* 1287 012029
[4] Yamanaka R and Faye D 2024 *Journal of the Japan Society for Aeronautical and Space Sciences* 72 5 pp 163-170
[5] Suliga A et al 2023 *IOP Conf. Series: Mater. Sci. Eng.* 1287 012026
[6] Shimamura H, Baba S and Miyazaki E 2011 Handbook of the Combined Space Effects Test Facility JAXA-RM-10-013

IOP Conf. Series: Materials Science and Engineering **1328** (2025) 012012 doi:10.1088/1757-899X/1328/1/012012

Laser cleaning of molecular contamination on optical components

Frank Wagner[1*], Amer Aoun[1,2], Richard-Nicolas Verrone[1] and Julien Eck[3]

[1] Aix Marseille Univ, CNRS, Centrale Med, Institut Fresnel, Marseille, France
[2] Centre National d'Études Spatiales, 18 Avenue É. Belin, 31401 Toulouse, France
[3] European Space Research and Technology Centre, Noordwijk, The Netherlands

*E-mail: frank.wagner@fresnel.fr

Abstract. Molecular contamination is a problem during long term storage of space instruments to be launched later. We investigate laser cleaning of molecular contamination on optical materials aiming to make a first step towards the development of a contactless remote cleaning method that allows the cleaning of critical optical components without disassembling the instrument. A nanosecond laser at 266 nm was used and, depending on the homogeneity of the beam profile, good results were obtained on ZnSe, CaF_2 and protected silver mirrors.

1. Introduction

Satellite constellations operating several identical or very similar satellites are more and more common. For best cost efficiency these satellites should be produced in parallel or at least one right after the other, even if their launch dates may be separated by several months or even years. However, after assembly, integration, testing and long term storage, even in clean room conditions, molecular contamination is observed on the surfaces. This molecular contamination originates from partial condensation of the outgassing species of organic materials in the storage conditions. Particularly for optical components, the layer or the droplets of molecular contaminants may cause important performance loss during operation of the satellite in vacuum. Thus cleaning the critical elements from molecular contamination is necessary before launch.

However, cleaning of these elements may not be possible as they may be difficult to reach (with a wiping tissue) after final system integration. This is the reason why in this study we explore the feasibility of non-contact cleaning by laser irradiation in the presence of oxygen and humidity.

Former studies worked mainly on laser cleaning processes for the microelectronics industries. In this context, mostly the cleaning of particulate contamination with UV gas lasers (Excimer lasers) was studied. From 1988 to 1995, a company, STI Optronics, developed an excimer laser based cleaning procedure for the semiconductor and aerospace industries [1] which has also been tested on telescope mirrors [2]. They are among the few people who tested and reported on cleaning of molecular contamination [1] in addition to particulate contamination. The most important results of their and preceding studies (see references in [1]) are that (i) UV lasers are better suited than IR lasers and (ii) that the samples need to be illuminated with a flat-top beam profile.

The choice of UV- rather than IR-lasers can be explained by the tendency of IR lasers to cause selective cleaning of some molecules [3]. The same fact is exploited advantageously in FTIR spectroscopy in order to identify / distinguish organic molecules.

The importance of a flat-top beam profile has been explained by Osiecki and Magee who stated that irradiation with small fluences needs to be avoided in order to avoid 'baking' of the contaminants. From today's point of view, one would rather say that we need to avoid the so-called 'laser-induced contamination' (LIC) effect which is mostly know from UV lasers operating in vacuum [4–6] but which can also cause typical LIC deposits in air [7].

It is also interesting to mention that cleaning of particulate contamination works much better if some humidity and at least one hydrophilic material (the substrate or the particles) is present [8,9]. Further, Grojo and co-workers showed that the fluence range for real laser cleaning, i.e. removal of contamination without causing damage to the substrate, may be rather narrow, at least for particle cleaning where the locally enhanced electric field may create a nanometric hole in the substrate right under the particle [9]. (The fluence of a pulsed laser beam is defined as the pulse energy per surface area in common units of mJ/cm^2.)

2. Experimental details

2.1 The choice of the laser

The UV lasers used in the literature are all Excimer lasers (with nanosecond pulse duration) which today became quite rare in research labs due to the environmental and safety regulations. These safety regulations cause high cost of ownership for such lasers that employ toxic gas mixtures including fluorine or chlorine. However, these lasers had the advantage of low spatial coherence which simplified the creation of flat-top beam profiles.

The modern UV nanosecond lasers are usually frequency converted Nd:YAG lasers emitting at 355 nm or 266 nm, which has to be compared to the Excimer laser wavelengths 351 nm, 308 nm, 248 nm, 193 nm and 157 nm. We thus chose the 4th harmonic at 266 nm to carry out our laser cleaning tests.

2.2 Beam homogenization

The near field beam profile at the 4th harmonic of our Innolas Spitlight 250 is shown in Figure 1a. It shows strong hot spots and is thus difficult to homogenize. Two approaches for the

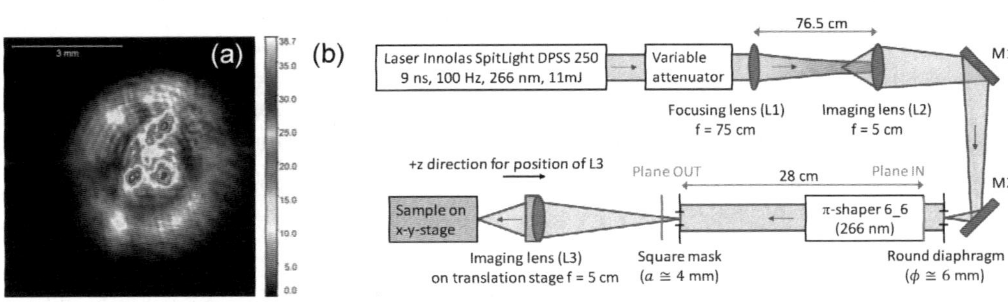

Figure 1. (a) The near field beam profile delivered by our laser at 266 nm. (b) A schematics of the setup using the *refractive* beam homogenizer. Its length is approximately 4 m and the transmission (pulse energy on the sample / pulse energy after the attenuator) is only 2%.

homogenization of UV laser beams are available: (i) refractive homogenization which needs a Gaussian beam profile as input and (ii) diffractive homogenization which can in principle deal with any input beam profile, if it was designed on purpose.

During this project we tested both types of homogenizers using a π-shaper from Edmund Optics and an off-the-shelf diffractive homogenizer from Holo/or.

In order to generate a beam profile that is acceptable for the refractive beam homogenizer a 4m long setup with low transmission was necessary (Figure 1b). We then placed a square stencil mask in the output plane of the homogenizer and imaged it onto the sample surface using a demagnification to obtain a fluence of up to 100 mJ/cm^2.

Using the diffractive beam homogenizer improved the transmission of the setup by a factor of 16, but provided a speckled beam with a strong zero order if used in the standard way with a focusing lens after the diffractive optical element (DOE). The conclusion of this trial was that one should never buy an off-the-shelf DOE if the laser has a complicated near field beam profile. We finally used the DOE in the non-standard way described in Figure 2a and obtained the beam profile in Figure 2b which should be compared to the beam profile obtained using the refractive homogenizer (Figure 2c).

In summary, the refractive beam homogenizer provided a much more homogeneous irradiation than the diffractive beam homogenizer. The latter suffered from speckles and zero order intensity. This observation may however not be generalizable to custom tailored diffractive optical homogenizers.

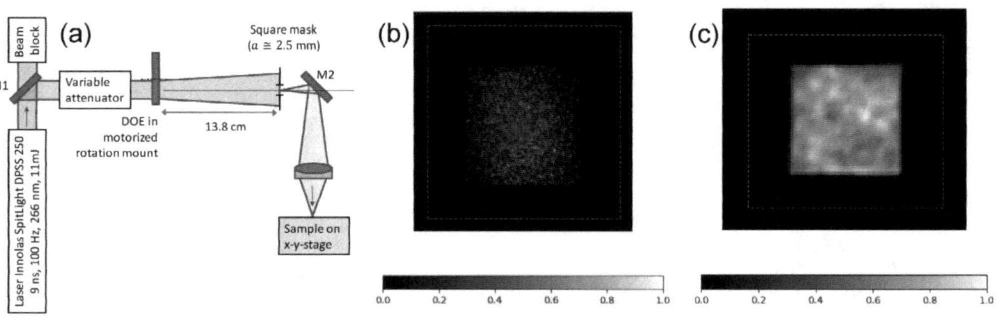

Figure 2. (a) A schematics of the setup using the *diffractive* beam homogenizer. Its length is approximately 1 m and the transmission (pulse energy on the sample / pulse energy after the attenuator) is 33%. (b) and (c) show typical beam profiles obtained with both setups. Both are normalized to the strongest pixel in the image. (b) shows the beam obtained with the *diffractive* homogenizer. The ratio of the peak intensity and the mean intensity is very high (≈8). (c) shows the beam obtained with the *refractive* homogenizer. The peak / mean ratio is less than 1.8.

2.3 Samples and characterization methods

As the standard characterization method for molecular contamination is transmission FTIR spectroscopy [10], we mainly tested contaminated FTIR samples made from ZnSe and CaF_2. These two materials can in fact be considered representative for many optical components as ZnSe absorbs and CaF_2 transmits the cleaning wavelength of 266 nm. We additionally used commercial off-the-shelf protected silver mirrors. These absorb the cleaning wavelength but are not

IOP Conf. Series: Materials Science and Engineering **1328** (2025) 012012 doi:10.1088/1757-899X/1328/1/012012

transmitting the IR radiation necessary for FTIR measurements. We thus characterized these samples simply by optical microscopy using differential interference and bright field contrasts.

All samples have been contaminated at ESA ESTEC and were then sent to the Institut Fresnel for cleaning. Only vacuum outgassing provided homogeneous contamination of reasonable strength. This method has been used to contaminate all three types of samples with the vapours stemming from an epoxy adhesive and (separately) from a silicone adhesive. However, solvent drop evaporation was also used to contaminate ZnSe samples using DC200 and paraffin. These samples were heavily and inhomogeneously contaminated

For quick feedback we defined the relative cleaning success, *s*, as the relative change in absorption of the strongest absorption line in the FTIR spectrum:

$$s = \frac{(1-T)_{\text{Before}} - (1-T)_{\text{After}}}{(1-T)_{\text{Before}}}$$

Here the suffixes "Before" and "After" refer to measurements before and after the cleaning test and T is the transmission of the sample at one of its absorption peaks. Defining the cleaning success (or the relative cleaning efficiency) by this expression provides results that are rather independent from the choice of the baseline for the FTIR spectra, provided that the same baseline parameters were used for the measurements before and after the cleaning test.

3. Results of the cleaning tests

3.1 Discussion of the irradiation parameters

The processes that occur on the sample when it is irradiated mainly depend on the fluence (=energy per pulse and per surface) of the irradiation (Figure 3). The full sample surface should be irradiated with fluences in the range where laser cleaning is obtained. This means that we need to irradiate at sufficiently high fluences (and small pulse numbers per site) to avoid cross-linking (LIC, baking), but we also need to irradiate at sufficiently low fluences to avoid laser induced damage.

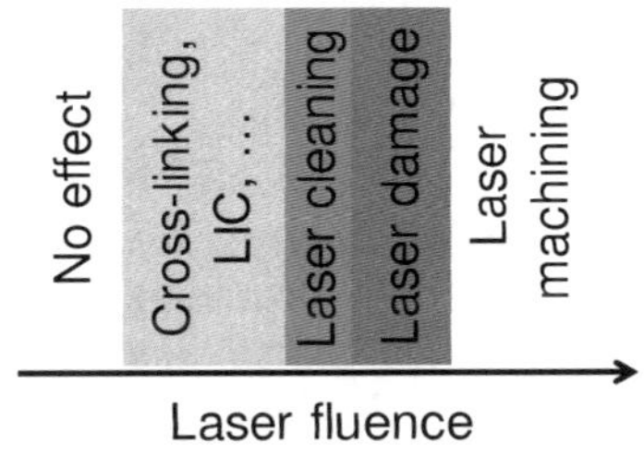

Figure 3. A schematics of the effects that increasing laser fluence has on a contaminated sample. The fluence range where laser cleaning is obtained may be narrow or may not even exist for some substrate-contamination combinations.

The number of pulses per site is the second parameter than can be adjusted during the irradiation. According to the literature, 4 – 16 pulses should be sufficient in case the fluence is well chosen [1]. In practice the number of pulses per site depends on the irradiated surface on the sample, the pulse repetition rate of the laser and the scanning speed of the sample which we move with a motorized x-y-table.

3.2 Results for the setup with the refractive homogenizer
With the refractive homogenizer we obtained a rather flat-top beam with a ratio between peak fluence and mean fluence being less than 1.8. The maximum fluence however was limited to approximately 100 mJ/cm^2 which was not sufficient for cleaning CaF_2 substrates that are transparent to the cleaning wavelength.

This setup was well suited for cleaning ZnSe substrates as long as they were not contaminated too heavily (Figure 4).

ZnSe contaminated with **paraffin** by drop evaporation:

➡ Rel. success = **96%**

Transmission FTIR at 2924 cm^{-1} on 4.5 x 4.5 mm²:
$(1\text{-}T)_{BEFORE}$ = 6.8 %
$(1\text{-}T)_{AFTER}$ = 0.3 %

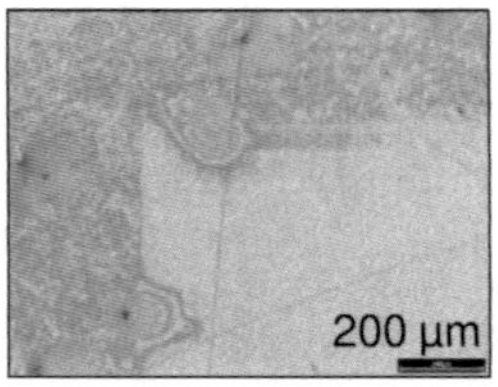

Figure 4. The setup with the refractive homogenizer is well suited for cleaning substrates absorbing the cleaning wavelength, like ZnSe. This sample was quite heavily contaminated with paraffin. We used 67 mJ/cm^2 and 4 pps (pulses per site).

The sample was irradiated with 4 pps (pulses per site). This means that the step between two laser pulses is half of the irradiated square surface in both directions. The center of the scanned area is thus exposed to 4 pps, but the frame of the width of the irradiated surface is exposed to less pulses. Due to this effect and due to locally stronger contamination, one can see an only partially cleaned band on the top of the cleaned surface in Figure 4.

Unfortunately, this setup was no longer available when we tried to clean the protected silver mirrors from adhesive outgassing. It would certainly have provided better results than the setup with the diffractive homogenizer.

3.3 Results for the setup with the diffractive homogenizer
With the diffractive homogenizer we obtained a speckled beam with a ratio between peak fluence and mean fluence being more than 8. The maximum fluence however was much higher than for the other setup. This allowed us to clean CaF_2 substrates that are transparent to the cleaning wavelength (Figure 5).

Due to the high laser damage threshold of CaF_2 the strong speckles did not damage the substrate as far as we can tell from our investigations on these FTIR samples that did not have good optical quality surfaces.

We also tried to clean the protected silver mirrors with this setup, but ran into problems as part of the cleaning light is absorbed in the metal and the bad homogeneity of the beam makes it difficult to find a fluence where the contamination is removed without damaging the mirror. Figure 6 shows the best compromise we found for a protected silver mirror that was contaminated with silicone adhesive outgassing.

IOP Conf. Series: Materials Science and Engineering **1328** (2025) 012012 doi:10.1088/1757-899X/1328/1/012012

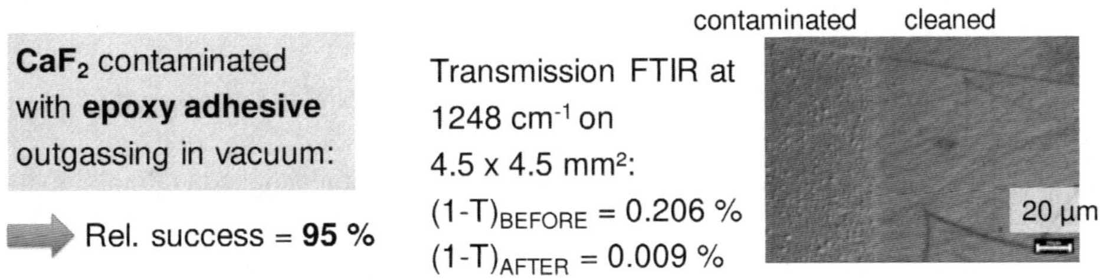

Figure 5. The setup with the diffractive homogenizer is suited for cleaning substrates that do not absorb the cleaning wavelength, like CaF_2. We used 9300 mJ/cm^2 and 4 pps.

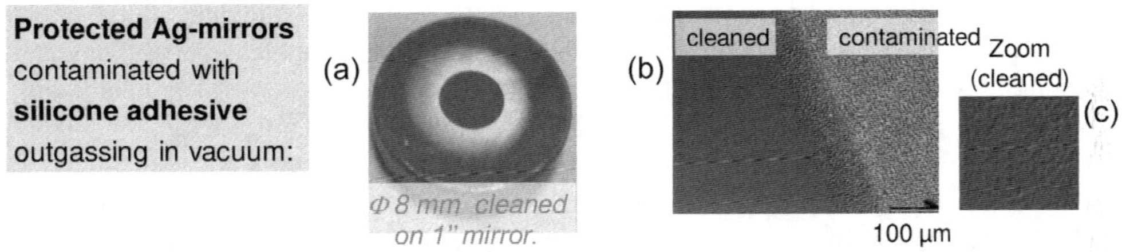

Figure 6. The setup with the diffractive homogenizer is not well suited for cleaning substrates that absorb the cleaning wavelength, like protected silver mirrors. This figure shows the best compromise that was found. We used 80.5 mJ/cm^2 and 6400 pps. (a) A photograph showing the diffusion of the contamination and the cleaned area. Some speckle damage is visible in the bigger microscope image (b), whereas some remaining roughness originating from the contamination droplets is visible in the smaller microscope image (c).

On the photograph (Figure 6a), the irradiated zone looks nicely cleaned as the diffusing effect of the contamination droplets vanished completely. However, the microscope pictures reveal some roughness in the cleaned area. Some speckle damage is visible in Figure 6b. We can recognize speckle damage due to their diagonal orientation, caused by the rotation of the DOE superposed with the horizontal scan of the sample (Figure 2a). On the smaller microscope image (Figure 6c) one can see some remaining roughness originating from the contamination droplets.

A consequence of using this not well suited setup for cleaning the mirrors is that we need many pulses per site (6400 pps) to achieve a reasonable cleaning effect. Using the setup with the refractive homogenizer would have been much better for this task.

Trying to clean epoxy adhesive outgassing from these mirrors was even more difficult. Significant speckle damage occurred before the contamination droplets were removed.

4. Summary and conclusions

In this project we investigated nanosecond laser cleaning at 266 nm wavelength as a means to remove molecular contamination from optical components. Laser cleaning would have the advantage of being a remote and contact less cleaning method and would thus ideally allow to clean the optical components without disassembling the instrument. Earlier laser cleaning tests on molecular contamination were carried out in the 1990s using excimer lasers. Today's standard

UV lasers have higher spatial coherence and are thus more difficult to homogenize. This is an important point as the laser cleaning fluence range may be quite narrow, making it necessary to irradiate the sample with a flat top beam profile in order to avoid laser damage (if the fluence is too high) and cross-linking of the contaminants (if the fluence is too low). The risk of degrading the surface is high if an inhomogeneous beam is used on an absorbing surface.

A refractive beam homogenizer was tested with success for cleaning of absorbing substrates. However, the near field beam profile of our laser was not well suited to the usage of the refractive beam homogenizer which limited us to fluences below 100 mJ/cm^2.

An off-the-shelf diffractive beam homogenizer was also tested but yielded a beam profile exhibiting strong speckles. In spite of this bad beam profile, it was possible to clean CaF_2 substrates that are transmitting the cleaning wavelength and have a high laser damage threshold.

Using the speckled beam on absorbing substrates (protected silver mirrors) could lead to a limited cleaning effect depending on the nature of the contaminant, but required using a high number of pulses per site and induced some damages of the size of a speckle.

If laser cleaning was to be used on satellite instruments one needs however to take this into consideration during the design phase of the instruments. Access paths for the cleaning beam and shields to protect possible neighbouring sensitive elements from the ablation products of the cleaning process need to be integrated in the instrument design.

References

[1] Osiecki R A and Magee T J 1990 Ultraviolet laser cleaning of mirrored surfaces *Proc. SPIE* **1329** 127–133.

[2] Kimura W D, Kim G H and Balick B 1995 Comparison of laser and CO_2 snow for cleaning large astronomical mirrors, *PASP* **107** 888.

[3] Mihesan C, Ziskind M, Chazallon B, Therssen E, Desgroux P, Gurlui S, Focsa C 2006 IR wavelength-selective laser desorption via OH and CH stretching modes *Applied Surface Science* **253** 1090–1094.

[4] Aoun A, Wagner F, Natoli J-Y, Gineste T, Faye D 2025 Laser-induced contamination: analyzing deposits from epoxy outgassing under 355 nm laser irradiation *IOP Conf. Ser.* **this volume**

[5] Riede W, Schroeder H, Bataviciute G, Wernham D, Tighe A, Pettazzi F, Alves J 2011 Laser-induced contamination on space optics *Proc. SPIE* **8190** 81901E.

[6] Wagner F R, Gebrayel El Reaidy G, Faye D, Natoli J-Y 2020 Laser induced deposits in contaminated vacuum environment: optical properties and lateral growth *Optics & Laser Technology* **122** 105889.

[7] Stehlik M, Wagner F, Zideluns J, Lemarchand F, Lumeau J, Gallais L 2021 Laser-induced contamination on dielectric coatings in sub-ps MHz regime at 515 nm *Proc. SPIE* **11910** 119101Q.

[8] Fourrier T, Schrems G, Mühlberger T, Heitz J, Arnold N, Bäuerle D, Mosbacher M, Boneberg J, Leiderer P 2001 Laser cleaning of polymer surfaces *Appl Phys* A **72** 1–6.

[9] Grojo D, Delaporte Ph, Sentis M, Pakarinen O H, Foster A S 2008 The so-called dry laser cleaning governed by humidity at the nanometer scale *Appl. Phys. Lett.* **92** 033108.

[10] European Cooperation for Space Standardization 2024 Detection of organic contamination of surfaces by infrared spectroscopy *European Cooperation for Space Standardization* **ECSS-Q-ST-70-05C Rev.2**.

Analyzing and cleaning laser-induced contamination deposits from epoxy outgassing under UV laser irradiation in vacuum

Amer Aoun[1,2*], Frank Wagner[1], Jean-Yves Natoli[1], Thomas Gineste[3] and Delphine Faye[2]

[1] Aix Marseille Univ, CNRS, Central Med, Institut Fresnel, Marseille, France
[2] Centre National d'Études Spatiales, 18 Avenue E. Belin, CEDEX 9, 31401 Toulouse, France
[3]Airbus Defence and Space SAS, 31 rue des Cosmonautes, 31402 Toulouse CEDEX 4, France

*E-mail: amer.aoun@fresnel.fr

Abstract. The mechanisms of deposition and removal of Laser-Induced Contamination (LIC) were studied in a vacuum environment using a combination of in-situ and ex-situ techniques. The optical properties of LIC deposits were analyzed in-situ through transmission measurements and Laser-Induced Fluorescence (LIF) and ex-situ using White Light Interference Microscopy (WLIM). Chemical analyses were conducted via in-situ gas-phase mass spectrometry and ex-situ Time-of-Flight Secondary Ion Mass Spectrometry (ToF-SIMS) measurements of the deposits. Correlations between the in-situ techniques demonstrated that our setup could detect different stages of both LIC deposition and cleaning processes. ToF-SIMS measurements revealed that LIC deposits undergo continuous chemical evolution under irradiation. These findings provide a deeper understanding of LIC, a phenomenon that degrades the performance of nanosecond UV lasers operating in vacuum environments and limits their lifetime.

1. Introduction

Laser-Induced Contamination (LIC) is a phenomenon in which, organic, nanometric, and light-absorbing deposits form on optical surfaces exposed to laser irradiation[1,2]. This effect is particularly frequent with UV lasers operating in space, where the deposition occurs due the interactions between the laser light, the optical component, and volatile organic molecules outgassed from organic components onboard the satellite. As such, the process is similar to Light-Induced (or enhanced) Chemical Vapor Deposition (LICVD)[3].

LIC deposits, due to their absorbing nature, were shown to pose a mission failure risk by significantly reducing the performance of optical components[4]. Since it is impossible to maintain the optics in space, the deposits created become permanent, and thus fall under the definition of Laser-Induced Damage: A permanent light-induced modification of an optical component that impacts its performance[5].

IOP Conf. Series: Materials Science and Engineering **1328** (2025) 012013 doi:10.1088/1757-899X/1328/1/012013

Due to its catastrophic consequences[6], understanding and mitigating LIC, become crucial tasks in ensuring the success of future space missions that rely on high-power laser-based technologies in vacuum.

Today, most LIC investigations are conducted in the framework of specific space missions [7,8]. This contribution, however, offers a systematic study aimed at broadening our understanding of the fundamental processes underlying this effect.

2. Experimental setup and methods

2.1 Experimental setup

The experimental setup used in this work is shown in Figure 1, and has been described in more detail elsewhere[9]. Briefly, it includes two vacuum chambers, the first chamber serves as the main, sample chamber, while the second smaller one, acts as the contaminant chamber. The two chambers are connected via a needle valve for contaminant flow regulation. The system is pumped with an oil free primary pump and a turbomolecular pump and can reach base pressures in the 10^{-6} mbar range. The setup is temperature controlled between room temperature and 180°C for the main chamber and 120°C for the contaminant chamber.

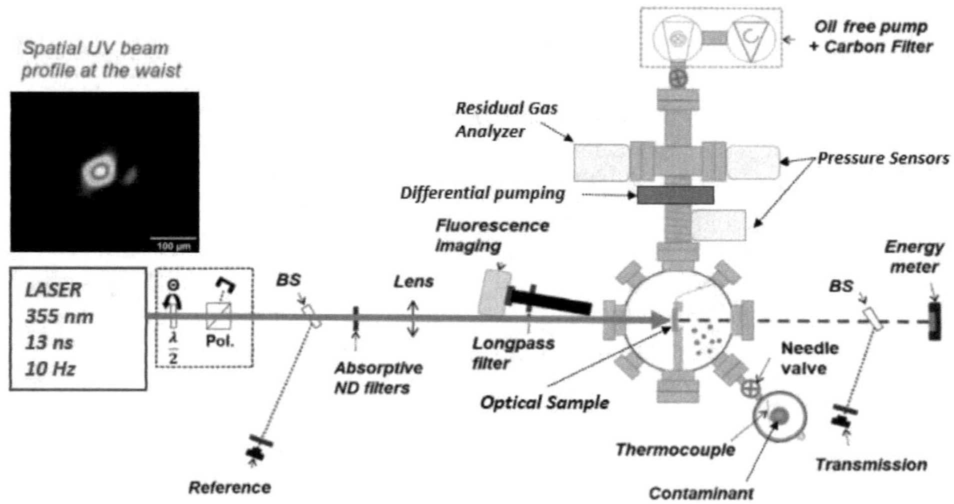

Figure 1. Vacuum LIC experimental setup.

The laser used in this work is a 10 Hz Nd:YAG laser, emitting 13 ns pulses at 1064 nm, which are then frequency tripled to 355nm. The UV beam passes through a rotating halfwave plate and a polarizer, allowing for adjustable attenuation. The beam is then focused by a UV fused silica lens onto the optical sample, which is placed inside the sample chamber. The beam diameter, corresponding to $1/e^2$ of the maximum intensity is approximately 60 μm at the waist, on the sample surface.

IOP Conf. Series: Materials Science and Engineering **1328** (2025) 012013 doi:10.1088/1757-899X/1328/1/012013

The optical samples used in this work were UV-fused silica windows coated with an SiO_2 layer, and the contaminant used was the outgassing at 60⁰C or 100⁰C of an epoxy-based adhesive (Scotch-weld 3M EC-2216 grey).

2.2 Transmission measurements
As mentioned earlier, LIC deposits are light-absorbing in nature[8,10,11]. The transmission variations induced by the deposits were monitored in-situ using energy meters and calibrated photodiodes placed before and after the chamber. After acquisition, the raw transmission data was smoothed using a 10-minute moving average to reduce the influence of the pulse-to-pulse laser energy fluctuations[9].

2.3 Laser-Induced Fluorescence (LIF)
Laser-Induced Fluorescence (LIF) is a technique widely used for LIC monitoring and risk assessment[8,12,13]. In our setup, the 355 nm laser serves as both the irradiation source and the fluorescence pump. At this wavelength and with the substrates that we used, LIF proved to be an exceptionally sensitive technique[9,13].

2.4 White Light Interference Microscopy (WLIM)
WLIM (optical profilometry) is also a commonly used technique for LIC risk assessment[2]. It is a quick and robust technique capable of detecting sub nanometric thicknesses, such as those of LIC deposits, that are usually a few nanometres thick.

2.5 Gas-phase mass spectrometry
Our setup is equipped with a Residual Gas Analyzer (RGA) that analyses the chemical composition of the contaminant atmosphere during the experiments. The RGA operates continuously throughout the whole outgassing period, and can detect ions with a mass-to-charge ratio (m/z) from 1 to 300 with a chosen resolution of 5 points per m/z. With these settings, a complete scan is performed in approximately 1 minute, which thus defines the time resolution of the data.

2.6 Time-of-Flight Secondary Ion Mass Spectrometry (ToF-SIMS)
To study the deposits' chemical evolution during irradiation, we performed a time (number of laser shots) series, where we created many deposits under identical irradiation and contamination conditions, with the only variable being the total irradiation time. Ex-situ ToF-SIMS measurements were then performed on these deposits, enabling quantitative characterization, and tracking of the chemical evolution due to irradiation.

3. Experimental Procedure

3.1 Chamber bakeouts and cleanliness
Before each LIC test, the chamber is baked out for at least 16 hours at 180⁰C for the main chamber, and 120⁰C for the contaminant chamber. The chamber's cleanliness is then assessed by performing a blank test with in situ LIF observation, followed by an inspection of the sample using the optical profilometer. If no fluorescence is detected during the blank test, and no resulting LIC

deposit is detected on the profilometer, the chamber is considered clean and ready for the upcoming LIC test.

3.2 Deposit cleaning conditions

It has been shown in other works, that the addition of oxygen to the contaminated atmosphere can inhibit LIC build-up or trigger the cleaning of an already made LIC deposit[4,8,14–16]. In this work, the outgassing conditions were maintained constant throughout both the deposition and cleaning phases, meaning that no additional oxygen leak was introduced to trigger the cleaning at the end. Nevertheless, due to the decaying outgassing rate of the contaminant, and the constant leak rate of air into the chamber, the air-to-contaminant ratio rises during the irradiation and reaches a point where the laser cleaning is triggered[14].

4. Results and discussion

4.1 In-situ tracking of EC-2216 LIC deposition and cleaning

Figure 2 presents the in-situ transmission and pressure data during deposition and cleaning as well as in situ fluorescence images of the deposit. A transmission loss is observed as the deposit grows, followed by a recovery near the end signalling that the deposit is being cleaned. An increase in fluorescence intensity during growth is observed. However, as the deposit starts getting cleaned and the transmission starts recovering, the fluorescence intensity starts dropping until it reaches residual levels. This correlation allows the identification and delimitation of the LIC deposition and cleaning processes. The pressure data shown in the figure is the pressure close to the sample and shows that no additional oxygen was introduced to trigger the cleaning.

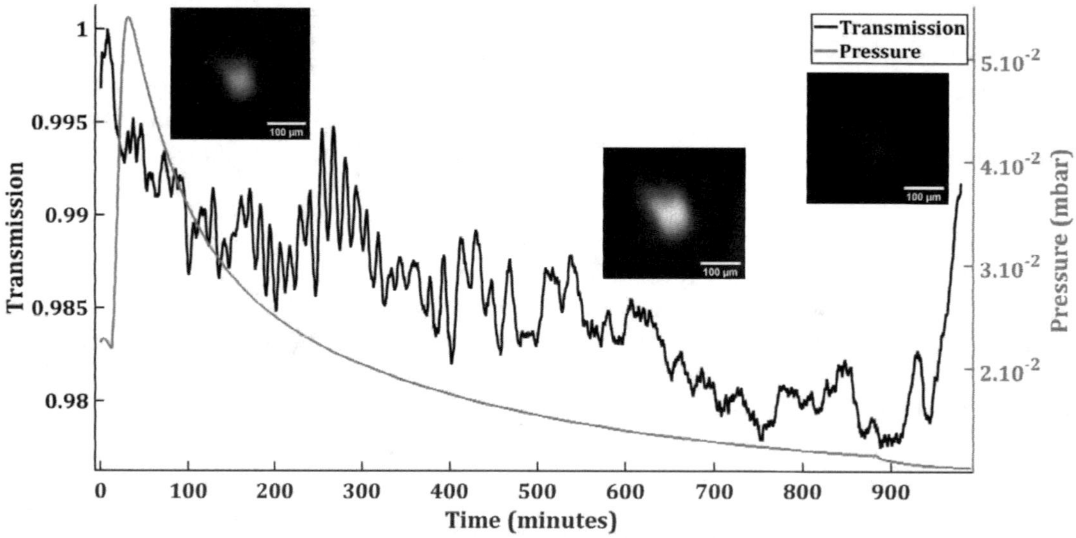

Figure 2. Plot showing the evolution of transmission, pressure, and fluorescence during the experiment. During deposition, the transmission decreases while the fluorescence intensity increases. When the cleaning starts, the transmission recovers, and the fluorescence starts decreasing until it reaches residual levels. The pressure curve shows the pressure near the sample and provides an outline of the outgassing conditions during the experiment.

4.2 ToF-SIMS and the chemical evolution under irradiation

Figure 3 shows profilometer images of two deposits: one formed with 90,000 laser shots and the other with 540,000 laser shots. Adjacent to the deposits, striped patterns of EC-2216 vapor condensation can be seen. These stripes are EC-2216 condensates and not LIC, as they are not formed due to the laser irradiation and are not irradiated during the experiment. However, their presence offers valuable insight and allows a comparison between the chemical composition of Laser-Induced Contamination deposits and contamination via condensation of EC-2216 vapor.

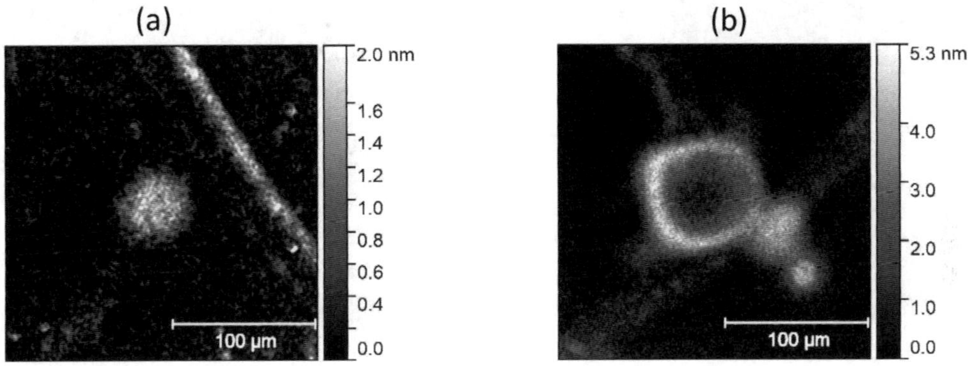

Figure 3. Profilometer images of LIC deposits at two different stages of deposition. (a) pancake shaped deposit formed after 90 000 laser shots, (b) donut shaped deposit formed after 540 000 laser shots. The stripes observed close to the deposits are EC-2216 vapor condensates.

Figure 4 presents ToF-SIMS images showing the distribution of CN^- around the same deposits and reveals that CN^- is present in both deposits and the condensation stripes surrounding them.

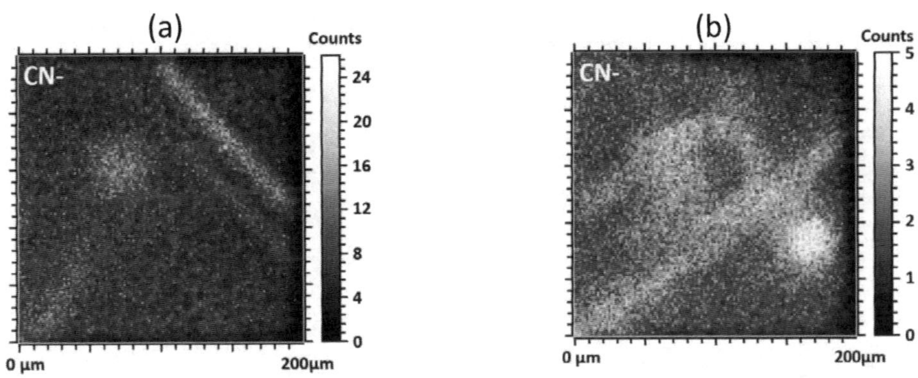

Figure 4. ToF-SIMS images showing the distribution of CN^- on the surface, (a) close to the 90 000 laser-shot deposit, (b) close to the 540 000 laser-shot deposit.

Figure 5 shows the distribution of a molecule with an atomic mass unit (AMU) of 356 and reveals that this molecule is absent in both the condensation stripes and the 90 000-shot deposit but appears only in the rim of the 540 000-shot deposit. This proves that the UV laser induces the formation of new molecules during deposition and that there is constant chemical evolution under irradiation, as the 356 AMU molecule does not appear before 90 000 laser shots, but somewhere between 90 000 and 540 000 laser shots.

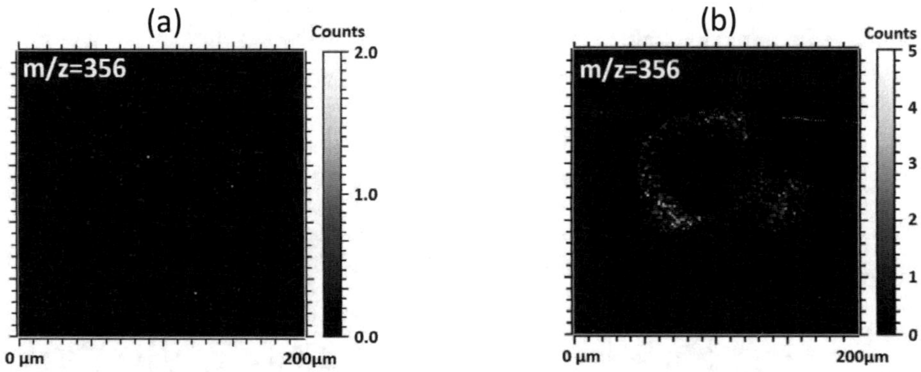

Figure 5. ToF-SIMS images showing the distribution of a molecule with an m/z (AMU) of 356. (a) close to the 90 000-laser-shot deposit, (b) close to the 540 000-laser-shot deposit.

4.3 In-situ gas phase mass spectrometry

Figure 6 shows the RGA signal recorded for Hydrogen gas (H_2) and a molecule with an AMU of 30. Initially both signals are decaying, but as soon as the cleaning starts, both signals significantly increase. Later, as the cleaning plasma decays, both signals start decaying at the same decay rate they had prior to the cleaning.

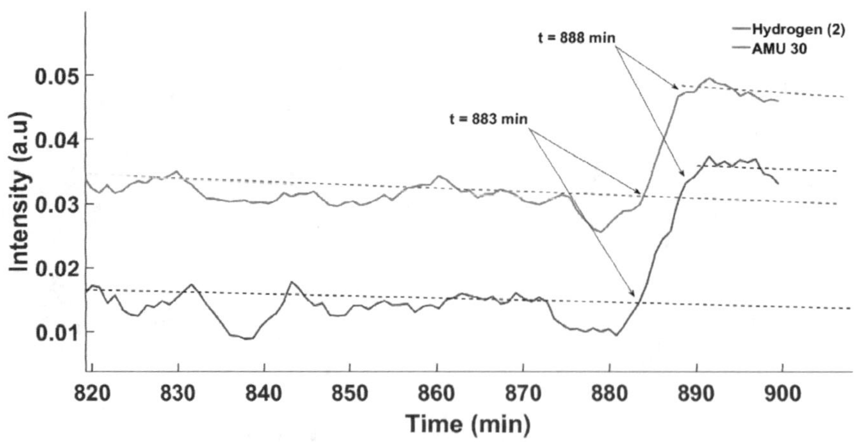

Figure 6. In-situ RGA data, showing an increase in the signal of hydrogen gas (H2) and a molecule with AMU 30 during the cleaning process. After the cleaning is done, the signals decay again at a similar rate to their decay prior to the cleaning.

This result proves that our setup, with the RGA, can potentially be used to identify chemical components of LIC deposits and identify chemical reactions underlying the deposition and cleaning processes.

Conclusion and perspective

This paper presents a systematic study, offering new insights into the fundamental processes underlying Laser-Induced Contamination (LIC), demonstrating that the chemical composition of LIC deposits differs from that of simply condensed contaminants. ToF-SIMS analyses of deposits at various growth stages provided evidence of a continuous chemical evolution of LIC under irradiation, as new molecules emerge at different stages of growth. Additionally, this work confirms that our experimental setup can track the deposition and oxygen-based cleaning of LIC deposits and can, using an RGA, potentially detect the chemical reactions underlying these processes.

Acknowledgments

This work is supported by the French space agency (CNES) and Airbus Defence and Space. The authors would also like to thank Jean-Paul Salvetat, PLACAMAT, Université de Bordeaux for the ToF-SIMS experiments and their discussion.

References

[1] F.R. Wagner, G. Gebrayel El Reaidy, D. Faye, J.-Y. Natoli, Laser induced deposits in contaminated vacuum environment: Optical properties and lateral growth, Optics & Laser Technology 122 (2020) 105889. https://doi.org/10.1016/j.optlastec.2019.105889.

[2] W. Riede, H. Schroeder, G. Bataviciute, D. Wernham, A. Tighe, F. Pettazzi, J. Alves, Laser-induced contamination on space optics, in: G.J. Exarhos, V.E. Gruzdev, J.A. Menapace, D. Ristau, M.J. Soileau (Eds.), Boulder, Colorado, 2011: p. 81901E. https://doi.org/10.1117/12.899190.

[3] E. Halary-Wagner, T. Bret, P. Hoffmann, Light-Induced CVD of Titanium Dioxide Thin Films I: Kinetics of Deposition, Chem. Vap. Deposition 11 (2005) 21–28. https://doi.org/10.1002/cvde.200306303.

[4] O. Lux, D. Wernham, P. Bravetti, P. McGoldrick, O. Lecrenier, W. Riede, A. D'Ottavi, V. De Sanctis, M. Schillinger, J. Lochard, J. Marshall, C. Lemmerz, F. Weiler, L. Mondin, A. Ciapponi, T. Kanitz, A. Elfving, T. Parrinello, O. Reitebuch, High-power and frequency-stable ultraviolet laser performance in space for the wind lidar on Aeolus, Opt. Lett. 45 (2020) 1443. https://doi.org/10.1364/OL.387728.

[5] ISO 21254-1:2011, ISO (n.d.). https://www.iso.org/standard/43001.html (accessed February 11, 2025).

[6] J.B. Abshire, X. Sun, H. Riris, J.M. Sirota, J.F. McGarry, S. Palm, D. Yi, P. Liiva, Geoscience Laser Altimeter System (GLAS) on the ICESat Mission: On-orbit measurement performance, Geophysical Research Letters 32 (2005) 2005GL024028. https://doi.org/10.1029/2005GL024028.

[7] N. Bartels, F. Hadinger, G. Taube, W. Riede, C. Dahl, K.-C. Voss, A. Ciapponi, R. Martins, L. Mondin, Molecular contamination testing with continuous-wave laser radiation at 1064 nm: de-risking activity for the LISA space mission, in: K. Minoglou, N. Karafolas, B. Cugny (Eds.), International Conference on Space Optics — ICSO 2022, SPIE, Dubrovnik, Croatia, 2023: p. 167. https://doi.org/10.1117/12.2690854.

[8] D. Wernham, J. Alves, F. Pettazzi, A.P. Tighe, Laser-induced contamination mitigation on the ALADIN laser for ADM-Aeolus, in: G.J. Exarhos, V.E. Gruzdev, J.A. Menapace, D. Ristau, M.J. Soileau (Eds.), Boulder, Colorado, 2010: p. 78421E. https://doi.org/10.1117/12.867268.

[9] G. Gebrayel El Reaidy, F.R. Wagner, D. Faye, J.-Y. Natoli, Study of the first stages of laser-induced contamination, Opt. Eng. 57 (2018) 1. https://doi.org/10.1117/1.OE.57.12.121903.

[10] D. Kokkinos, H. Schroeder, K. Fleury-Frenette, M.P. Georges, W. Riede, G. Tzeremes, P. Rochus, Laser optics in space failure risk due to laser induced contamination, CEAS Space J 9 (2017) 153–162. https://doi.org/10.1007/s12567-016-0137-1.

[11] H. Schröder, P. Wagner, D. Kokkinos, W. Riede, A. Tighe, Laser-induced contamination and its impact on laser damage threshold, in: G.J. Exarhos, V.E. Gruzdev, J.A. Menapace, D. Ristau, M. Soileau (Eds.), Boulder, Colorado, USA, 2013: p. 88850R. https://doi.org/10.1117/12.2030002.

[12] H. Schröder, W. Riede, E. Reinhold, D. Wernham, Y. Lien, H. Kheyrandish, In situ observation of UV-laser-induced deposit formation by fluorescence measurement, in: G.J. Exarhos, A.H. Guenther, K.L. Lewis, D. Ristau, M.J. Soileau, C.J. Stolz (Eds.), Boulder, CO, 2006: p. 64031K. https://doi.org/10.1117/12.695347.

[13] G.G. El Reaidy, F.R. Wagner, D. Faye, J.-Y. Natoli, Laser-induced contamination deposit growth mechanisms on space optics, Opt. Eng. 61 (2022). https://doi.org/10.1117/1.OE.61.9.096110.

[14] M. Hippler, P. Wagner, H. Schroeder, W. Riede, Laser-induced contamination of space borne laser systems: impact of organic contamination and mitigation by oxygen, in: J. Egges, C.E. Soares, E.M. Wooldridge (Eds.), San Diego, California, United States, 2016: p. 99520N. https://doi.org/10.1117/12.2236897.

[15] W. Riede, P. Allenspacher, H. Schröder, D. Wernham, Y. Lien, Laser-induced hydrocarbon contamination in vacuum, in: G.J. Exarhos, A.H. Guenther, K.L. Lewis, D. Ristau, M.J. Soileau, C.J. Stolz (Eds.), Boulder, CO, 2005: p. 59910H. https://doi.org/10.1117/12.638765.

[16] A.P. Tighe, F. Pettazzi, J. Alves, D. Wernham, W. Riede, H. Schroeder, P. Allenspacher, H. Kheyrandish, Growth mechanisms for laser induced contamination on space optics in vacuum, in: G.J. Exarhos, D. Ristau, M.J. Soileau, C.J. Stolz (Eds.), Boulder, CO, 2008: p. 71321L. https://doi.org/10.1117/12.804423.

IOP Conf. Series: Materials Science and Engineering **1328** (2025) 012014 doi:10.1088/1757-899X/1328/1/012014

Can thin films deposited by ALD reduce the outgassing? A case study by dynamic outgassing on 3D-printed polyether ether ketone (PEEK)

Théo Henry[1], Johanna Harpur[1,2], Mircea Alexandru Helici[1], Ricardo Martins[1], Ugo Lafont[1], Kaisa Aab[3], Aivar Tarre[3], Kaupo Kukli[3], Maido Merisalu[3], Orcun Ergincan[1,*], Malgorzata Holynska[1,*]

[1] European Space Agency (ESA), European Space Research and Technology Centre (ESTEC), Keplerlaan 1, Postbus 299, 2200 AG Noordwijk, The Netherlands
[2] Airbus Netherlands B.V. Mendelweg 30, 2333 CS Leiden, The Netherlands
[3] Institute of Physics, University of Tartu, Ostwaldi 1, Tartu, Estonia

[*]E-mail : orcun.ergincan@esa.int, malgorzata.holynska@esa.int

Abstract. The use of thin films as a barrier to prevent outgassing from the source is proposed as a strategy rather than focusing on reducing contamination after outgassed molecules are released. To explore this concept, Atomic Layer Deposition (ALD) was utilized to apply a thin oxide coating, approximately 110 nm thick, onto a 3D-printed polyether ether ketone (PEEK) sample. Dynamic outgassing experiments were performed according to the ECSS standard to study and to compare the kinetics of outgassing for a coated and a uncoated sample. The obtained results suggest that PEEK materials have favourable outgassing properties with negligible collected volatile condensable material (CVCMs). In addition to the latter, ALD coating reduces the rate of outgassing and slows the reabsorption of moisture. These results are encouraging, when water or ice can be problematic for space missions.

1. Introduction

Contamination-related risks include outgassed deposits condensing on sensitive surfaces. This can alter their thermo-optical properties (e.g., coatings, optics) or impair performance (e.g., solar arrays, radiators), ultimately diminishing the effectiveness of key instruments. In extreme cases, may even result in the loss of the spacecraft [1]. When exposed to a vacuum, volatile chemicals within various spacecraft components can gradually migrate from the contamination source to surfaces through diffusion, eventually escaping via desorption or evaporation. These processes are the primary contributors to a spacecraft's molecular contamination budget. Surfaces subjected to cold temperatures require special consideration, as molecules tend to remain on them for extended periods—or even indefinitely—at sufficiently low temperatures in a vacuum such as the case for ESA's Euclid mission [2].

To minimize contamination risks, several common practices are implemented. include selecting low-outgassing materials, improving material formulations, and enforcing strict inspections and control procedures during Manufacturing, Assembly, Integration, and Testing (MAIT). Additional measures include purging certain areas with nitrogen gas to reduce moisture and contamination ingress, optimizing the view factor between contamination sources and sensitive spacecraft components, and conducting in-ground or in-flight bake-outs. Novel approaches are also being explored including anti-contaminant coating designed to repel [3] or decompose the outgas species by photocatalysis [4], others such as adsorber coatings [5] and honeycomb traps are also used. These approaches are promising but are complex to develop and their efficiency in-orbit is difficult to validate. Despite all of these, contamination remains an active topic because none of these approaches fully eliminate it.

The search for novel cross-contamination risk mitigation techniques remains an area of interest in overcoming the limitations of material selection for satellite design. For this reason, atomic layer deposition (ALD), a thin-film deposition technique based on the sequential use of a gas-phase chemical process, presents a promising approach to reduce the impact of outgassing products from materials [6]. In addition to its general applications, ALD has demonstrated potential for space applications, particularly protective functions against atomic oxygen [7-10]. ALD is capable of coating wide range of materials, including the possibility of coating an entire cavity in a single step or for 3D printed materials having complex shapes. The advantage of this technique could also be leveraged to reduce the contamination risks and thereby enhance the protective capabilities of such coatings.

Rather than addressing outgassed molecules after their release, the focus is placed to directly limiting outgassing by using thin films that acts as an outgassing barrier. Previous studies have shown promising results [6]. To investigate this approach, ALD was used to apply with a thin oxide coating (about 110 nm) to coat a 3D-printed polyether ether ketone (PEEK) sample. Dynamic outgassing experiments were conducted to study and compare the outgassing kinetics of the coated and uncoated sample. A special attention was given to the mass recovery of the samples, concerning the moisture exchange.

1.1. Atomic layer deposition (ALD)

ALD consists of sequential reactions of chemical precursors with the surface of the material being treated (see Figure 1). ALD is a subclass of Chemical Vapor Deposition (CVD) process. The potential applications of ALD are vast, due to its numerous advantages. The deposition pressure (~1 mbar) is relatively high compared to most Physical Vapor Deposition (PVD) methods. Unlike traditional CVD techniques, ALD offers precise control over thickness, enabling the deposition of ultra-thin coatings just a few nanometers thick while maintaining substrate properties such as flexibility. Furthermore, ALD facilitates the deposition of both amorphous materials and their crystalline phases at lower temperatures than conventional CVD techniques.

One of ALD's primary strengths is its exceptional coating conformity, even on complex 3D structures at both micro and macro scales. Its high aspect ratio coverage reduces the risk of masking effects, which are often encountered in PVD methods. Furthermore, ALD is compatible with a broad range of substrates and precursors, making it a versatile option for various applications [11]. Under optimal deposition conditions, the coating exhibit excellent adhesion, optical and electrical/dielectric properties and with a high density. A key consideration of the ALD technic is the need to carefully control the ALD window — the temperature range required for an optimum deposition (e.g. avoid incomplete reaction, decomposition) — to obtain high quality coatings. Moreover, the injection and purging sequences can also have a significant impact (e.g. too short purging times) on the outcome. The main drawbacks of ALD include mainly its cost, due to low deposition rate and the limited volume capacity of most reactors. These challenges can be mitigated through spatial atomic layer deposition (SALD), which operates sequentially in space rather than time [12]. The size of reactors can scale up to 1300 mm x 2400 mm [13]. Further introductive content on the ALD can be found on the following references [11,14-17].

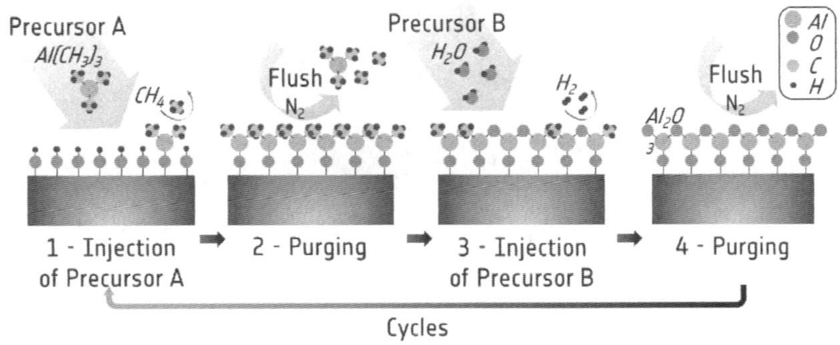

Figure 1. Schematic representation of the ALD process. Credit: ESA

2. Experimental

2.1. Materials

The polyether ether ketone (PEEK 4000 from Evonik) is used for this study. The are printed using an APIUM P220 printer with a 0.4 mm nozzle. This printer is equipped with an extrusion head capable to be heated up to 500 °C, enabling processing of various types of thermoplastics, including PEEK. The slicer software Simplify 3D is used to generate the corresponding gcode. The printing process was performed with a + 45/- 45 raster layer orientation and a 100 % infill for flat sample, while for the honeycomb/hexagonal structure is printed with a specific wall thickness. Dimafix liquid adhesive is applied on the print bed to ensure good adhesion during printing. After printing, the sides of the samples are roughly polished to remove the residual printed layer. Following up, they are cleaned with ethanol by flushing and gently rubbing the sample with the aid of an ISO wipe. To ensure a proper cleaning, the samples are then soaked in a beaker of ethanol placed in an ultrasonic bath for 10 minutes. The printed structure is a honeycomb shape composed of nineteen alveoli for a total size of 55 x 55 x 70 mm to optimize the effective surface area (> 450 cm^2). Two different samples are made, one to remain as pristine, and the other one to be coated by ALD.

2.2. ALD Coating

Prior to the coating process, the sample is cleaned using pressurized air to remove loose dust particles. Additional cleaning is carried out with isopropanol, followed by drying in a laminar flow cabinet. The ceramic coating is then applied by thermal ALD in a commercial PICOSUN® R200 reactor at a deposition temperature of 125 °C. The heating conditioning phase lasts about 1 h to reach target temperature. The deposition occurs under a partial pressure of 1 mbar. The nanolaminate consists of ten alternating layers of amorphous Al_2O_3 and TiO_2 with a total thickness of 110 nm (see Figure 2). The first layer Al_2O_3 is deposited to a thickness of 20 nm, while all subsequent layer of Al_2O_3 and TiO_2 have a thickness of 10 nm. For the synthesis of TiO_2, water and $TiCl_4$ are used. The precursors for Al_2O_3 are water and TMA (trimethylaluminum).

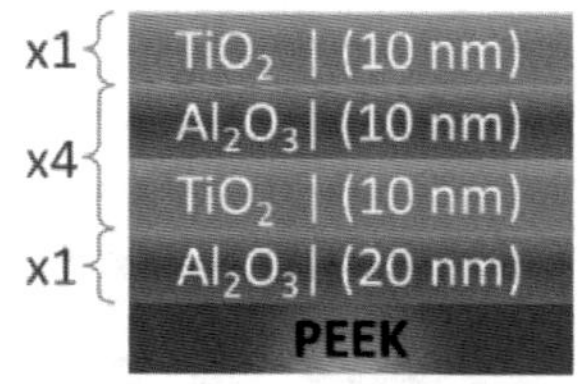

Figure 2. Schematic representation of the ALD dielectric multilayer coating with the detailed stack and its nominal thicknesses. Credit: ESA

2.3. Dynamic outgassing test

The test setup consisted of a vacuum chamber, a vacuum interlock chamber, a temperature control system, a Knudsen Cell, four QCMs and a data acquisition system. The cryogenic QCM (CQCM), located on a cryogenic plate, was operating at -170 °C to measure the Total Mass Loss (TML) parameter. The QCM measures the deposition in terms of frequency change, and it is converted to mass using the mass-sensitivity of the QCM and the geometrical view factor. The remaining three QCMs, which monitor the deposition of outgassing condensable at various temperatures (CVCM), were set to operational temperatures of -75 °C, -50 °C, and -25 °C. The standard dynamic outgassing test included five temperature steps, each with a delta temperature of 25 °C, from 25 °C up to 125 °C, with each step lasting 24 hours. The test was limited to 125 °C to avoid going beyond the glass transition temperature of PEEK being around 140 °C. Following each test step, QTGA was performed on the CQCM at a heating rate of 1 °C·min^{-1} up to -75 °C to analyse the thermal behaviour of the condensable. At the end of the dynamic outgassing test, QTGA was also performed sequentially on four QCMs. Further details on the test setup and the analysis methods are provided in ECSS-Q-TM-70-52A. The pressure throughout the entire test's duration was below $1.8 \cdot 10^{-6}$ mbar. Prior to the test, an in vacuum bake-out of the facility was performed at 450 °C. A blank test was performed at 150 °C for 24 h to verify the cleanliness of the facility under actual test conditions. The background levels of the QCMs were found to be negligible. In addition, and prior to the dynamic outgassing experiments, the samples were conditioned for 24 h at 22 ± 3 °C and 55 ± 10 %RH. After the tests, the samples were placed in a conditioning area at 22 ± 3 °C and 55 ± 10 %RH until the end of the mass recovering test. An external balance was used for the ex-situ sample mass measurements (Sartorius RC2105 with an accuracy of 10 µg). In addition, CaF_2 window was installed in the chamber to measure potential condensable after the experimentation according to the ECSS-Q-ST-70-05C.

3. Results and discussions

3.1. PEEK and ALD coating

The Figure 3 shows the samples after being printed and coated by ALD. The ALD coating gives a goldish reflection to the sample. This is attributed to the multilayered structure modifying the overall optical properties of the sample by constructive and destructive interferences with respect to the PEEK substrate. The pristine and coated samples weight 27.13722 g and 27.23022 g respectively. The slight mass difference is essentially attributed to small variations during the printing process and the polishing of the residual layer.

Figure 3. Pictures of the samples used for the outgassing tests. Credit: ESA

3.2. Dynamic outgassing test results

The total mass loss (TML) during the experiments is presented in Figure 4, relative to the weight of the samples measured before the experiments. The pristine and the coated samples exhibit distinct behaviours. For the pristine sample, TML reaches approximately 0.20 %. After the second step at 50 °C, the maximum TML is nearly reached and is completed during the first part of the third step at 75 °C. The two last steps do not show any additional mass evolution suggesting that all the species from the sample were outgassed at the beginning of the third temperature step. In contrast, for the coated samples, each step shows a continuous mass evolution, with the TML rate increasing as the temperature raising. Moreover, the amount of TML released during each step is also increases with increase of temperature, which contrasts the trend observed for the pristine sample. By the end of the test, at the fifth step, the TML has not yet plateaued and reaches approximately 0.12 %. This suggest that not all the species from the sample were outgassed, implying that the coating drastically reducing the outgassing rate. Additionally, during the conditioning phase in vacuum (<20 °C) before the temperature steps, the frequency changes reached 6200 Hz for the pristine sample but only 70 Hz for the coated sample. This further supports the idea that the coating effectively limits the outgassing. Since the TML did not reach saturation for the coated sample, it is not possible to determine if the coating permits to reduce the maximum TML.

The CVCMs were also measured by the other QCMs (Figure 5). The results show no significant difference between the two samples, with values remaining significantly low, below $2 \cdot 10^{-4}$ %. Therefore, it is concluded that the printed PEEK sample does not outgas CVCMs and that all the collected TML originates from non-condensable species. As a result, it is not possible to assess the coating's ability to block or reduce CVCMs. However, this suggests that PEEK is a material outgassing-friendly material due to low CVCM and low TML. This absence of CVCM was further confirmed at a later stage during the QTGA of the QCMs at -25 °C, -50 °C and -75 °C. Additionally, no peaks were detected by FTIR on the CaF_2 witness placed inside the chamber prior to the test.

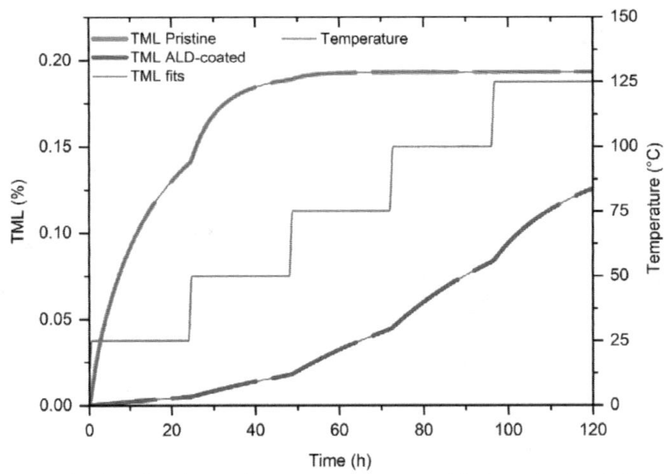

Figure 4. TML evolution during the dynamic outgassing test with the obtained fits. The temperature steps are displayed in grey. Credit: ESA

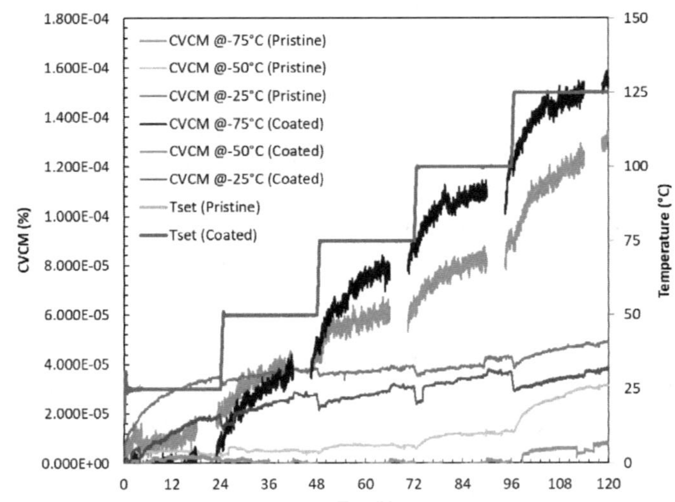

Figure 5. CVCMs collected on the QCMs operated at - 25 °C, - 50 °C and - 75 °C during dynamic outgassing test of the pristine and ALD-coated PEEK. The temperature steps are displayed in green. Credit: ESA

Predictions can be made from the TML data by applying a mathematical treatment to the outgassing data. This involves extracting parameters including the acceleration factor, the apparent activation energy and the residence time–temperature dependency coefficient associated in the outgassing process. Although the robustness of this approach is limited due to early saturation of the TML in the pristine samples, which prevents reliable extraction of variables, it was still used to perform trend analysis at ambient temperatures. The results, presented in Figure 6 suggest that outgassing can be delayed by approximately two orders of magnitude in this case. This result is coherent with the notable TML slope differences between the pristine and coated sample (Figure 4).

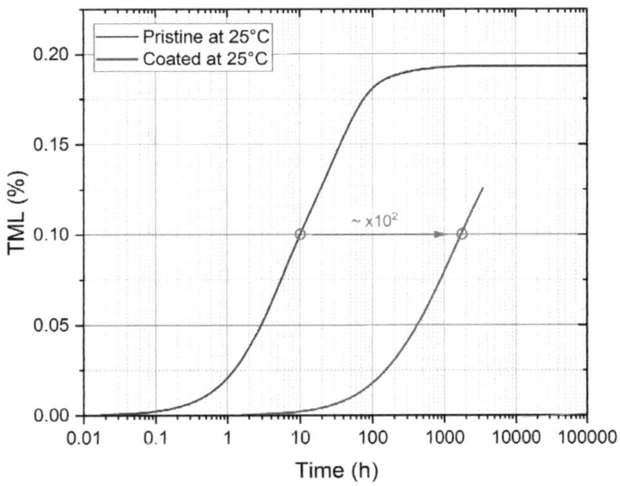

Figure 6. Prediction of the TML in time according to the experimental data and their modelling for the pristine PEEK and its ALD-coated version. Credit: ESA

To provide a complete overview, the QTGAs performed after each temperature step on the CQCMs are shown in Figure 7. The derivative data reveals a single desorption peak at approximately -100 °C for both samples, with a captation temperature located around -120 °C. This suggest that there is either one type of molecule which condensed on the CQCM or multiple molecules but with very similar vapor pressure characteristics. Interestingly, this temperature is typically associated with water [18, 19]. Therefore, it is assumed that observed features represents water. Due to thermal inertia, the centre of the peak shifts to higher temperatures as the deposited mass increases (indicating higher frequency difference). It is also worth noting that the frequency differences are vary with the temperature steps, showing opposite trends for the pristine and the coated sample. This is consistent with the TML measurements discussed earlier, where the pristine sample releases more content in its initial steps, while the ALD coating behaves in the opposite manner.

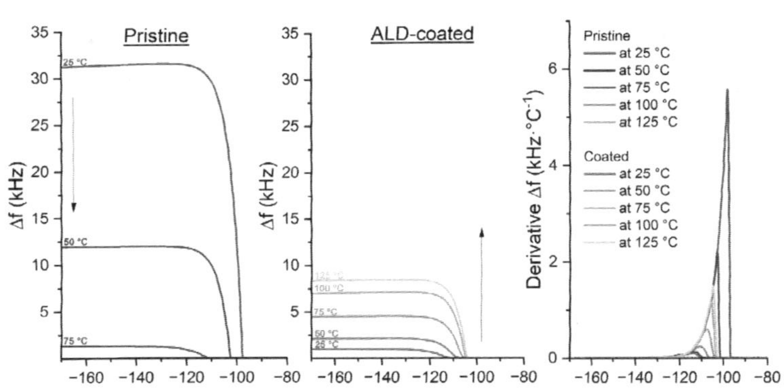

Figure 7. QTGA curves at 1 °C·min-1 of the condensed material on CQCM, kept at -170 °C during each temperature test step of the dynamic outgassing test. Credit: ESA

3.3. *Mass recovered results*

After the outgassing test, the mass of the coated sample was found to increased gradually, while the mass recovery for the pristine sample was reached after approximately 4 days (Figure 8). After 200 days, the mass of the coated sample due to moisture reabsorption is still increasing slightly and at this rate will reach saturation after a year. This difference is attributed to the coating acting as a buffer, reducing moisture diffusion into the PEEK bulk, and consequently showing the mass regain. To assess the diffusion through the coating, a 1D diffusion approach was considered. This model assumes a uniform initial distribution of species which might diffuse, with different surface concentrations at the extremes of the one dimension. This corresponds to a flow through a membrane and can be modelled as it follows [20] with $\frac{M_t}{M_\infty}$ the relative mass evolution (here in g), n an integer, D the diffusion coefficient (here in $cm^2 \cdot day^{-1}$) t the time (here in day), l the thickness of the layer (in here in cm):

$$\frac{M_t}{M_\infty} = 1 - \frac{8}{\pi^2}\sum_{n=0}^{\infty}\frac{1}{(2n+1)^2}\cdot e^{-\frac{D(2n+1)^2\pi^2 t}{l^2}} \tag{1}$$

Hence, by considering a total mass loss Δm_{max}, the mass variation Δm equals:

$$\Delta m = \Delta m_{max}\cdot\left[1 - \frac{8}{\pi^2}\sum_{n=0}^{\infty}\frac{1}{(2n+1)^2}\cdot e^{-\frac{D(2n+1)^2\pi^2 t}{l^2}}\right] \tag{2}$$

This model was applied to the mass recovery results of the coated sample. An additional y_0 parameter was applied to add a slight flexibility to the fitting assuming that the sample weighted before the outgassing experiment might have not totally recovered moisture up to saturation after the ALD process. The fit reported in Figure 8 was obtained with a $R^2 = 0.999$, $y_0 = -0.0050\ g$, $D = 6.09 \cdot 10^{-13}\ cm^2 \cdot day^{-1}$ and $\Delta m_{max} = 0.0727\ g$ considering the nominal thickness of 110 nm. Interestingly, the saturation value of the recovered mass is similar for both the coated and uncoated samples. This suggests that, while the coating does not fully block moisture exchanges (e.g. adsorption, diffusion) with the atmosphere, but it significantly reduces the rate of moisture uptake. It is worth noting that this observation is not entirely consistent with the different TMLs measured *in-situ*. The cause of this discrepancy has not been identified but may be attributed to rapid moisture desorption during the pumping phase or additional outgassing of the coated sample after the phase at 125 °C. Although this study focused on ALD coating to limit water outgassing, the results obtained are also promising for further demonstrating a reduction in CVCM outgassing.

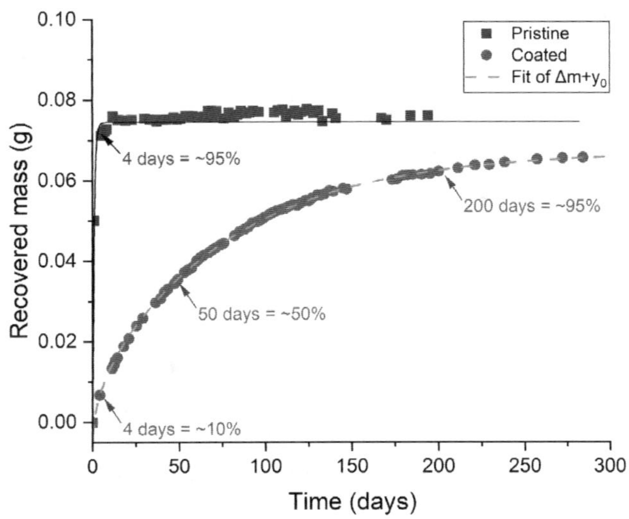

Figure 8. Mass measurements of the re-adsorbed mass after the outgassing test of the PEEK samples. Credit: ESA

4. Conclusions and perspectives

This study demonstrates that 3D-printed PEEK exhibits very low outgassing properties, with negligible CVCM and minimal moisture content, making it highly recommended for application requiring low outgassing behaviour. When coated with an ALD layer of dense dielectric oxides, both the outgassing and the re-adsorption kinetics are significantly reduced, at least in terms of water. This is true even though the coating thickness is thin (few tens of nanometers). However, due to absence of CVCM, it is not possible assess the coating impact on larger molecules. Nevertheless, it is expected that the hindrance would be greater for larger molecules. Therefore, it is believed that this coating could also reduce the kinetics of CVCM outgassing. Additionally, the kinetic exchange limitations could also be improved with thicker coating or denser materials.

This work also demonstrate that the conditioning steps are essential to properly determinate the RML, and that in some cases, the duration of 24 hours referred in the ECSS-Q-TM-70-52A might be too short. It is therefore recommended to extend the mass measurement until it stabilizes to improve the results reliability.

From an application perspective, the use of such coating might slow down and reduce the risks of water re-adsorption. This is relevant during various mission phases, such as, MAIT, transportation (if not under N_2) and pre-launch phases, off-times or between purging phases (e.g. storage/de-storage). The coating might also mitigate the risks of outgassing during commissioning, sun exposure or hot phases. If the present results can be extended to CVCMs, the overall outgassing and contamination risks might be reduced, particularly if the coating can block the bigger molecules which are assumed to be responsible to the molecules yellowing under sun exposure, and reducing (thermo-) optical properties of optics, mirrors, paints or reflectors. This could also possibly offer the opportunity to use materials having interesting properties but discarded due to their too high outgassing characteristics.

Due to the flexibility of ALD, this method could be tailored for numerous applications. For example, leveraging its CVD principle and excellent coverage for high-aspect-ratio surfaces, it could coat an entire cavity or CubeSat at once. The low deposition temperature could permit to coat plastics and polymers. Finally, the capability of ALD to deposit precise thicknesses and various oxides and fluoride materials could permit to adapt the multilayer to improve the thermo-optical properties of a surface while simultaneously limiting outgassing.

Disclaimer

This work shows an example of investigation with trendlines. It does not aim to qualify or precisely quantify properties nor promote or demote any product.

Fundings

This research was supported by ESA/ESTEC (research fellowship). The views expressed herein can in no way be taken to reflect the official opinion of the European Space Agency (ESA) and are not intended to endorse technologies, companies, or products.

References

[1] A Tribble et al 1996 *1996 Space Programs and Technologies Conference*, Contamination Control Engineering Design Guidelines for the Aerospace Community
[2] M Schirmer et al 2023 *Astron. Astrophys.* **675** A142
[3] M Holynska et al 2019 *J. Spacecraft Rockets* **56** 1358
[4] N Shimosako et al 2023 *Acta Astronaut.* **213** 29
[5] N S Abraham 2015 *Contamination, Coatings, Materials and Planetary Protection Workshop, NASA Applications of Molecular Adsorber Coatings*, NASA Goddard Space Flight Center
[6] J V Anguita et al 2019 *Nat. Mater.* **19** 317
[7] R Cooper et al 2008 *Thin Solid Films* **516** 4036
[8] I Gouzman et al 2010 *ACS Appl. Mater. Interfaces* **2** 1835
[9] T K Minton et al 2010 *ACS Appl. Mater. Interfaces* **2** 2515
[10] W Huang et al 2023 *IOP Conf. Ser. Mater. Sci. Eng.* **1287** 012014
[11] V Miikkulainen et al 2013 *J. Appl. Phys.* **113** 023507
[12] P Poodt et al 2012 *J. Vac. Sci. Technol. A: Vacuum, Surfaces, and Films* **30** 010801
[13] Anon Beneq releases the P1500: World's Largest ALD Reactor, accessed on 8.04.2025
[14] M Ritala and M Leskelä 2002 *Handbook of Thin Films* 103
[15] M Leskelä and M Ritala 2003 *Angew. Chem. Int. Ed.* **42** 5548
[16] S M George 2010 *Chem. Rev.* **110** 111
[17] R W Johnson et al 2014 *Mater. Today* **17** 236
[18] A Suliga et al 2020 *Proc. SPIE.* **11489**, *Systems Contamination: Prediction, Control, and Performance* 55
[19] A Suliga et al 2023 *Contamination, Coatings, Materials and Planetary Protection Workshop*, Identifying and Quantifying Water Species in Spacecraft Materials Using QTGA Analysis: Implications for Molecular Contamination Modelling and Bake-Out Efficiency, NASA Goddard Space Flight Center
[20] J Crank 1979 *The Mathematics of Diffusion*

IOP Conf. Series: Materials Science and Engineering **1328** (2025) 012015 doi:10.1088/1757-899X/1328/1/012015

Ground-based experimental study on the contamination potential of a freely expanding 10 N bi-propellant thruster plume

L. J. Buntrock[1*], M. Grabe[1], L. Schmidt[1], C. Jéger[2], J. van den Eynde[2]

[1] German Aerospace Center (DLR), Institute of Aerodynamics and Flow Technology, Göttingen, Germany
[2] European Space Agency (ESA), ESTEC, Noordwijk, The Netherlands

*E-mail: leonie.buntrock@dlr.de

Abstract. Satellite thruster plume constituents pose a challenge when it comes to estimating their potential to degrade or contaminate spacecraft surfaces. Therefore, ESA has commissioned a study that covers the characterization of two European bi-propellant attitude control thruster plumes and their potential contamination impact. In this regard, a freely expanding 10 N MMH/MON bi-propellant thruster plume is investigated in this ground-based experimental campaign. Methods of investigation include quartz-crystal microbalances (QCM), material sample impingement, high-speed imaging of the nozzle exit and mass spectrometry (MS) measurements. QCM experiments supply data on mass flux as well as desorption rates at temperatures ranging from 80 to 300K, which help to estimate the impact of plume gas on surfaces of different temperatures. The optical and chemical analysis of actual spacecraft surface material witness coupons that were exposed to the plume under a wide range of angles, spanning from plume centreline to the backflow region, complements this study. Furthermore, high-speed imaging of the thruster nozzle supports the interpretation of results with information on ejected droplet velocities, trajectories and size distribution, especially regarding potential droplet impact features which were identified on witness materials in the first part of this on-going study.

1. Introduction

One of the main goals of this study is the identification of plume species and their general capability for redeposition on spacecraft surfaces, as well as their specific behaviour when impacting surface material types that are typically present on a spacecraft. Bi-propellant thrusters are of high interest in this regard, as they are widely used and require shielding mechanisms or sensible material selection in their vicinity due to the emission of gaseous and liquid species. These emitted species bare the potential to alter or damage sensitive surfaces by mere condensation, mechanical impact or chemical reactions. A thorough understanding of the contaminant's composition, quantity, spatial distribution and surface behaviour is therefore crucial to satellite design.

State of the art European plume expansion models and contamination models are based on past experimental findings which were all limited in their ability to replicate space-like vacuum conditions throughout the entire length of a pulse. Technically speaking, the DLR CCG facility

IOP Conf. Series: Materials Science and Engineering **1328** (2025) 012015 doi:10.1088/1757-899X/1328/1/012015

sampled by Dettleff rises in background pressure up to the 10^{-1} mbar range for typical 500 ms pulses [1] and the Hamburg-Chamber-1 used by Trinks at Hamburg University of Technology was able to sustain a background pressure of 10^{-4} mbar only for pulses as short as 50 ms [2]. Especially the backflow region could not be well represented, as a built-up of chamber background pressure during thruster firing produces a so-called barrel shock, a compression shock system which confines the plume gases to a region in the near-field around the plume axis [3]. Collected data were therefore only covering a freely expanding plume within a confined region, mostly within 60° off-axis or less [1, 4].

In contrast to the aforementioned limitations, the experimental work in this project is performed at the DLR high-vacuum plume test facility Göttingen for chemical thrusters (STG-CT) [5], which is able to sustain a background pressure level of 10^{-6} mbar during firing of approx. 500-5000 ms pulses, depending on thruster type. This paper focuses on plume composition analysis and material sample impact under free expansion conditions realized with a 10 N MMH/MON bi-propellant thruster at STG-CT. Gathered data help to improve plume expansion models as well as contamination models in the future, especially with regard to novel backflow information.

2. Facility

STG-CT is a cryogenic high-vacuum facility that replicates space-like conditions for plume testing of small reaction control thrusters (<20 N). It is unique in the sense of creating vacuum conditions for free plume expansion which allow for representative plume parameter collection, even in the backflow region. The operating principle of STG-CT is based on a cryogenic, LHe cooled copper cylinder, Fig.1, which instantly freezes out any impacting plume molecules and therefore prevents the chamber background pressure from rising above 10^{-6} mbar. Given that the surface is at a temperature below 4.5 K, even hydrogen, a plume exhaust gas present in most combustion reactions, is captured in this process.

Figure 1. High-vacuum plume test facility for chemical thruster (STG-CT) at DLR Göttingen. The LHe-cooled copper shroud encloses the test section and is the innermost cylinder, thermally protected from the outer stainless-steel vessel by a LN_2-cooled radiation shield. Image credit: DLR. All rights reserved.

The longest feasible single pulse duration at STG-CT is dictated by the heat load that the cryopump of the facility can take up without an excessive throughput of evaporating helium and always depends on the type and size of thruster in use. Moreover, the accumulated firing time for

IOP Conf. Series: Materials Science and Engineering **1328** (2025) 012015 doi:10.1088/1757-899X/1328/1/012015

one test cycle is governed by the total available amount of liquid helium, typically in the range of 20-40 pulses.

3. Experimental setup

3.1 Thruster operation

A 10 N bi-propellant reaction control thruster operated with MMH (mono-methyl hydrazine) and MON3 (mixed oxides of nitrogen containing 3% of nitric oxide) as propellant and oxidizer respectively. A constant regulation of feedline pressures ensures a stable mass flow rate to operate at nominal conditions for all test cases of this experimental campaign.

In the interest of improving numerical modelling of thruster plume flow fields, the project covers near-field plume characterisation with a pitot probe for impact pressure and a thin film thermometer for heat flux estimations [6]. The aforementioned methods are not discussed in this paper.

As the thruster is sitting on a rotatable mount (+/-120°), angular profiles can be achieved for all below mentioned methods. Additionally, thruster pulses modes are varied from single pulse durations between 20 and 500 ms, to pulse trains with 100 ms and five repetitions. However, with material samples installed, 26 pulses of 500 ms, identical to the best possible extend with regard to feedline pressure and thruster temperature, are fired while the thruster is held in a fixed position. This allows for the deduction of an average mass deposit or damage per pulse from the post-test analysis of materials. Thruster temperatures are recorded from a Pt200 platinum resistance thermometer mounted to the outer surface of the combustion chamber, along with a type K thermocouple (Ni-Cr-Ni) attached to outside of the nozzle lip.

3.2 High-speed imaging

A high-speed camera (Phantom UHS Series, Model v1210) equipped with an Infinity K2 DistaMax Long-Distance Microscope and a tele-lens (AF-S DX NIKKOR 55–200 mm) images the thruster nozzle from an angled viewport (55°) on top of STG-CT. A DC operated LED light blurred by a diffusor plate is positioned beneath the nozzle, producing a backlight for shadowgraphy imaging at frame rates between 100 and 6000 frames per second (fps).

3.3 Material samples

A multitude of different material samples were distributed inside the chamber for plume impingement across a wide angular range from +124° to -66°, including centreline (3°) and two backflow positions (114° and 124°), Fig.2. Samples were of circular shape with a diameter of

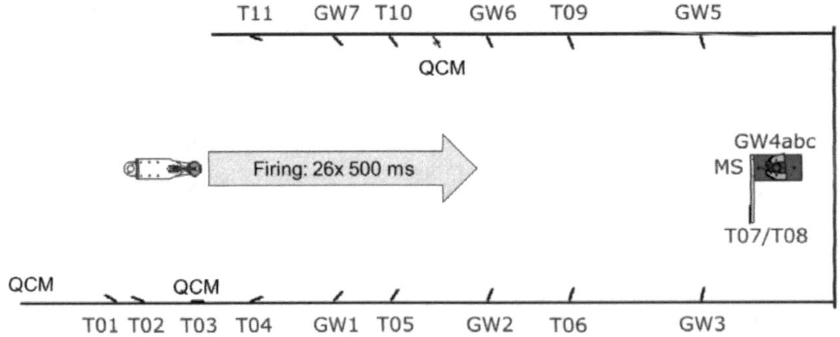

Figure 2. Schematic setup of material testing, including three QCMs and a mass spectrometer (MS). Author's own work.

25,4 mm and were held on a supporting ring by a retainer ring in order to facilitate contactless pre- and post-processing, Fig.3. All specimens were assembled in a clean-room tent and protected with a Teflon® cover during their installation inside the chamber. After plume exposure, the vacuum chamber was slowly vented once back at room temperature and the samples were recovered, transported to the clean-room tent in closed boxes, briefly inspected and photographed and then stored inside ESD-bags purged with argon.

Most material samples were either directly procured from satellite surface material manufacturers, e.g. black and white paints, yellow Kapton®, ESD (electro-static discharge) coatings, SSM (secondary surface mirror), OSR (optical solar reflector), or representative of materials used on satellite surfaces, such as anti-reflective coatings, gold mirror, aluminium and other coatings.

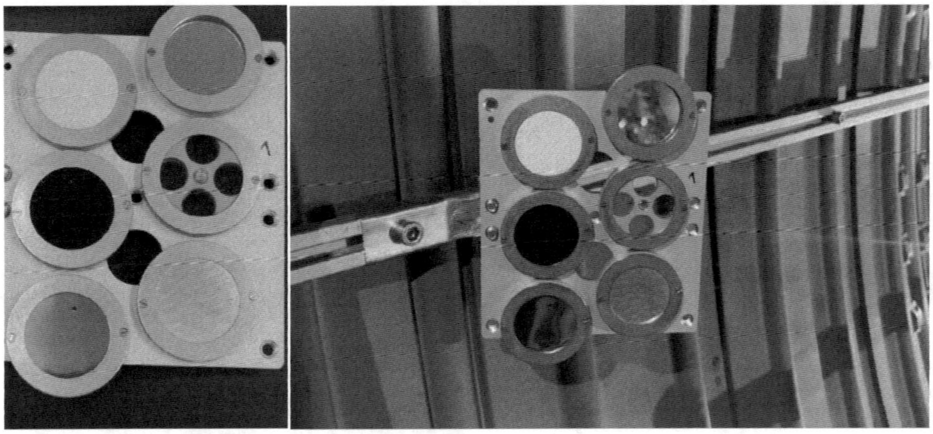

Figure 3. Exemplary material samples (white paint, black paint, yellow Kapton©, OSR/metal-coated fused silica, ESD/ITO on glass, copper) pre-test as prepared in the clean room tent (left) and installed inside STG-CT (right). Author's own work, photos taken at DLR Göttingen.

In addition, pristine silicon wafers were added to the set in the angular range of 7°-114° for a SEM-EDS (scanning electron microscopy with energy dispersive X-ray spectroscopy, SEM *SU3900*, Hitachi High Technologies and EDS detector *Ultim Max*, Oxford Instruments) analysis of permanent deposits independent of large-scale surface structures of the substrate, as they are present on surface materials like Kapton®, paints, etc. Silicon wafers will be discussed in this paper as the chemical and optical analysis of the other sample types is still on-going. Wafers were not sputter-coated due to time and technical constraints, which is, however, sought for future test.

3.4 Mass spectrometry

STG-CT features a quadrupole mass spectrometer (MS, Hiden Analytical Ltd., model "HAL PIC 101") sitting at plume centre at about 3 m downstream. It records the gaseous plume species at a resolution of integer mass-to-charge ratios (m/z = 1-100) in-situ, giving direct insight into the completeness of the chemical reaction between the two propellants and the ratio of the detected reaction products. This paper focussed on permanent deposit observations and is limited to small remarks regarding this technique.

3.5 Quartz-crystal microbalance (QCM)

Three QCMs (QCM Research, Mark18) can be positioned at desired locations across STG-CT to measure mass deposit per solid angle at temperatures as low as 80 K with a resolution down to an atomic layer. In this project 28°, 90° and 142° are probed, Fig.2, with additional angles covered by the rotation of the thruster in the first part of the test campaign. Once the plume exposure is complete and the chamber is still under vacuum and cryogenic conditions, a QCM-thermogravimetric analysis (QTGA) is conducted. The different vapour pressures of the deposited molecular species enable their consecutive, although potentially reciprocatively affected, evaporation when the QCM surface is heated at 2 K/min.

4. Intermediate results

4.1 High-speed imaging

The images recorded with lower frame rate (100 fps) show cooling film droplets running down the nozzle (v = 0.05 m/s), starting towards the middle of the pulse, Fig.4 left. As soon as the pulse ends, the droplets start to evaporate, indicated by the slow blurring of their contours. Right before the next pulse, i.e. approx. 5 min later, the droplets cannot be seen any longer in the nozzle. After approx. 25 pulses, droplets sit on the nozzle lip, Fig.4 right, as they advance further towards the lip with each firing.

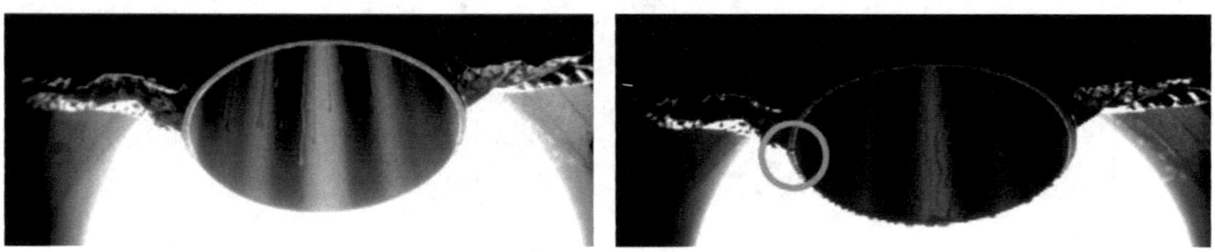

Figure 4. Cooling film advancing towards the nozzle lip (left) and cooling film liquid sitting on the nozzle lip after approx. 25x 500 ms pulses (right). Author's own work, photos taken at DLR Göttingen.

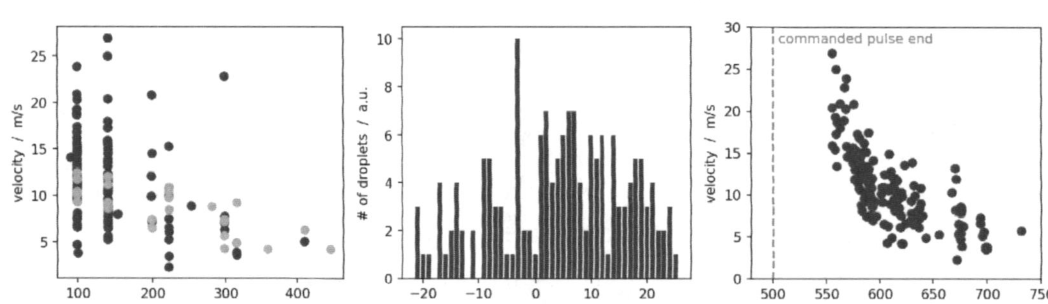

Figure 5. Droplets statistics as observed after the commanded end of a 500 ms pulse: Larger droplets tend to be expelled at slightly slower velocities (left), the ejection angle spans over a region of +/-21° (centre) and representation of decreasing droplet velocities after pulse end which corresponds to the collapsing combustion reaction and hence overall reduced gas flow velocities (right). Author's own work.

In all recordings with higher frame rate, no droplets were observed during the pulse firing. Potentially they could be below the camera's resolution limit (< 98 µm). However, between 55 and 140 ms after the commanded pulse end, droplets exit the nozzle in each pulse, mainly with velocities between 5 and 20 m/s and a size smaller than 400 µm, Fig.5. Larger droplets tend to be slightly slower compared to smaller ones. The droplets follow straight trajectories at angles up to 21° from the nozzle axis. The number of visible droplets varies for each pulse, from only one visible at all to several within milliseconds. A correlation between appearance, size and trajectory is not yet identified.

Only the very first cold pulse featured a spray ejection approx. 31 ms after the pulse trigger. During this cold start-up, the spray expels at an angle up to 20° from the nozzle axis. The event lasts for less than 2 ms. While the spray is barely visible by eye, Fig.6 left, details of the spray are uncovered by subtracting the images from a reference image, Fig.6 right.

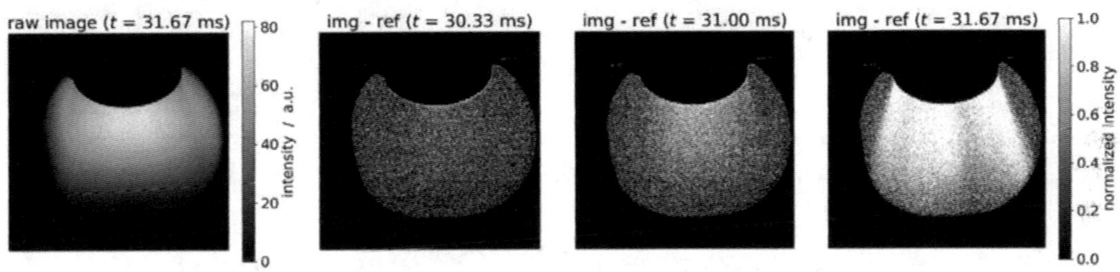

Figure 6. Fine spray observed during first cold-start pulse at about 31 ms after valve opening (left), enhanced by image subtraction for better visibility (right). Author's own work.

4.2 Material samples

When opening the chamber after plume exposure of the samples under constant thruster conditions, some materials appeared unaltered at first sight, while some had a milky appearance (glasses) and on several others (silicon wafers, OSR, gold mirror) droplet patters where identified by the unaided eye, Fig.7. All samples except for silicon wafers were sent to the chemistry lab of University Leipzig where they are currently being analysed, which is why this paper focusses on analysis results obtained for silicon wafers using a SEM with EDS detector in-house.

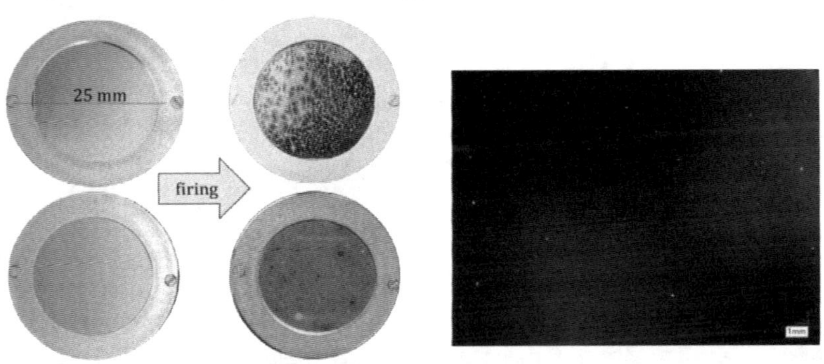

Figure 7. Exemplary samples (OSR, top and gold mirror, bottom) in comparison from pristine (left) to exposed state after 26 pulses of 500 ms (centre), together with an optical image of a backflow silicon wafer (114° position) under lateral lighting (right). Author's own work, photos taken at DLR Göttingen.

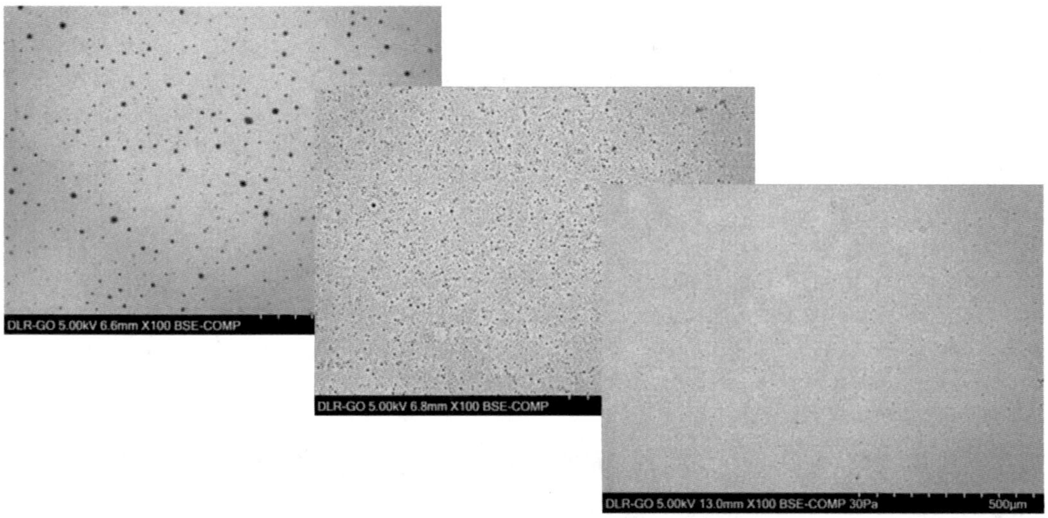

Figure 8. SEM images of silicon wafers with permanent deposit patterns decreasing in size and significantly increasing in number of what appears like droplets from 7°, to 19° and to 23° position (top left to bottom right). Author's own work, photos taken at DLR Göttingen.

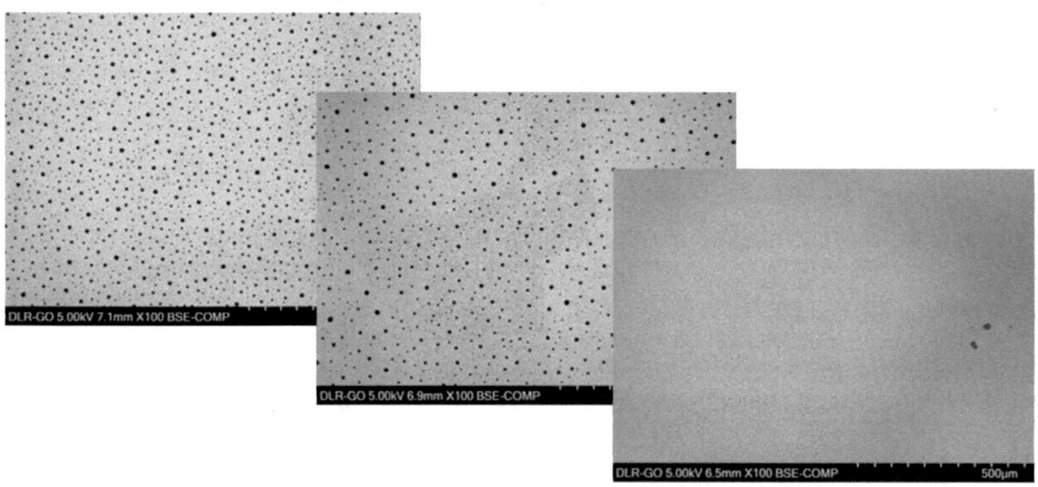

Figure 9. SEM images of silicon wafers with similar permanent deposit patterns at 33° (top left) and 42° (centre), but considerably higher droplet density and smaller particle size distribution at 66° position (bottom right). Author's own work, photos taken at DLR Göttingen.

SEM measurements of permanent deposit patterns were analysed on representative wafer areas shown in Fig.8 and Fig.9 after a general inspection with optical microscopy. The perfectly spherical appearance of nearly all features suggests that they are droplets or at least created by the influence of droplets. The droplet size evaluations in areas of even deposit results in diameters ranging from 0,1 – 25 μm. It is remarkable that both the 23° and 66° position have upper diameter values limited to about 3,5 μm. Another anomaly differing from the aforementioned results was observed on both the 19° and 23° position wafers. At each wafer an impact site of what appears to have been a larger droplet dispersed into smaller ones on impact was detected, Fig.10.

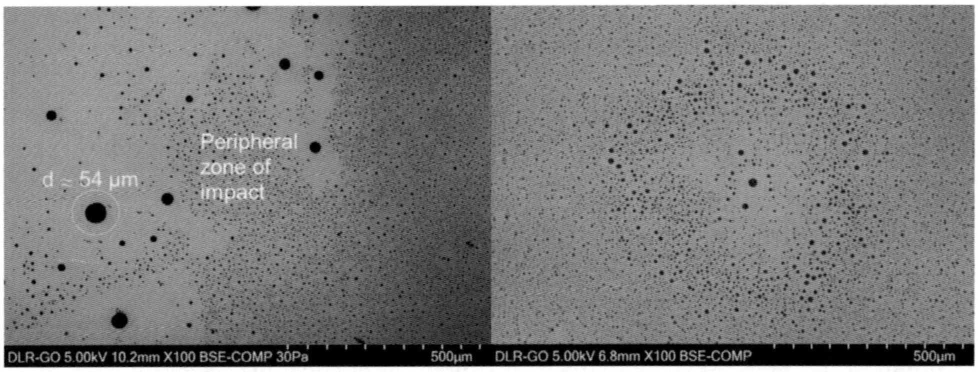

Figure 10. Exceptional findings: Presumable impact sites of initially larger droplets supposedly dispersed on impact on silicon wafers located at 23° (left, diameter of entire impact zone about 4 mm, not represented in its entirety) and 19° (right, diameter of impact zone about 750 µm). Author's own work, photos taken at DLR Göttingen.

Remaining feature sizes in the range of 50 µm are twice as big as the largest findings within regions of uniform deposit.

The backflow position exhibits a minor amount of permanent deposits, which is why an optical microscopy image is used in the area of highest droplet number density, Fig.7, right, in order to obtain a peak value representing this sample.

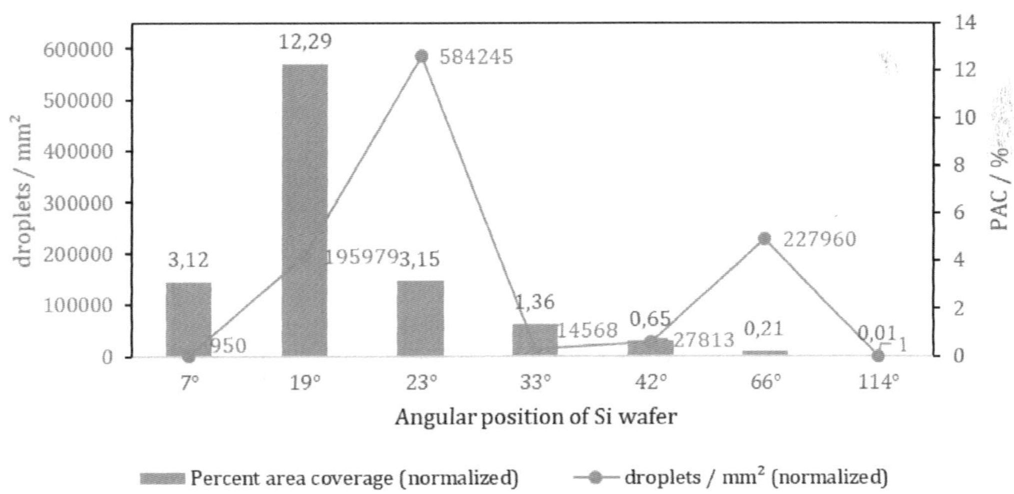

Figure 11. Droplet statistics analysed from SEM imaging and normalized over the thruster to sample distance: The total amount of droplets per mm² is shown in orange (line) and the percent area coverage shown in blue (bars). The value for the 114° position were obtained by optical imaging and represent maximum values found on that sample. Author's own work.

A statistical examination of droplet number densities and percent area coverage values, both normalized to the thruster sample distance, reveals two major insights, Fig.11. Among the sampled positions, 23° and 66° feature two maxima in droplet number density, along with their deviating particle size distribution describe above. The highest percent area coverage was found at the 19° position.

An EDS analysis of various droplet features across all samples revealed the presence of nitrogen, oxygen and carbon (descending in percentage), expectable from the propellants in use.

4.4 Quartz-crystal microbalance (QCM)

Presented QCM results are based on two different test cycles. Firstly, angular mass deposition data where gathered by the rotation of the thruster, Fig.12, left. Secondly, for the sake of better comparability, pulse length dependencies were recorded when firing onto two QCMs at fixed positions (90° and 142°), Fig.12, right, equal to QTGA data presented in Fig.13, which are based on QCM deposits accumulated from 26 thruster firings under constant angular positions.

Mass deposition at different angles is logged on a scale of μm/cm² and decreases by approximately three orders of magnitude when looking at values near plume centreline up to the backflow region. The latter exhibits a non-negligible mass deposition that was repeatedly recorded for the default QCM2 position of 142° off centreline.

Varying the on-time of the thruster from 20, 50, 100, 200 to 500 ms leads to an increase in deposited mass for both QCM1 fixed at 90° (T=80 K) and QCM2 at 142° (T=110 K), however, the trend is not linear, showing a declining increase for longer pulse durations.

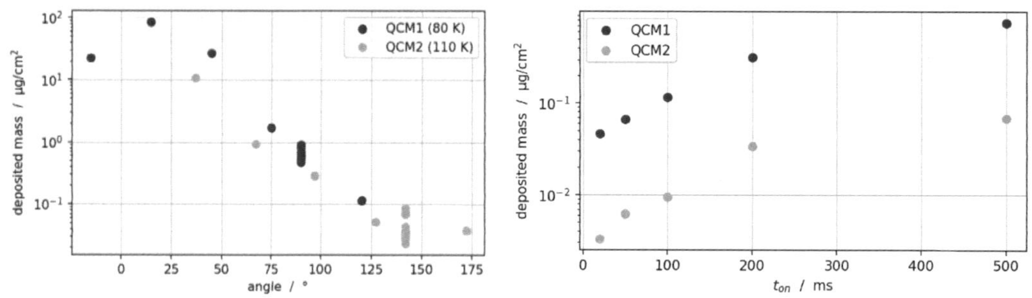

Figure 12. QCM deposit decreases with increasing angle off plume centreline (left). On-time variations of the thruster from 20 to 500 ms lead to a non-linear increase in mass deposited for QCM1 fixed at 90° (T=80 K) and QCM2 at 142° (T=110 K) (right). Author's own work.

Despite adsorbing species at a higher substrate temperature of 120 K, QCM2 experienced a mass deposition exceeding the one of QCM1 at 100 K by an order of magnitude: QCM2 at the 28° position recorded 1-10 μg/cm² on average per pulse, while QCM1 at the 90° angle recorded about 0,25-0,5 μg/cm², which is coinciding with higher desorption rates of QCM2 during the QTGA process, Fig.13.

TGA results from QCM1 positioned at 90° (T=100 K) and QCM2 at 28° (T=120 K), Fig.13, both exhibit a desorption peak near 200 K. Additionally, a minor desorption process is observed around 215 K in both cases. Both peaks presumably originate in unburnt or partially burnt, heavier propellant species, as all other regular plume exhaust products like N_2, H_2, CO_2, CO, NO, etc. would desorb at lower temperatures due to their lower vapour pressures.

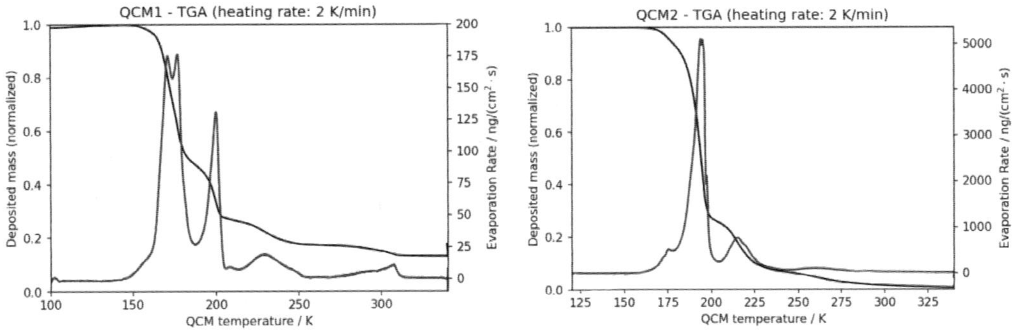

Figure 13. QTGA data after 26 pulses of 500 ms fired onto QCM1 positioned at 90° (T=100 K) (left) and QCM2 at 28° (T=120 K) (right). Both sensors exhibit a desorption peak that corresponds to water close to 200 K and a minor peak around 225 K, while QCM1 shows two major overlaying desorption peaks starting at around 160 K. QTGA was carried out at a chamber background pressure of 10^{-5} mbar. Author's own work.

At the 90° position two additional, overlapping desorption processes take place between 160-180 K, which cannot easily be matched to a plume species. Part of that mechanism can supposedly be linked to the desorption of water with a corresponding vapour pressure at the present vacuum level in the low 10^{-5} mbar range.

5. Discussion

The present study confirms the finding of droplets being expelled from the nozzle of a 10 N bi-propellant reaction control thruster operating on MMH/MON under nominal conditions as found in previous studies conducted in conventional vacuum chambers [3]. This result highlights the need to understand the plume and all its components, including gaseous and liquid (and for other thruster types solid) constituents, for a complete assessment of its contamination potential.

An operation of the thruster was realized under constant nominal conditions, excluding the first cold-start pulse of each testing sequence as the structure of the thruster has to heat up with the help of the combustion process, as no heating unit is provided on this kind of thruster.

High-speed imaging of the nozzle exit had the potential to reveal droplets larger than 98 micrometres i.e. larger than the ones imaged by SEM analysis of material samples. This implies that either numerous smaller droplets were present in the plume and deposited onto the materials but could not be filmed due to limited resolution, or larger droplets tend to break apart when impacting a surface, leaving behind a finer dispersion of droplets which are then found ex-situ. Regarding the origin of droplets, it is important to note, that there exist different sources: First, the cooling film made of pure MMH advances to the nozzle lip where droplets are ripped off by the plume flow or evaporate in a yet undefined manner. Second, incomplete evaporation of injected liquid propellants inside the combustion chamber during, especially likely during start-up and shut-down phases due to lower temperatures, can lead to the direct emission of these very droplets [7]. The latter is the phenomenon that was presumably observed with our high-speed camera.

Droplets that were identified on silicon wafer substrates across a wide angular range, indicate an order of magnitude for percent area coverage established after 26x 500 ms firings

IOP Conf. Series: Materials Science and Engineering **1328** (2025) 012015 doi:10.1088/1757-899X/1328/1/012015

under space-like vacuum conditions. However, we believe that an extrapolation of these findings should be examined with attention to detail, as the plume composition varies with thruster operational points, as shown by QCM mass deposition rates of different pulse lengths. Especially the backflow area should be treated with great care, as this study has so far only investigated QCM data together with a single material sample. Already one impact feature as seen on two other wafers - which has a non-negligible probability of impacting a surface anywhere in the backflow when considering the cooling film advancing to the nozzle lip - could significantly alter the observations made so far.

In addition, the behaviour of droplets during the venting of our chamber with ambient air remains an open question, as the water introduced during this process would have a strong tendency to react, should the droplets be composed of a MMH-derivative, as all hydrazine are readily soluble in water. A chemical analysis of exposed samples promises to help understand whether chemical reactions with substrate materials took place in addition to the aforementioned processes.

QCM measurements supply data on how much deposit a surface impacted by the plume would take up depending on its angular position with respect to plume centreline and, equally important, related to a certain surface temperature. The latter makes a difference when considering if a gaseous species will stick to a satellite surface at all. Furthermore, QTGA measurements reveal another crucial information for satellite design or operation: How much does a surface need to be heated in order to re-evaporate the deposited mass? The observation of permanent deposit ex-situ at room temperature is supported by the remaining mass on QCM1 after heating up to 330 K, Fig.13 left.

6. Outlook

In summary, this paper gives a first insight into the results of an overall larger study on the contamination potential of a 10 N bi-propellant MMH/MON thruster plume, investigated under free stream conditions.

A remaining item is the detailed chemical analysis of a multitude of different material samples that were exposed to the plume, which will help to improve statistical conclusions and to better differentiate between findings that are broadly applicable across material's categories, or else others that are highly depended on the type of substrate material. Furthermore, the analysis of optical properties of the samples before and after plume exposure will allow for a rating of the sensitivity of different materials to plume exposure.

A better understanding of the exact chemical species evaporating during a QTGA could be achieved by the combination with a parallel MS measurement. The results of in-situ MS plume analysis are still under investigation but first analysis indicate an incomplete combustion process containing, amongst others, CH_4 as one of the exhaust gases. This finding will further be studied but corresponds to a numerical study on a similar type of 10 N thruster [8].

The overall goal of this investigation is an improved database for well-founded recommendations during the design and operational phase of a satellite. Another part of this study complements these findings with gas dynamic plume characterisations which in turn serve as points of reference for improved numerical plume modelling. The latter will focus on plume expansion and contaminant deposition and their comparison to experimental results, e.g. gas species distribution, QCM data, etc. Future research should be devoted to the in-situ observation of deposits on surface samples in order to better understand effects during impact, mechanisms occurring under vacuum conditions and facility effects during warm-up and venting. A higher camera resolution could potentially reveal a broader spectrum of droplets being ejected from the nozzle.

Acknowledgements

This work is funded under ESA Contract No. ESA AO/1-11328/22/NL/MG. The authors gratefully acknowledge the work of the technical team of STG-CT, DLR (Julian Ammer, Günter von Roden, Christopher Geile) for their essential work on the test setup and their dedication to maintain and operate the facility.

References

[1] Dettleff G 1991 *Prog. Aerosp. Sci* **28** pp 1-71 (10.1016/0376-0421(91)90008-R)
[2] Trinks H 1990 Gas species separation effects in exhaust plumes, *AIAA/ASME 5th Joint Thermophysics and Heat Transfer Conference* p 1734
[3] Dettleff G and Grabe M 2011 Basics of Plume Impingement Analysis for Small Chemical and Cold Gas Thrusters, *Models and Computational Methods for Rarefied Flows* pp 1-40
[4] Trinks H 1990 Exhaust Plume Data Handbook *ESA Progress Report contract 7510/87/NL/PH*
[5] Grabe M 2016 *JLSRF* **2** A86 (10.17815/jlsrf-2-139)
[6] Buntrock L J, Grabe M, Schmidt L, Jéger C and van den Eynde J 2024 *9th Space Propulsion Conference*
[8] Hoffman R J, Kawasaki A, Trinks H, Bindemann I and Ewering W 1985 *AIAA 20th Thermophysics Conference*
[9] Ecker T, Fechter S and Grabe M 2024 *9th Space Propulsion Conference*

IOP Conf. Series: Materials Science and Engineering **1328** (2025) 012016 doi:10.1088/1757-899X/1328/1/012016

Temporal evolution of the morphology of contaminants deposits

Aurélie Zamo[1], David Lansade[1*], Jean-François Roussel[1*], Guillaume Rioland[2] and Delphine Faye[2]

[1] DPHY, ONERA, Université de Toulouse, Toulouse, France
[2] CNES, Toulouse, France

*E-mail: david.lansade@onera.fr / jean-francois.roussel@onera.fr

Abstract. The effects of contamination on space systems have been studied for many years. However, the effects especially due to the morphology of contaminant deposits have not been studied extensively. ONERA has developed a vacuum chamber for studying molecular contamination in which experiments aimed at studying changes in morphology. Ex-situ optical microscopy analyses revealed temporal maturation phenomena: coalescence and solidification of the deposit. A systematic study of contaminant/substrate/pressure conditions combinations allowed a first understanding of the influence of the pressure on changes in deposit morphology after contamination. These results show the importance of identifying the morphology of contaminant deposits within a short delay in order to anticipate changes in morphology.

1. Introduction

Exposed to space environment, some materials that make up satellites outgas molecules, known as contaminants. These contaminants can condense on sensitive surfaces of the satellite, significantly altering their properties. The main consequences of this contamination are loss of transmittance in optics [1], reduced solar panel power output [2], and changes in solar absorptance of thermal coatings [3].

In-flight measurements showing differences in aging behaviour of various contaminated materials were previously poorly understood. Recent work suggests differences in morphology (cluster vs. film) of the contaminant deposits may be responsible for these effects. The way the contaminant deposits age under ultraviolet (UV) irradiation is greatly influenced by their morphology (film vs clusters). For example, the increase in solar absorptivity of a contaminated substrate can be highly important in the case of a film deposit. Despite the interest in contamination morphology [4], the physical processes are little studied and, moreover, difficult to predict.

One issue for studying the morphology of contaminant deposits when there is no in-situ measurement possible is the need to restore atmospheric pressure to characterize the samples. The experiments presented in this paper aim at determining the influence of the pressure (vacuum breakdown) on potential changes in deposit morphology after contamination.

In order to establish the effects of atmospheric pressure on contaminated surfaces, a study was carried out comparing the pressure conditions under which the samples were held. To achieve this, a systematic study of contaminant/substrate/deposition temperature combinations and characterizations of contaminated surfaces are used to understand the phenomena governing the morphology of a deposit, thus refining predictions.

2. Experimental

2.1 Vacuum contamination chamber

ONERA has developed a vacuum test facility for studying contamination mechanisms. The facility is equipped with an effusion cell in which a certain quantity of contaminant is introduced. Once the effusion cell reaches a specific temperature, a gaseous flow of contaminant is generated. The presence of a temperature-regulated copper envelope (known as shroud) enhances the transport of molecules (contaminant gas flow) towards various targets: substrates and quartz crystal microbalances (QCMs). These targets, positioned on a sample holder, can also be temperature-controlled to study contaminant deposition according to various parameters.

QCMs enable real-time monitoring of contaminant deposition and detection of any reemission of deposits. QCMs used in this work are equipped with quartz crystals vibrating at a resonant frequency of 6 MHz. The crystal is a quartz disk 14 mm in diameter, with a gold coating on both electrodes faces. The quartz crystals used are "AT-cut" and have were cut according to a particular crystallographic direction (Neyco® AUDA 6).

Figure 1 shows the positioning of the various elements described above.

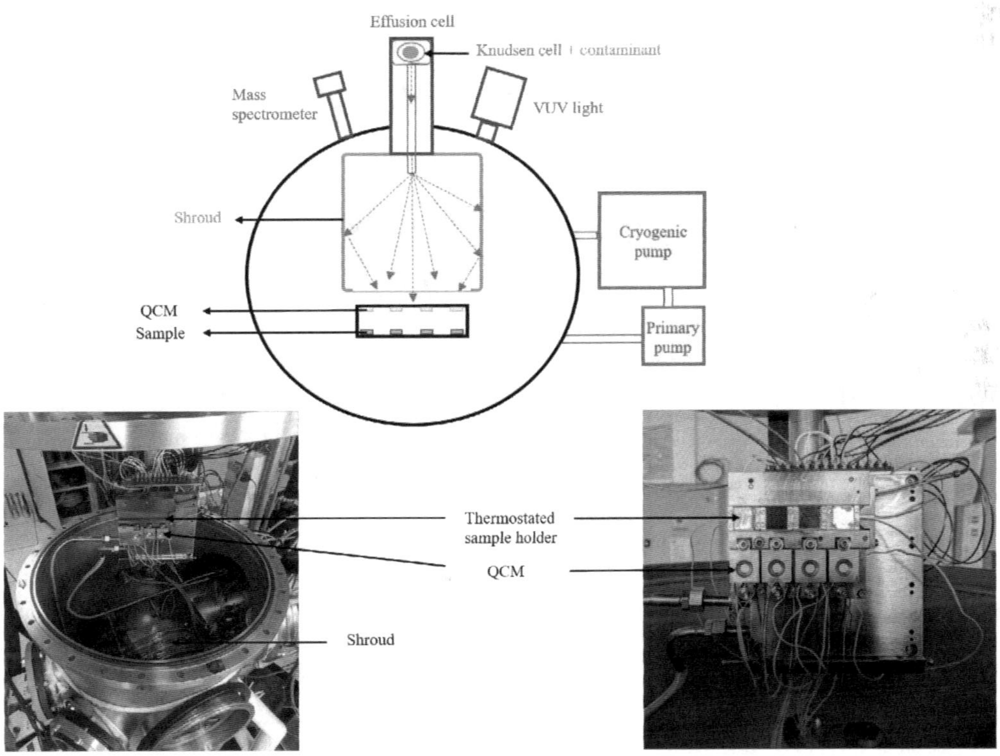

Figure 1. COPHOS test equipment (Credit: ONERA).

2.2. Contaminants and materials

When a material outgases, a multitude of contaminants are emitted and can contaminate a satellite surface [5]. Given the diversity of outgassing species, it is complex to determine the impact of each species on the morphology of the deposit. We have therefore focused on studying

IOP Conf. Series: Materials Science and Engineering **1328** (2025) 012016 doi:10.1088/1757-899X/1328/1/012016

the morphology of the deposits of a pure compound: diphenyl phthalate. This contaminant is representative of molecules outgassed in flight as it belongs to a group of molecules often found as plasticizers. Phthalates can be found from polymers such as polyesters and polyurethanes. These polymers are used in adhesive tapes, resins or thermal control coatings (MLI). Contaminated substrates are materials that can be used for optical surfaces, thermal control or solar panels, like: fused quartz, polyimide (Kapton tape), black silicone paint.

2.3. Characterization of contaminant deposits: optical microscopy

Deposits need to be accurately characterized in order to understand their behavior on various substrates and consequently determine the physico-chemical mechanisms responsible for changes in morphology. Thus, various characterization techniques have been employed.

To characterize the deposits, an optical microscope (Nikon LV 100 Eclipse) coupled to a camera (Imaging Source – DKF330X365) was used. By modulating the contrast or the angle of inclination of the light, it is possible to observe micrometric deposits. This makes it possible to determine the morphology of deposits (film or droplets), droplet diameters and sphericity. The morphology of the studied deposits can also be observed at different times.

3. Results

3.1. Test plan

Samples were contaminated with diphenyl phthalate. QCMs and samples surfaces were kept at a constant temperature of 5 °C for the whole deposition time under secondary vacuum ($P \sim 1x10^{-6}$ mbar). Such temperature inhibits contamination reemission under vacuum, detrimental to the study. When the QCMs and samples holder are stable, the effusion cell is heated and maintained at 190 °C. The temperature of the effusion cell is relatively high, which is unusual for spacecraft. This choice of temperature was motivated by the objective of minimizing the time required for the deposit. As soon as the effusion cell is empty (no more incident molecular flow), the pressure in COPHOS is:

- Case 1: maintained for 3 days under dry nitrogen at 1 bar
- Case 2: maintained for 3 days under secondary vacuum ($P \sim 1x10^{-6}$ mbar)

The mass of diphenyl phthalate in the effusion cell is 10.2 mg and 10.5 mg for case 1 (dry nitrogen) and case 2 (secondary vacuum), respectively.

3.2. QCM measurement of deposit mass

Despite targets' temperature maintained at 5 °C, a reemission phenomenon is detected in the case of a deposit maintained under secondary vacuum, whereas it does not appear under dry nitrogen at 1 bar (Figure 2). This difference is explained by the fact that when maintained under secondary vacuum, the reemission flow from the targets surface is in free molecular regime. After reemission none of the molecules collide, so no return to the contaminated surface is possible. On the opposite, a flow of reemitted molecules under dry nitrogen has a high probability of colliding with nitrogen molecules close to the contaminated surface due to the fact that the mean free path at such pressure is in the order of 100 nm [6], [7]. As a result, almost every reemitted molecule return to the surface and the reemitted flux becomes so weak that it is not measurable on QCMs.

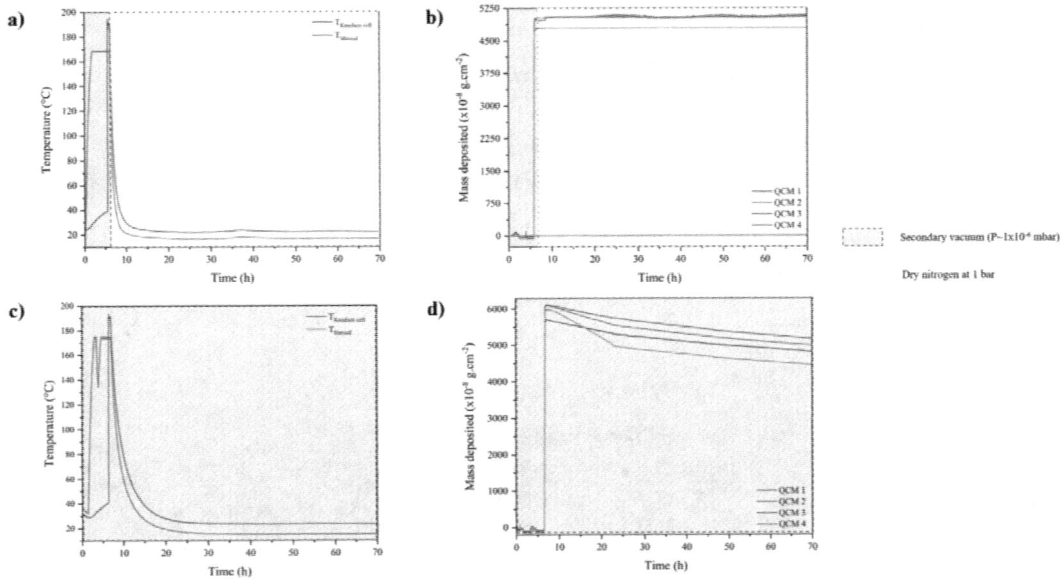

Figure 2. Temporal evolution of Knudsen cell and shroud temperatures for a) case 1 maintained under dry nitrogen at 1 bar and c) case 2 maintained under secondary vacuum at 1x10-6 mbar.

Temporal evolution of diphenyl phthalate deposited and re-emitted masses measured on QCM à 25 °C for b) case 1 maintained under dry nitrogen at 1 bar and d) case 2 maintained under secondary vacuum at $1x10^{-6}$ mbar (Credit: ONERA).

Note: QCM 1 malfunctioned during the experiment. Due to the geometry of the contamination system, the deposited mass is assumed to be close to that measured on QCM 4.

3.3. Morphology of contaminant deposits

It is important to note that the contaminated samples presented below were maintained for 3 days under dry nitrogen at 1 bar or under secondary vacuum before being characterized by optical microscopy (at ambient pressure and room temperature). It was possible to observe different morphologies depending on the nature of the contaminated substrate, regardless of the pressure conditions in the chamber (Figures 3 to 5). Observations of every samples as a function of time were also carried out in order to consider possible changes in morphology.

3.3.1. Polyimide

At first sight, it seems difficult to detect any major differences between contaminated polyimide samples maintained under dry nitrogen or secondary vacuum: the deposits form droplets (Figure 3). However, the surface area covered by diphenyl phthalate molecules appears to be greater in the case of deposits maintained under dry nitrogen at 1 bar. This phenomenon results from the reemission of deposit under secondary vacuum. Smallest droplets evaporate more easily from the surface and cannot return to the surface (free molecular flow).

In the case of contaminated samples maintained under dry nitrogen, numerous droplets are in contact with each other, but it is still possible to observe isolated droplets. For the most part, the droplets show an approximately central zone with no apparent deposit. The droplets are between 20 and 30 micrometers in size. Deposition on polyimide maintained under secondary vacuum for 3 days also shows droplets between 16 and 25 micrometers, with a central zone without apparent substance.

Temporal analysis of the samples was carried out 1 month after they were exposed to air (samples are stored in closed boxes at 20% humidity and 22°C.). The results of these observations confirm changes in morphology. However, these temporal maturation phenomena are not identical between the two experimental conditions. In the polyimide sample maintained under dry nitrogen for 3 days, droplets solidification appears, with a reduction in size (ten micrometers) and also the formation of dendrites. In the experiment maintained under secondary vacuum, the clusters also solidify and the smallest clusters seem to have evaporated.

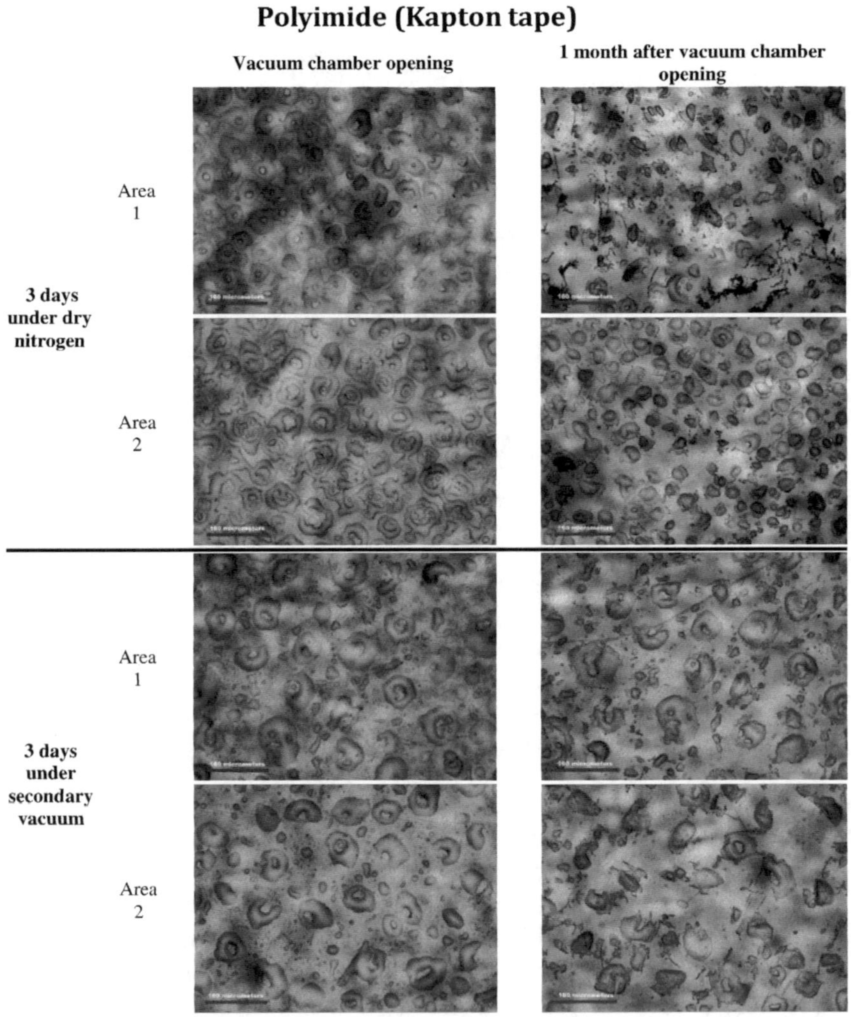

Figure 3. Optical microscopy imaging of the temporal evolution of diphenyl phthalate deposits on polyimide samples for different pressure conditions (Credit: ONERA).

3.3.2. Fused quartz

Figure 4 shows that diphenyl phthalate is deposited on a fused quartz surface in the form of droplets (maintained under secondary vacuum and dry nitrogen).

Although the deposit is spread over a large part of the sample surface, in the case of the deposit maintained under dry nitrogen, the larger droplets are surrounded by smaller droplets (~5 µm). Large areas without apparent deposit structure are also visible over a large part of the surface. In the case of deposits maintained under secondary vacuum for 3 days, the droplets show a less well-structured external structure than in the previous case. The number of small droplets also appears to be reduced compared with the case maintained under dry nitrogen at 1 bar. The absence of smaller droplets can be explained by the reemission of the deposit under secondary vacuum. As the ratio surface/volume is higher for small structures, their evaporation from the surface is accentuated.

Observation of the fused quartz samples 1 month after opening COPHOS reveals that maturation phenomena are visible at ambient pressure and room Indeed, it is possible to distinguish the effects of coalescence between droplets, crystallization and dendritic growth. However, these phenomena only seem to be visible when the sample was kept under dry nitrogen at 1 bar for 3 days. In fact, when the sample was kept under secondary vacuum for 3 days before being returned to air, the coalescence process seemed weaker. This slower maturation may be due to the greater distance between droplets, which prevents them from interacting with each other.

IOP Conf. Series: Materials Science and Engineering **1328** (2025) 012016 doi:10.1088/1757-899X/1328/1/012016

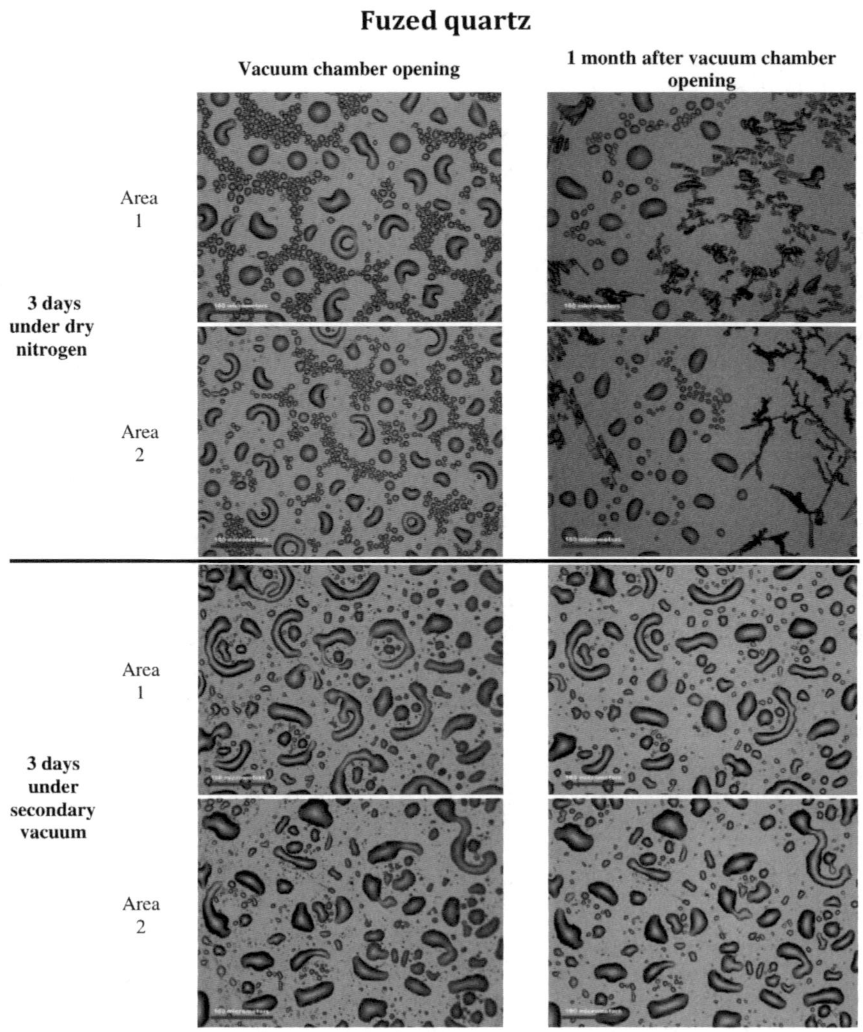

Figure 4. Optical microscopy imaging of the temporal evolution of diphenyl phthalate deposits on fused quartz samples for different pressure conditions (Credit: ONERA).

3.3.3. MAP PNC, silicone-based black paint

No matter what pressure conditions were applied to the silicone-based black paint samples (kept under 1 bar dry nitrogen or under secondary vacuum), surface coverage by the diphenyl phthalate deposit was heterogeneous. Figure 5 shows a few areas presenting diphenyl phthalate crystals. However, a difference in crystal shape is visible 1 month after the samples were re-exposed to the air. In the case where the deposit was kept under dry nitrogen for 3 days, the crystals are concentrated together. On the contrary, when maintained under secondary vacuum for 3 days, the deposits observed is in the form of needles, when it initially appeared to be in the form of small clusters.

The composite nature of this silicone-based black paint and its roughness probably influence strongly the adsorption process of molecules on the surface.

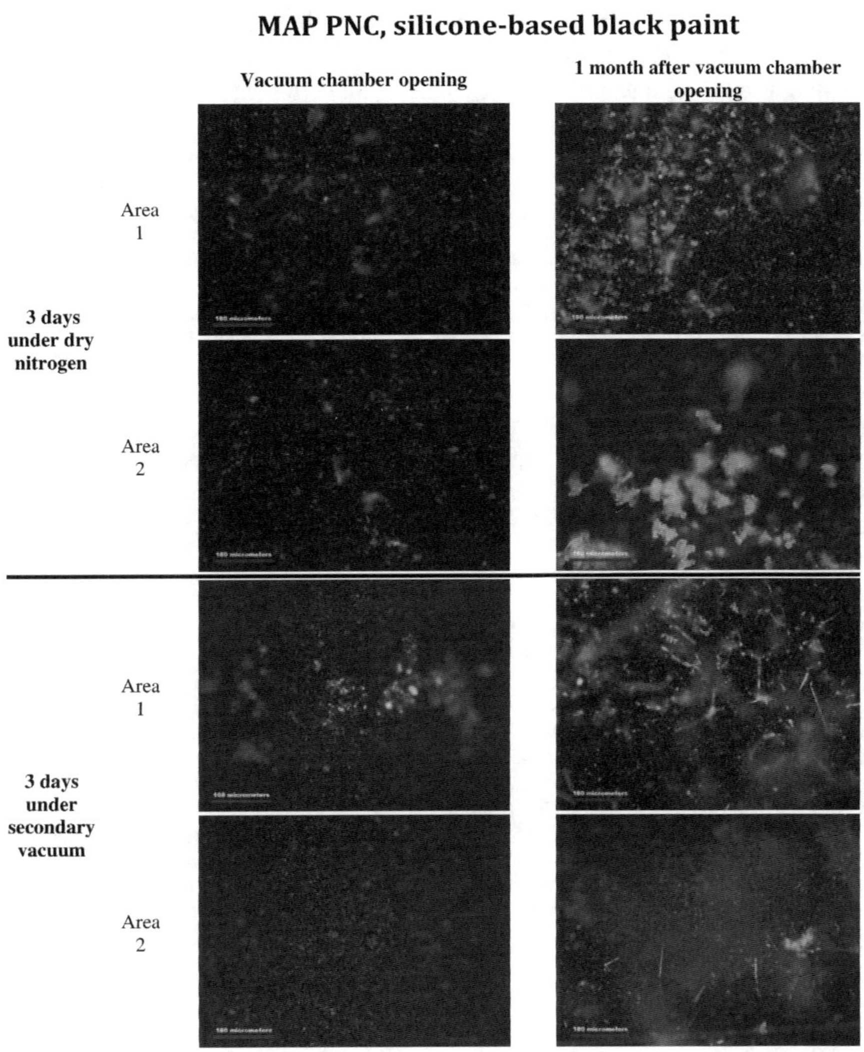

Figure 5. Optical microscopy imaging of the temporal evolution of diphenyl phthalate deposits on silicone-based black paint samples for different pressure conditions (Credit: ONERA).

4. Conclusion

The effects of pressure on morphological changes in diphenyl phthalate deposits were investigated. For this purpose, diphenyl phthalate deposits were studied on various substrates (fused quartz, polyimide and silicone-based black paint) with a deposit maintained under dry nitrogen at 1 bar and another one under secondary vacuum.

When contaminants are deposited in the gas phase, some of the molecules adsorbed on a surface may reemit. However, this reemission flux can be modified according to the conditions under which the deposit is maintained. When the deposit is maintained under secondary vacuum, molecules are reemitted in free molecular flow and the associated flux is large. As a result, the smallest droplets evaporate and the quantity of molecules present on the surface lowers. When the deposit is kept under pressure at 1 bar, the molecular flux reemitted is drastically reduced as the reemitted molecules are subject to numerous collisions and may therefore return to the contaminated surface. These returns modify the morphology of the deposit: the smallest droplets are maintained on the contaminated surface and may interact one with each other.

The temporal analysis reveals a maturation process for all deposits and substrates. In-depth observations using optical microscopy (*ex situ*) revealed coalescence phenomena between several droplets and the transformation of droplets into crystals/dendrites. These phenomena, observed at ambient temperature and atmospheric pressure, occur as soon as the deposit is formed under secondary vacuum.

This study revealed that contaminant deposition morphology is dependent on a number of parameters: the nature of the contaminated surface, surface roughness and deposition pressure conditions. Further work is currently underway to determine the physico-chemical processes responsible for the morphology of contaminant deposits and to establish a means of in-situ characterization.

References

[1] Rioland G., Shimazaki K., Miyazaki E., and Kimoto Y., 2020 Transmittance measurements and predictions of optics contaminated with space materials in the ultraviolet–visible–near-infrared range *JATIS* **6**(1) p. 018004, p1.
[2] Liu D., Liu S.H., Panetta C.J., Olson K.R., Hong S.M., Alaan D.R., Mann Colin J., and Luey K. 2010 Effects of contamination on solar cell coverglass *35th IEEE Photovoltaic Specialists Conference* pp. 002563–002568.
[3] Remaury S., Nabarra P., Bellouard E., and d'Escrivan S. 2011 In-Flight Aging of Thermal Coatings: THERME Experiment *Journal of Spacecraft and Rockets* **48**(1) pp. 27–33.
[4] Luey K. T., Olson K. R., and Coleman D. J., 2018 Optical system contamination: formation of films and droplets *JATIS* **4(3)** p. 036001
[5] Lansade D., Roussel J.-F., Faye D., Rioland G., Suliga A., and Ergincan O., 2023 *IOP Conf. Ser.: Mater. Sci. Eng.*, vol. **1287**(1) p. 012022.
[6] Al-Dmour E. 2020 Fundamentals of Vacuum Physics and Technology *2017 CERN–Accelerator–School course on Vacuum for Particle Accelerators.*
[7] Franchetti G. 2013 Vacuum I *CAS - CERN Accelerator School: High Power Hadron Machines* pp.309-326.

IOP Conf. Series: Materials Science and Engineering **1328** (2025) 012017 doi:10.1088/1757-899X/1328/1/012017

New Materials for LEO, GEO and Planetary Environments: Preliminary Results from MISSE-17 Experiment

J. Kleiman*, Z. Iskanderova, R. Ng, A. Tang

Integrity Testing Laboratory, Markham, Ontario, Canada
jkleiman@itlinc.com

Abstract. ITL developed coatings and surface treatments that include prototypes of multifunctional coatings with charge-dissipative and dust mitigation properties, including Lunar dust Work Function Matching Coating (WFMC) that can be applied to surfaces of various space materials. Selected samples were flown on MISSE-17 Material Experiment, to evaluate their resistance to space environmental factors like thermal cycling, vacuum, VUV and atomic oxygen.

Two thermal control materials, i. e. a 1 mil KaptonHN with a doped diamond-like coating (DLC) on it and an aluminized 1 mil KaptonHN with a DLC film on it were selected for the ram direction. For the wake position, a doped DLC on a 1 mil aluminized KaptonHN and a doped DLC on a silver-coated Teflon FEP were selected, to check the influence of both the thermal cycling and UV exposure. In addition, two samples of Lunar dust Work Function Matching Coating (WFMC) on Kapton HN (WFMC/Kapton) were prepared and added, one in ram direction and one in wake direction.

The MISSE-17 Experiment was launched from Kennedy Space Centre on March 15, 2023 and returned on SpX-29 flight in December 2023. Samples faced atomic oxygen fluence of ~8.0E+20 atoms/cm^2 in ram direction and 3.76E+18 atoms/cm^2 in wake direction over 145 days and 157 days, respectively.

The results discussed in this paper include weight and surface morphology changes, the thermal optical properties and surface analysis.

1. Introduction

With an ever-increasing range of activities in space that extend from the very low Earth orbit (VLEO) to the low Earth orbit (LEO), Geostationary orbits (GEO) and planetary environments like Lunar and Martian environments, the need in materials and coating that will withstand the harsh space conditions is growing.

For each operating environment the requests for material properties can be different. Thus, for materials flown in VLEO and LEO missions, the atomic oxygen remains the major hazard [1-6].

The VLEO satellites, that are orbiting at an altitude of 200-300 km and that provide advantages for high-resolution Earth observation and atmospheric science, are becoming important not only for science, but also for Earth observations and for military and national security. The performance and the erosion behaviour for various space hydrocarbon, fluorinated, and perfluorinated polymers that are candidates for use in VLEO orbits needs to be studied [1,2]. While for systems operating in GEO orbits the charging effects are one of the major hazards [7-9], because the Electric Orbit-Raising (EOR) launches to bring spacecrafts to GEO are being used increasingly, as they require less fuel than conventional chemical rockets, the influence of atomic oxygen for such missions should also be considered [1]. With electric propulsion, the thrust is weaker than chemical rockets, but this small thrust can be provided for long duration. Due to the significant amount of time (months) the spacecraft remains at lower altitudes before it reaches GEO, the external materials are therefore exposed to some AO (exposure to AO was previously negligible for a "fast" transfer to GEO with chemical rockets). The AO exposure remains much less than at a typical LEO mission, but is not negligible with a typical estimated AO fluence around 6E19 at/cm2 or less. Materials with very low thickness such as thin film coatings can therefore be impacted by this AO fluence [6]. For system operating on Mars and Moon surfaces, the presence of dust represents one of the major challenges, in addition to the charged particles and low temperatures [10].

ITL developed coatings and surface treatments [3,4,9,10] that include prototypes of multifunctional coatings with charge-dissipative and dust mitigation properties, including Lunar dust Work Function Matching (WFM) coating that can be applied to surfaces of various space materials. Selected samples were flown on MISSE-17 Material Experiment, to evaluate their resistance to space environment factors like thermal cycling, vacuum, VUV and atomic oxygen.

2. Experimental

A total of six samples were prepared for inclusion in the MISSE-17 experiment (Table 1). Two thermal control materials, i. e. a 1 mil KaptonHN with a Si-doped diamond-like coating (DLC) on it and an aluminized 1 mil KaptonHN with a Si-doped DLC film on it, both with a surface resistance 2·E8 [Ω/□], and one work function matching coating, consisting of a (Si+O)-doped DLC coating on an aluminized Kapton, with surface resistance of 1·E9 [Ω/□], were selected for the ram direction.

For the wake position, an N-doped DLC on a 1 mil aluminized KaptonHN with a surface resistance 7.5·E7 [Ω/□], and a N-doped DLC on a silver-Inconel coated Teflon FEP with a surface resistance 1·E8 [Ω/□], and one work function matching coating, consisting of a (Si+O)-doped DLC coating on an aluminized Kapton, with surface resistance of 1·E10 [Ω/□], were selected, to check the influence of both the thermal cycling and UV exposure. Figure 1 and Table 1 provide a summary on the samples flown on MISSE-17.

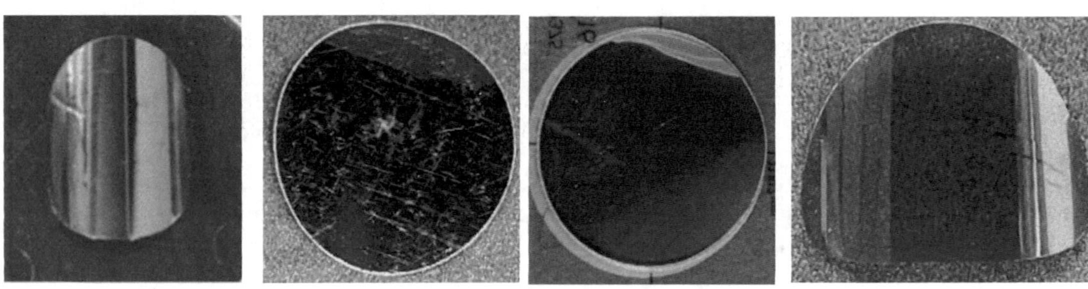

CK CT Ag CK Al WFM CK

Figure 1. Samples prepared for MISSE-17 mission. See Table 1 for sample identification. Credits: own work.

Table 1. Summary of information on the flown samples

Position on MISSE	Sample ID		Surface resistance (SR) [Ω/□]
WAKE	CT Ag	N-doped DLC on Teflon FEP/Ag/Inconel	1·E8
WAKE	C K Al	N-doped DLC/Kapton/Al	7.5·E7
WAKE	WFM C K	(Si+O)-DLC/Kapton /Al	1·E10
RAM	WFM C K	(Si+O)-DLC/Kapton /Al	1·E9
RAM	C K Al	Si-doped DLC/Kapton/Al	2·E8
RAM	CK	Si-doped DLC/Kapton	2·E8

The flown samples represent three novel materials that serve as charge -dissipating coatings for use in GEO and planetary environments, as thermal protection materials to maintain the thermal optical properties of structures exposed to the sun radiation and also as coatings with work function matching the lunar dust work function to mitigate the dust accumulation on the surfaces.

The MISSE-17 Experiment was launched from Kennedy Space Center on March 15, 2023 [11] and returned on SpX-29 flight in December 2023 [12]. The flight conditions are provided in **Table 2** [12].

Table 2. Summary of flight conditions in the MISSE-17 experiment

	Positioning in Flight	
	RAM	WAKE
Fluence of atomic oxygen (at/cm^2)	8.0E20	3.76E18
Duration	145 days 4 hrs. 36 min	157days 4 hrs. 22 min
Flux of atomic oxygen (at/cm^2/s)	6.38E13	2.77E11

All samples were analyzed before and after the flight using a number of analytical techniques. The surface morphology and the elemental composition of the samples were studied by scanning electron microscopy (SEM) combined with energy dispersive x-ray microanalysis (EDS). An environmental SEM was used in the studies mostly in secondary electron mode. The EDS system is equipped with a light element detector that allowed to detect all elements from carbon and up.

The changes in the surface elemental composition were also analyzed using the X-ray photoelectron spectroscopy (XPS). The spectra were obtained on the ThermoFisher Scientific K-Alpha. Monochromatic Al K-alpha X-ray used with a spot size of 400 micron. Survey spectra were obtained (pass energy - 200 eV), which was followed by the elements of interest obtained at low resolution (PE - 150 eV) from which the relative atomic % were taken. The C 1s peak was also collected in high resolution (PE - 25 eV). All data work-up was performed using software supplied with the instrumentation (Avantage v5.926 - ThermoFisher Scientific, E. Grinstead UK)).

The thermal optical properties of all samples were measured before and after the exposure. For the solar absorptance, firstly, the total reflectance spectrum was obtained in accordance with ASTM E903 [14] using an Agilent Cary 5000 UV-VIS-NIR spectrophotometer equipped with an integrating sphere. The instrument recorded reflectance data from 250 nm to 2500 nm. Then, the solar reflectance (ρ_s) was obtained by calculating a weighted average with an extraterrestrial solar spectral irradiance at air mass zero, specified in ASTM E490 [15], as the weighting function. The solar absorptance (α_s) was then calculated by subtracting the solar reflectance from 1, with solar transmittance (τ_s) = 0 in the Kirchoff relationship:

$$\alpha_s + \tau_s + \rho_s = 1 \qquad (1)$$

The total normal emittance (emissivity, ε) of the samples was measured with reference to the ASTM E408, Test Method A [16] using an infrared reflectometer manufactured by Gier-Dunkle Instruments, Model DB-100. This method is mainly for opaque samples.

3. Results and Discussion

The samples were photographed before the flight and at monthly intervals during the flight with the pictures being provided. Figure 2 shows the images of the three samples mounted in RAM direction right after opening the MISSE-17 tray and after 5 months of exposure to the FAO in LEO.

provides the optical images of the samples after they were delivered back to Earth.

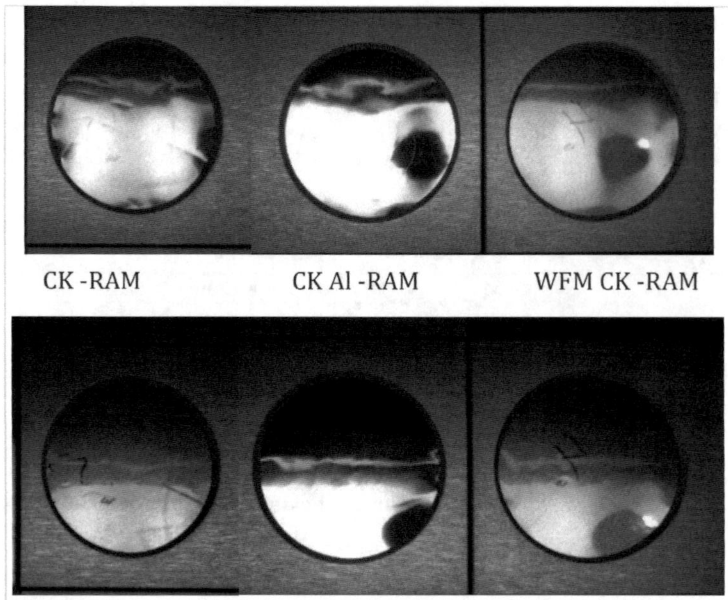

Figure 2. Images of samples mounted on the MISSE sample rack in RAM direction. Upper row: preflight picture; Lower row: after 5 months in flight (Images provided by Aegis as part of contract #21-17-0037 between Aegis Aerospace and Integrity Testing Laboratory, Inc).

Figure 3. Samples positioned in RAM direction after return to Earth and removal from the sample cassette. Credits: own work.

As can be seen from the images above, the samples flown in RAM direction while remaining shiny and without major observable defects in the pictures from the flight, demonstrated some surface changes in images taken upon their return. Table 3 presents the summary of data on mass change of the flown samples. All flown samples, positioned in RAM and WAKE directions, experienced very small mass losses, demonstrating erosion yields of the order of 1E-28 at/cm^3 in the RAM and 1E-29 at/cm^3 in the WAKE directions.

The samples in the RAM direction had shown a ~ 0.5% - 1.0% weight loss. Analyzing the images in Figure 4 through Figure 6 that present the SEM/EDS analysis of the flown samples that were positioned in the RAM direction, it can be suggested that the surfaces of all three samples remained smooth and not etched, demonstrating good AO erosion protection.

Table 3. Summary of data on weight changes of the MISSE-17 ITL flight samples

Sample ID	Position on the ISS	Weight (g)				Comments
		Before Flight (g)	After Flight (g)	Weight Change (μg)	Weight Change (%)	
C T Ag	WAKE	0.140000	0.139901	-99.0	-0.07	N-DLC Teflon FEP/Al/Inconel
C K Al	WAKE	0.017750	0.017730	-20.3	-0.11	N-DLC/Kapton/Al
WFM C K	WAKE	0.017230	0.017211	-19.3	-0.11	(Si+O)-DLC/KaptonHN
WFM C K	RAM	0.017190	0.017715	-85.5	-0.48	(Si+O) -doped DLC/Kapton HN
C K Al	RAM	0.017800	0.017033	-156.7	-0.91	Si-DLC/Kapton/Al
C K	RAM	0.017260	0.017092	-168.0	-0.97	Si-DLC/Kapton

Selectively etched regions were seen on all three samples, with small, but distinct pits developed. The formation of such selectively eroded pits is usually associated with surface defects like scratches, dents, contaminants present in the material, etc..

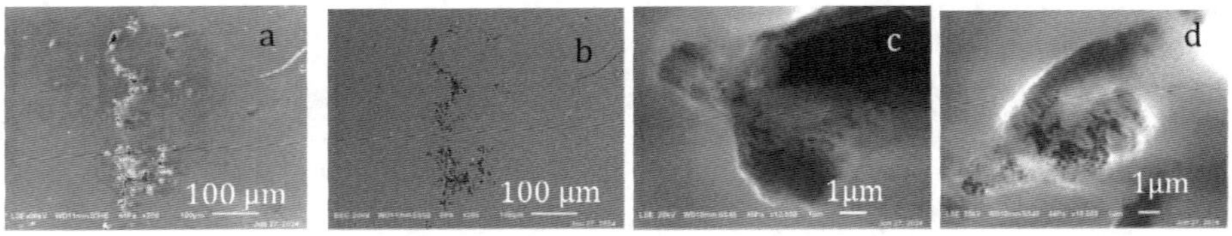

Figure 4. Scanning electron microscopy analysis of sample RAM-CK Al (N-doped DLC/Kapton/Al back). Credits: own work.

a) Enlarged secondary electron SEM image showing some erosion features on the surface.
b) Backscattered electron SEM image showing the same region of the surface as in a).
c) and d) Further enlarged secondary electron SEM images showing the erosion features.

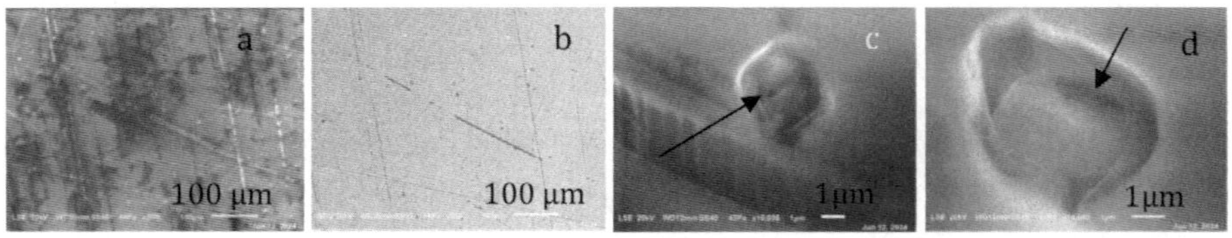

Figure 5. Scanning electron microscopy analysis of sample RAM-CK (Si-doped DLC/Kapton). Credits: own work.

a) Enlarged secondary electron SEM image showing some erosion features on the surface.
b) Backscattered electron SEM image showing the same region of the surface as in a).
c) and d) Further enlarged secondary electron SEM images showing the erosion features. Most of the particles (such as marked with arrows) were identified as CaP particles.

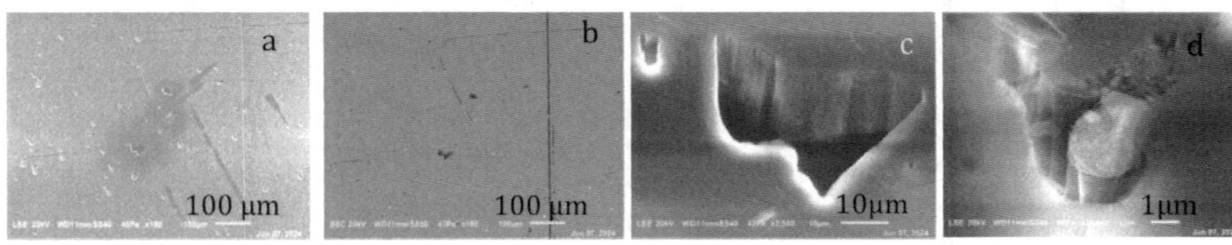

Figure 6. Scanning electron microscopy analysis of sample RAM- WFM C K ((Si+O) -doped DLC/Kapton HN). Credits: own work.

a) Enlarged secondary electron SEM image showing some erosion features on the surface.
b) Backscattered electron SEM image showing the same region of the surface as in a).
c) and d) Further enlarged secondary electron SEM images showing the erosion features.

All sample positioned in the WAKE direction exhibited negligible ~ 0.1% weight loss, with the surfaces showing practically no erosion features. The EDS analysis indicated that some contamination with Na, K, Ca and Si happened (Figure 7 through Figure 9).

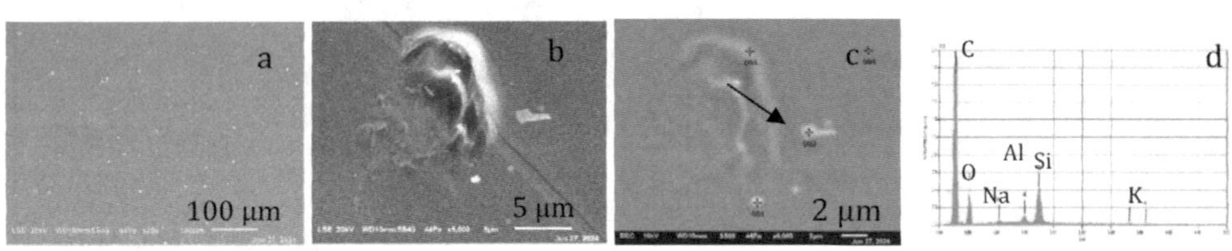

Figure 7. Scanning electron microscopy analysis of sample WAKE-C K Al (N-doped DLC/Kapton/Al back).

a) Enlarged secondary electron SEM image showing the surface morphology. Credits: own work.
b) Further enlarged secondary electron SEM image of a different area showing the surface. No erosion was noticed on the sample.
c) Backscattered electrons compositional SEM image of a surface region with particles.
d) EDS Spectrum collected from the white-contrast particle marked with the arrow.

IOP Conf. Series: Materials Science and Engineering **1328** (2025) 012017 doi:10.1088/1757-899X/1328/1/012017

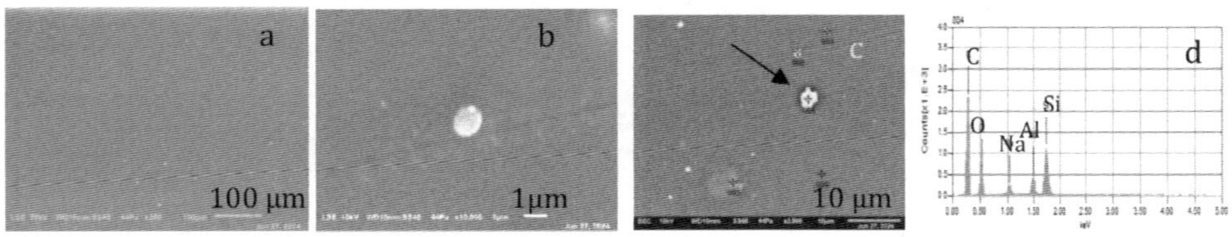

Figure 8. Scanning electron microscopy analysis of sample with a work function matching coating WAKE-WFM C K ((Si+O)-DLC/Kapton HN). Credits: own work.

a) Enlarged secondary electron SEM image showing the surface morphology.
b) Further enlarged secondary electron SEM image of a different area showing the surface. No erosion was noticed on the sample.
c) Backscattered electrons compositional SEM image of a surface region with particles.
d) EDS Spectrum collected from the large white-contrast particle marked with the arrow.

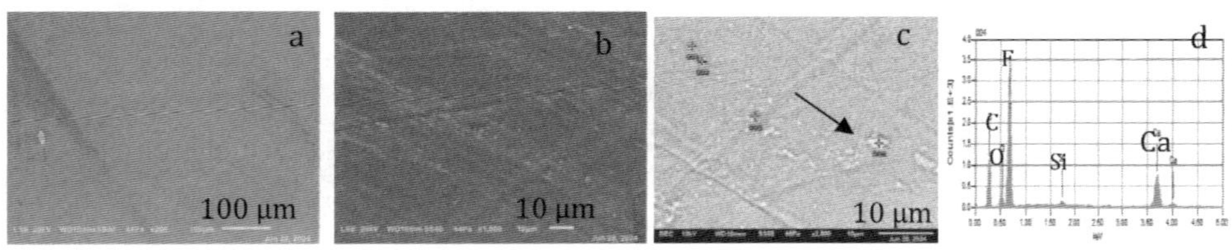

Figure 9. Scanning electron microscopy analysis of sample WAKE-CT Ag (N-doped DLC on Teflon FEP/Al/Inconel). Credits: own work.

a) Enlarged secondary electron SEM image showing the surface morphology.
b) Further enlarged secondary electron SEM image of a different area showing the surface. No erosion was noticed on the sample.
c) Backscattered electrons compositional SEM image of a surface region with particles.
d) EDS Spectrum collected from the large white-contrast particle marked with the arrow.

The thermal optical properties of all flown samples that are important for thermal control materials, were evaluated before and after the flight. Table 4 summarizes all measurements. The solar absorptance, in most samples, increased slightly, with only exception being samples CK (Si-doped DLC on Kapton) and WFM CK in RAM direction ((Si+O) -doped DLC/Kapton). The thermal emittance of all samples in RAM and WAKE positions decreased slightly. The α/ ε ratio that is a very important parameter in controlling the temperature of an object for all materials [17], increased somewhat, with only exception being the WFM CK sample in RAM direction.

Table 4. Summary of data on thermal optical properties' changes of the MISSE-17 ITL flight samples

Sample ID	Position on the ISS	Thermal Optical Before Flight			Thermal Optical After Flight			Comments
		α	ε	α/ε	α	ε	α/ε	
C T Ag	WAKE	0.233	0.811	0.29	0.263	0.808	0.33	N-DLC Teflon FEP/Al/Inconel
C K Al	WAKE	0.443	0.654	0.68	0.465	0.637	0.73	N-DLC/Kapton/Al
WFM C K	WAKE	0.373	0.647	0.58	0.419	0.632	0.66	(Si+O)-DLC/KaptonHN
WFM C K	RAM	0.370	0.649	0.65	0.367	0.637	0.56	(Si+O) -doped DLC/Kapton HN
C K Al	RAM	0.350	0.646	0.57	0.355	0.633	0.58	Si-DLC/Kapton/Al
C K	RAM	0.423	0.654	0.54	0.384	0.639	0.60	Si-DLC/Kapton

XPS analysis was conducted on the samples before the flight and after they were returned. As it is well known, the XPS is a surface analysis technique that samples the outmost 3-6 nm of the surface. Table 5 and

Table 6 show the XPS survey data collected from the samples before and after the flight, respectively. The data on pristine Kapton and Teflon FEP is also included in Table 5 for comparison. The DLC layers on the three samples flown in RAM direction were all, either Si or Si+O doped. The Si concentration in the upper layers, in "as prepared samples" varied from ~9% to 27% (see Table 5). Analyzing the XPS data for the samples flown in RAM direction after their return (Table 6), it can be suggested that an SiO_x layer formed on all three samples with x close to 2. Such a layer enhances the protection of the underlying Kapton from the atomic oxygen in LEO orbit. Future work on these samples will include evaluation of the formed SiO_x layer thickness and a more detailed investigation of its surface properties.

Table 5. Summary of sample surface composition from XPS survey scans before flight

Peaks	Si2p	C1s	N1s	O1s	F1s	
Position	Rel.At.%	Rel.At.%	Rel.At.%	Rel.At.%	Rel.At.%	Comments
Kapton 500HN pristine	0	77	6	17	0	Ref [18]
Teflon PTFE pristine	0	33-35	0	0-1	65-67	Ref [19]
C T Ag -WAKE	0.5	77.5	7.2	10.3	4.5	N-DLC Teflon FEP/Al/Inconel
C K Al -WAKE	0.3	85.7	6.0	7.7	0.3	N-DLC/Kapton/Al
WFM C K -WAKE	11.8	62.3	2.1	23.6	0.2	(Si+O)-DLC/Kapton HN
WFM C K -RAM	13.43	53.28	2.97	30.01	0.21	(Si+O) -doped DLC/Kapton HN
C K Al -RAM	27.0	42.9	1.7	28.4	0.1	Si-DLC/Kapton/Al

CK -RAM	8.4	75.7	1.8	14.0	0.0	Si-DLC/Kapton

Table 6. Summary of sample surface composition from XPS survey scans after flight

Peaks	Si2p	C1s	N1s	O1s	*F1s*	
Position	Rel.At.%	Rel.At.%	Rel.At.%	Rel.At.%	Rel.At.%	*Comments*
C T Ag -WAKE	15.8	28.8	1.5	36.4	17.4	N-DLC Teflon FEP/Al/Inconel
C K Al -WAKE	9.3	51.8	2.0	33.2	3.7	N-DLC/Kapton/Al
WFM C K -WAKE	9.1	49.9	2.1	34.2	4.8	(Si+O)-DLC/Kapton HN
WFM C K -RAM	29.1	11.1	0.5	59.2	0.0	(Si+O) -doped DLC/Kapton HN
C K Al -RAM	29.7	9.7	0.6	60.0	0	Si-DLC/Kapton/Al
CK -RAM	29.6	8.92	0.42	61.1	0.0	Si-DLC/Kapton

The initial analysis of the XPS survey data of samples mounted in the WAKE direction (Table 6) suggested that the sample surfaces were contaminated with Si and F. Thus, samples CT Ag and CK Al that originally did not contain Si, had ~9 -15% silicon in the surface layers after the flight (Table 6). Also, the CK Al and the WFM CK samples that did not contain any F, had shown ~4%-5% of F after the flight (compare the data in Table 5 and Table 6). The sources of these contaminants are not clear presently. A more in-depth analysis will be conducted on the wake-flown samples. The depth distribution of the elements detected in the surface layers will be evaluated using high resolution XPS to understand better the nature of the formed structures.

4. Conclusions

In summary, six samples representing different functional solutions developed at ITL for use in various space and planetary environments like VLEO, LEO, GEO, Lunar were flown on MISSE-17 mission for a period of 6 months and returned to ITL. The initial analysis of the flown samples indicated that they retained all basic functional properties, with very small changes in surface morphology, weight and thermal optical properties. Formation of a SiOx layer on the surfaces of the samples flown in RAM directions was suggested, based on the XPS analysis. Cross-contamination with Si and F of samples flown in the WAKE direction was observed. Due the small pool of samples flown and incomplete analysis of the data, because of the short time available for the analysis, it is difficult to make definite conclusions at this point on the nature of the observed contamination patterns. Future work is planned to continue the analysis of the samples using high-resolution XPS to explain better the obtained results.

Acknowledgments

We acknowledge the financial support of Canadian Space Agency (CSA) [20STDP007]. The help from Aegis Corporation in providing the information on the MISSE-17 flight details is acknowledged with thanks. We also acknowledge the help in XPS analysis from Dr. Peter Brodersen from the Ontario Centre for the Characterization of Advanced Materials (OCCAM).

References

1. Iskanderova Z, Tagawa M, Kleiman J, Nishioka S and Horimoto S 2023, Space polymers erosion in simulated LEO and VLEO environments, 2023 *IOP Conf. Ser.: Mater. Sci.*, *Eng.* 1287 012039, **DOI** 10.1088/1757-899X/1287/1/012039, IOP Conference Series: Materials Science and Engineering, Volume 1287, A joint conference on Materials in the Space Environment (ISMSE 15 & ICPMSE 13) 18/09/2022 - 23/09/2022 Leiden, Netherlands.
2. Tagawa M, Yokota K, Minton T et al 2013, Survivability of Silicon-doped diamond-like carbon films in energetic atomic/molecular oxygen beam environment, Aerospace and Space Science Proceedings 32, 547.
3. Iskanderova Z, Kleiman J, Gudimenko Y, Cool G, Tennyson R 1997, Surface modification of polymers and carbon-based materials by ion implantation and oxidative conversion, US Patent number: 5683757, November 4, 1997.
4. Kleiman J I 2010, Surface Modification Technologies for Durable Space Polymers, MRS Bulletin, Volume 35, Issue 1: Materials Issues in the Space Environment , 55 – 65, DOI: https://doi.org/10.1557/mrs2010.617.
5. Schäff S and Jürgens M 2016, Advanced Electric Orbit-Raising Optimization and Analysis with LOTOS 2, Astos Solutions GmbH, Stuttgart, Germany, 6th International Conference on Astrodynamics Tools and Techniques, 17 March 2016, Darmstadt, Germany.
6. Iskanderova, Z, Kleiman, J., Bussieres, F., Tagawa, M., Ng, R., Sodhi, R. 2018, Enhancement of atomic oxygen resistance of charge dissipative ion-beam treated surfaces of space polymers. Presented at Joint 14th ISMSE/12th ICPMSE Conference, Biarritz, France, 1–5, Oct 2018, http://www.ismse2018.com/wp-content/uploads/80.pdf .
7. Iskanderova, Z., Kleiman, J., Issoupov, V. and Bussieres, F. 2009, Surface Treatment Technology for Changing, Turning, and Adjusting Surface Resistivity of Space Polymers , Proceedings of ICPMSE-8, Ed. J. Kleiman, Toronto, Canada, 317-326.
8. Kleiman, J., Iskanderova, Z., Bussieres, F., Grigorevskiy, A. and Sodhi, R. 2011, Ion Beam Treatments for Enhancement of Surface Conductivity and Durability of Space Polymers: Results, Analysis, Mechanisms, Aerospace and Space Science Proceedings 32, 547, Eds. J. Kleiman, M. Tagawa, Y. Kimoto, 317-326.
9. Iskanderova Z, Kleiman J, Bussieres F 2013, Method of Making Charge Dissipative Surfaces of Polymeric Materials with Low Temperature Dependence of Surface Resistivity and Low RF Loss, US Patent 9174396B2, 2013-07-03.
10. Kleiman J, et al 2022, Regolith adherence characterization (RAC) experiment on the Moon and it's ground-based simulation: Materials Issues, Joint International Conference on Protection of Materials from Space Environment, ISMSE15/ICPMSE13 , 18 -23 September, 2022 Leiden, The Netherlands, https://iopscience.iop.org/issue/1757-899X/1287/1 .
11. https://news.satnews.com/2023/04/11/aegis-aerospace-celebrates-commercial-dod-payloads-in-space/
12. https://www.linkedin.com/posts/jeff-buell-724029aa_another-misse-mission-completed-activity-7111508332183486464-i_a-/?trk=public_profile_like_view .
13. Personal communication, Jeff Buell (AEGIS) <jeffrey.buell@aegisaero.com>

IOP Conf. Series: Materials Science and Engineering **1328** (2025) 012017 doi:10.1088/1757-899X/1328/1/012017

14. Annual Book of ASTM Standards, "Standard Test Method for Solar Absorptance, Reflectance, and Transmittance of Materials Using Integrating Spheres", ASTM E903-20
15. Annual Book of ASTM Standards, "Standard Solar Constant and Zero Air Mass Solar Spectral Irradiance Tables", ASTM E490-22.
16. Annual Book of ASTM Standards, "Standard Test Methods for Total Normal Emittance of Surfaces Using Inspection-Meter Techniques", ASTM E408-13(2019).
17. Henninger J H 1984, Solar Absorptance and Thermal Emittance of Some Common Spacecraft Thermal Control Coating, NASA Reference Publication 1121, 1, April 1984
18. Kleiman J I 2004, Surface Modification of Polymers, Paints and Composite Materials Used in the Low Earth Orbit Space Environment. MRS Online Proceedings Library 851, 462–474. https://doi.org/10.1557/PROC-851-NN8.6
19. https://chatgpt.com/c/68141e1f-c818-800b-9603-d2a5a9a3b8d9

IOP Conf. Series: Materials Science and Engineering **1328** (2025) 012018 doi:10.1088/1757-899X/1328/1/012018

Material recycling and hardware re-use for Moon and Martian settlement

Francesco Caltavituro[1*], Dr. Marco Berg[1], Alexandra Hornicar[1], Dr. Manuel Ortega[2], Matvei Andreev[2], Frank Koch[3], Dr. Barbara Imhof[4], Chris Gilbert[4], René Waclavicek[4], Dr. Daniel Schubert[4], Dr. Advenit Makaya[5]

[1] Human Spaceflight Department, OHB System AG, Bremen, Germany
[2] Azimut Space GmbH, Berlin, Germany
[3] Orbit Recycling, Berlin, Germany
[4] LIQUIFER Systems Group GmbH, Vienna, Austria
[5] ESTEC, European Space Agency, The Netherlands

*E-mail: francesco.caltavituro@ohb.de

Abstract. The next big leap in institutional and commercial space exploration is inextricably linked to the human desire to reach the Moon and Mars and to establish a permanently off-Earth inhabited colony [1], [2]. In the last decades, this has been one of the major, more demanding challenges for industry and agencies, since this imposes a creative re-thinking of previous mission approaches. Earth dependency has always represented the bottleneck for freely conceiving a human outpost on the Moon and Mars. It is widely recognized that a key enabler to any sustainable presence in space is the ability to manufacture necessary structures and spare parts in-situ and on-demand by recycling and re-using the available resources [3]. This will reduce cost, volume, and up-mass constraints, being also in line with the ESA space debris mitigation policy towards environmentally sustainable space activities. In this frame, the scope of this paper is to present the concepts investigated during ESA's (European Space Agency) HARMONISE activity concerning both in-situ materials recycling and partial or complete re-use of end-of-life hardware to serve different purposes during Moon/Mars exploration missions. The investigation approach was threefold: recycling of polyethylene Ziplock® bags into 3D-printable filament; melting and casting of scrap aluminium for tools fabrication, and partial re-utilisation of rack blind panels for habitat furniture design. For each of these strategies, a dedicated demonstrator has been designed, manufactured, and tested with respect to pre-defined success criteria, to fulfil functional requirements for both Earth and lunar/Martian scenarios. The HARMONISE (Hardware Recycling for Moon and Martian Settlement) activity will contribute to ushering a new, more sustainable space exploration era, supporting an emerging circular economy through in-orbit servicing and off-Earth manufacturing by 2050.

1. Introduction

1.1 Up-front analysis

To establish a baseline for the analysis carried out on recycling technologies and the potential benefits, the activity began with a survey of past, present and future missions to the Moon and to Mars. The survey provided an overview of all known landing missions to the Moon and to Mars. These missions have, to date, delivered 188.6 T of material to the Moon and 10.3 T to Mars. From this starting point, macro estimates were made of the nature of the material which could in principle be available for processing.

Supplementary, less relevant details such as mission cost, scientific goals and payloads were included where they were easily available. Factors influencing missions expected to take place in the foreseeable future were discussed, and an overview of future missions was presented. Human lunar landing missions were included where planning dates have been announced, but the current published schedule is optimistic. In compliance with the guidance provided by ESA, this survey was based primarily on information publicly available.

1.2 Project scope

The scope of this paper is to present the concepts investigated during the ESA-funded HARMONISE activity concerning both in-situ materials recycling and partial or complete re-use of end-of-life hardware to serve different purposes during Moon/Mars exploration missions.

Three demonstrators were realized addressing material recycling and partial or complete reuse of no longer useful hardware: recycling of polyethylene Ziplock® bags into 3D-printable filament; melting and casting of scrap aluminium for tools fabrication, and partial re-utilization of rack blind panels for habitat furniture design.

Each demonstrator has been designed, developed, manufactured and tested against prescribed functional requirements and predefined success criteria considering both Earth and lunar/Martian scenarios.

At the end of the activity, a comprehensive analysis has also been performed to optimize material usage efficiency and hardware reusability, along with a critical evaluation of the manufactured demonstrators. This also allowed for relevant improvements of them based on observations and lessons learned.

Finally, the HARMONISE team was also able to define innovative guidelines and recommendations for the equipment brought for exploration missions, which pointed out how crucial it is that the design for recycling is already considered from the early definition phase of a mission.

2. Payload rack blind panels reused as furniture

2.1 Initial idea and concept trade-offs

Within the first demonstration project, an overall system analysis and open concept brainstorming was conducted. The main paradigm for this initial task was to elaborate which kind of furniture elements can be built out of payload rack blind panels to fulfil basic living needs of astronauts inside a future habitat. A starting point for the initial investigation were examples for achievable geometric forms, as can be seen in Figure 1. Baseline for the panel connection were dedicated 3D- printed connectors that connect either two or three panels with each other.

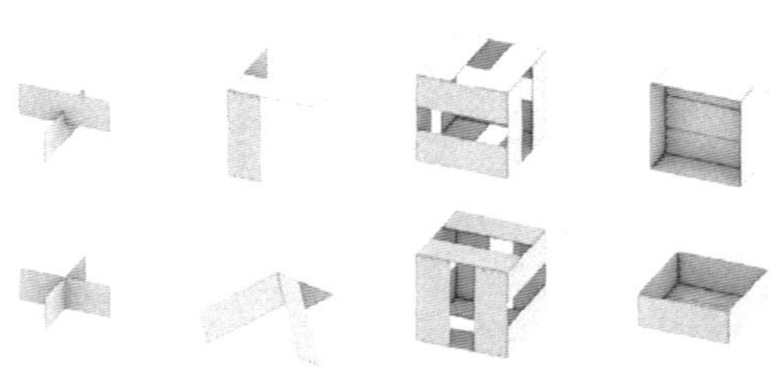

Figure 1. Examples for some achievable geometric forms, created with close-out panels [Source: LSG].

Starting from here, different activities and functional areas within a future habitat were analysed, and preliminary furniture concepts were created, such as tables, chairs, bunk beds, divider walls, blinds, adjustable ladders, and various shelving systems. Figures 2-4 present examples of this initial concept analysis. The LIQUIFER team decided to further investigate the design of an archetypical chair as an example demonstrator.

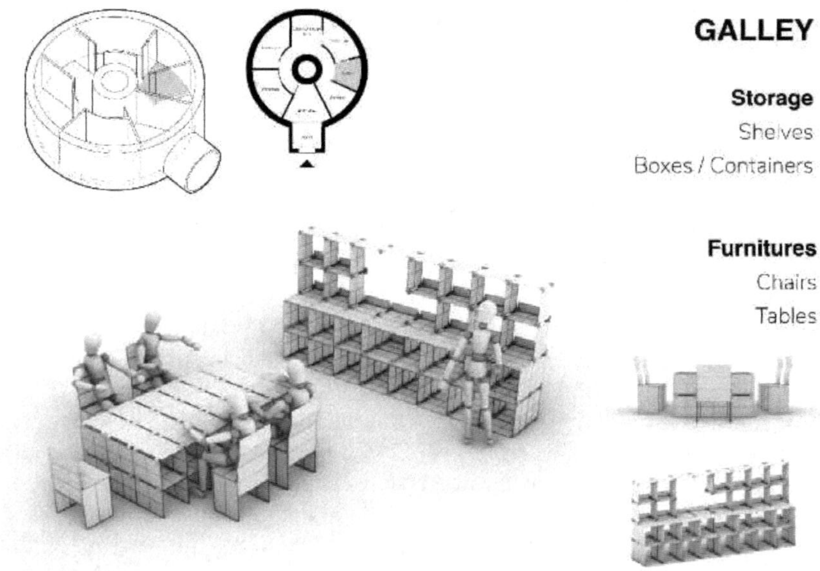

Figure 2. Concept for the galley/social area incl. chair/ table combination and storage capability [Source: LSG].

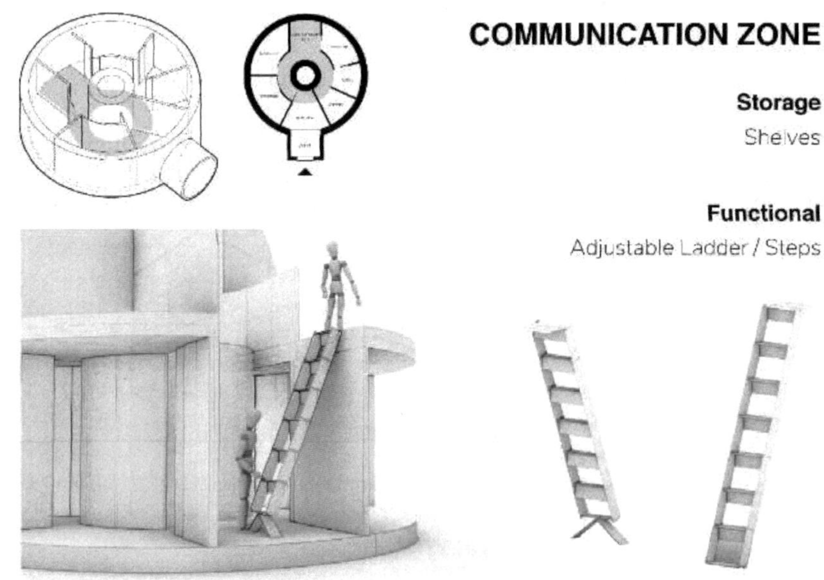

Figure 3. Concept for a ladder system [Source: LSG].

IOP Conf. Series: Materials Science and Engineering **1328** (2025) 012018 doi:10.1088/1757-899X/1328/1/012018

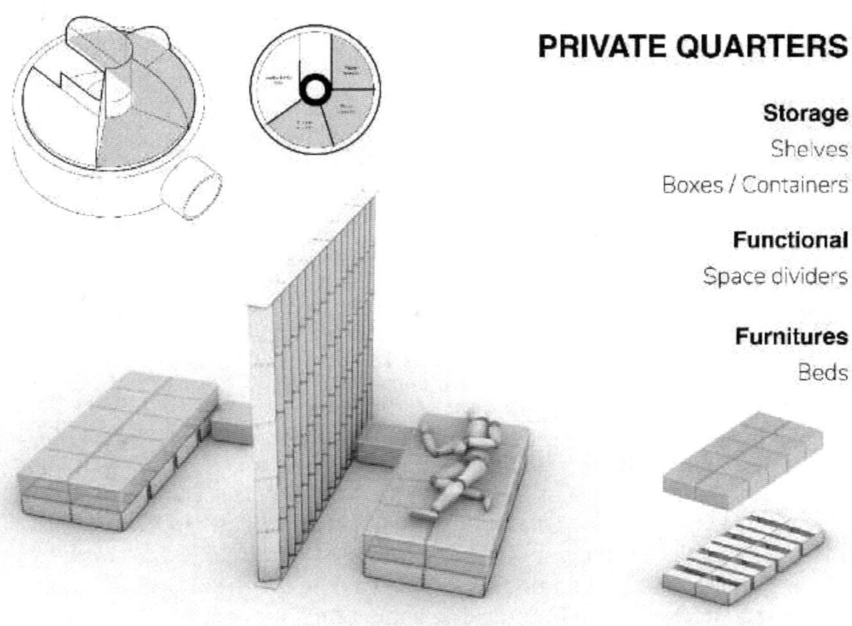

Figure 4. Concept for a space divider in combination of a bed system [Source: LSG].

Generally, a chair represents a common and often used furniture element. Further, the chair design enables the demonstration of multiple aspects of the envisioned recycling approach (connector design, achievable stiffness, general utility), but also allows the easy performance of a solid testing and validation phase. Further, the chair appears to be the most generic object of application that could possibly be needed in each of the notional functional areas of a future planetary human outpost. It is a simple, but challenging example for an improvised structure under both static and dynamic stresses.

2.2 Used material and recycling concept

For the creation of the demonstrator (here: chair design no. 1) three main components were selected:

- Different dimensional variants of close-out panels;
- cargo transfer bag dividers;
- 3D-printed connectors.

The different close-out panels (see Figure 5) used for the demonstrator are made of aluminium and include a set of pre-drilled holes, originally used to install the panel in the experiment rack, in order to structurally secure the experiment facility during the launch phase.

The close-out panels represent the main component element for the structural integrity of the chair. The main paradigm here was to limit the adjustment processing to the panels (e.g., drilling of additional holes, or other fixation adjustments such as nuts), since these are not easily doable on an extra-terrestrial environment especially in the early settlement phase.

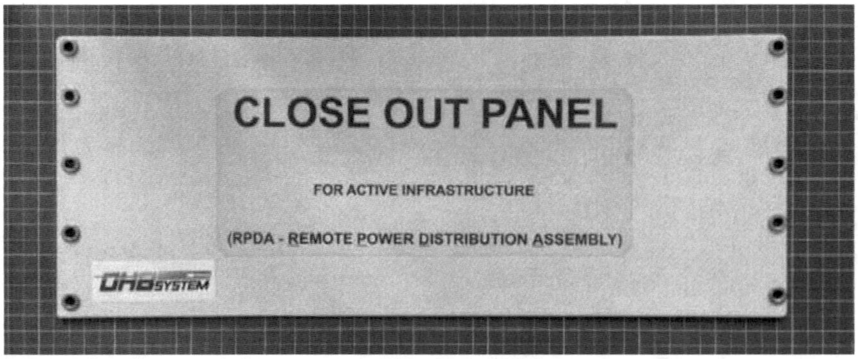

Figure 5. Example of a typical close-out panel [Source: OHB].

In addition to the close-out panels, a combination of several different types of cargo transfer bag dividers (see Figure 6) were used for the upholstery of the demonstrator seating area, and the backrest area. The attachment of the dividers to the close-out panels was performed by attaching individual self-adhesive hook and loop strips, which will later interlock with the already existing Velcro® strips of the cargo transfer bags dividers. Optionally, a second layer of cargo transfer bags dividers was considered initially, but postponed for the actual testing phase.

For connecting the close-out panels dedicated connectors are needed. These connectors shall make use of the already existing drill holes of the close-out panels.

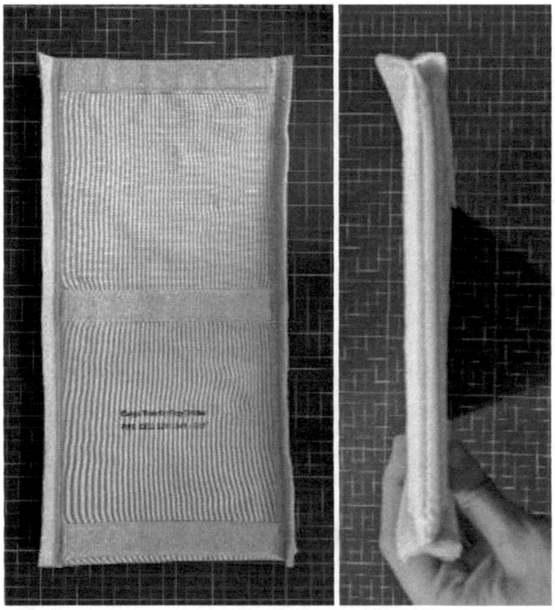

Figure 6. Example of cargo transfer bags dividers [Source: OHB].

Initial material assessment concluded that the use of a dedicated aluminium-based casting process would be too complex and would not provide sufficient accuracy for the subsequent assembly of the geometric demonstrator form. Therefore, the use of 3D-printed plastic connectors was used as first choice of manufacturing (see Figure 7). Additionally, the main design logic followed the approach to limit the necessary material consumption and at the same time to provide enough structural integrity and stiffness to the chair demonstrator.

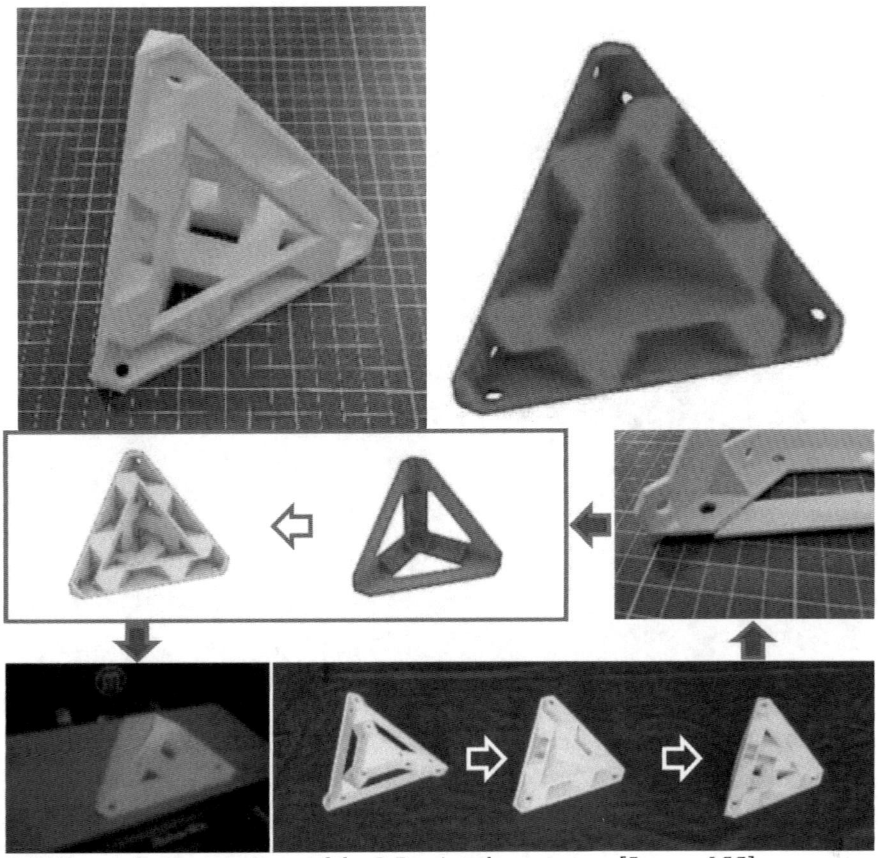
Figure 7. Different design iterations of the 3-D printed connectors [Source: LSG].

While the material for the later system shall be a result of internal in-situ recycled plastic, the actual foreseen demonstrator of this project will rely on 3D-printed PLA (Polylactic Acid) components. Various design iterations were performed in order to investigate and find out the best suitable layout for connecting the close-out panels.

2.3 Demonstrator design and recycling process

From the different demonstrator candidates, a chair was initially chosen. Figure 8 shows the final test design of the demonstrator. The manufacturing process of the chair demonstrator was conducted by printing the connectors and assembly of the different components.

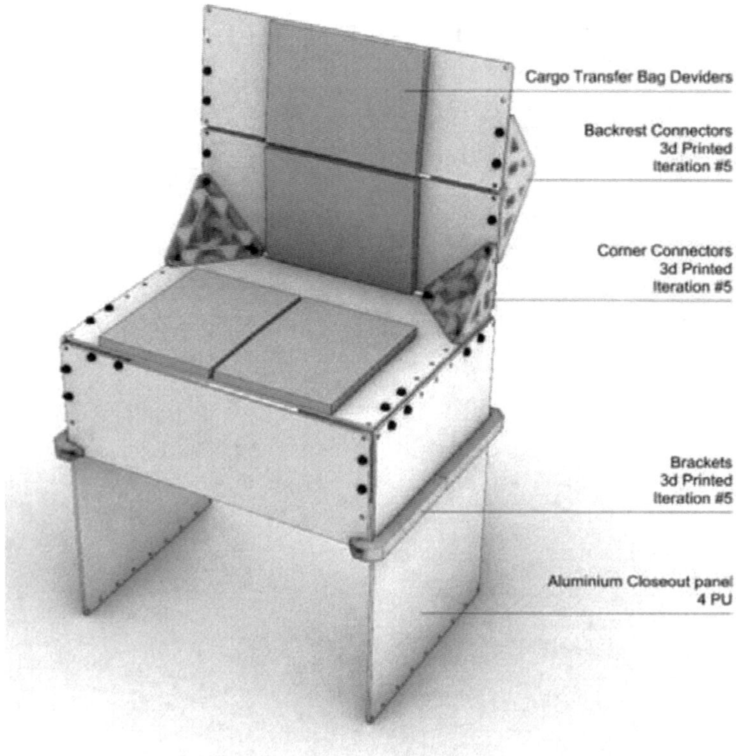

Figure 8. CAD drawing of developed chair (type: one) [Source: LSG].

The following parameters were collected for the production of the demonstrator:

- Printing Time: Total: 94 hours;
- Filament material: PLA (needed mass: 1,25 kg);
- Tool: Creality Ender-3 v2 (Total peak power: 350 W);
- Screws (type: M6/20mm Stainless Steel Hex Socket screw, number: 48);
- Nuts (type: M6 Stainless Steel Hex nut, number: 48);
- Velcro® band (Type: 20 mm black hook end loop; total needed lengths: 120 cm);
- CTB Divers: (Type: Half Size, needed number: 2; Type: Full Size, needed number: 1);
- Close-Out Panel 1: (Type: 4U, number: 4, single mass: 1,1 kg);
- Close-Out Panel 2: (Type: 8U, number: 3, single mass: 2,2 kg).

The final demonstrator (see Figure 9) resulted in a total mass of 13,3 kg, including all 3D-printed connectors, screws & nuts, and upholstery elements. Sitting height is 50,3 cm, the back rest assembly adds an additional 34,9 cm so that the total height of the chair is 85,2 cm. The width of the chair is 52,9 cm. The chair provides a seat area of 1828 cm^2.

2.4 Demonstrator testing phase

The full test campaign consisted of several phases which tested various aspects of the demonstrator. Investigation factors such as external environmental factors (e.g., radiation, vacuum, micro debris/ meteoroids, etc...) were considered negligible, as the architectural elements and furniture will only be used inside the habitat infrastructure.

The following requirements were addressed during the test phase of the demonstrator #1.

Table 1: Demonstrator #1 requirements.

Requirement ID	Functional requirement	Test Procedure
HAR-LSG-REQ-FUNC-0004	The demonstrator (here: furniture, such as chair, bed, bench, or ladders), when assembled, shall withstand its own weight and further the weight of a 98.5 kg person (95-percentile of the whole-body mass; Male; American; according to NASA-STD-3000, Man-System Integration Standards [RD 01]).	Test procedure as follow: • The prototype will be tested with certified weights. • The procedure will foresee to load the prototype with weights (15 kg steps) until ~ 100 kg will be reached.
HAR-LSG-REQ-FUNC-0005	The demonstrator (here: furniture such as chair, bed, bench) shall allow comfortable usage in the range of human ergonomics according to NASA-STD-3000, Man-System Integration Standards [RD 01].	Test procedure as follow: • The prototype design will be reviewed according to the human factors for adequate usage of a chair. • NASA-STD-3000, Man-System Integration Standards will be considered for evaluating the proper usage. Further testing w.r.t. ergonomics: • Three different persons (main human distinguishing element: height and weight) will sit on the prototype, performing nominal work procedures (e.g. writing/ office work, performing light work) • Each test will be last for 60 min. • Person's details to be reported anonymously • Concluding ergonomic parameters will be evaluated and noted. Q1: General usability is within acceptable range? (A: Low, medium, high) Q2: Prototype's stiffness is within acceptable range? (A: low, medium, high) Q3: Movability of the prototype is within acceptable range? (A: low, medium, high) Q 4: Sitting comfort level (soft padding) is within acceptable range? (A: low, medium, high)

HAR-LSG-REQ-FUNC-0010	The geometric structure of the demonstrator shall allow easy transport though the interjoining module hatches and internal doors with the dimensions 800 mm x 2000 mm. If this requirement cannot be met, the demonstrator shall be able to be partly disassembled (and later been reassembled) in order to meet this overall requirement.	Test procedure as follow: • The prototype will be handled by a human in order to perform several handling tests such as: Task 1: Lifting prototype (to a height of 1.2 m above ground), Task 2: Moving prototype from location 1 to location 2, Task 3: Manoeuvring prototype through a door (hatch), and Task 4: Carrying prototype upstairs (~25 steps). • The test procedure will be video recorded in order to demonstrate the different handling procedures.
HAR-LSG-REQ-FUNC-0018	The connectors of the demonstrator shall not be permanent and shall allow easy dissemble or reconfiguration procedures.	Test procedure as follow: • The prototype will be assembled. The time for the construction will be recorded. • The prototype will be disassembled. The time for the dissemble will be recorded as well • In case of additional design options, the time to reconfigure the prototype from design option one (here: chair) into a different design option (if applicable) will be recorded. • The assembly, disassembly, and reconfiguration tests will be video recorded.
HAR-LSG-REQ-FUNC-0022	An adequate attachment mechanism shall prevent slipping or detaching of the soft padding (cargo transfer bag dividers) from the close-out panels, which occurs by normal usage of the demonstrator (e.g. frequently seating/ lying on its surface, incorporating normal body movements).	Test procedure as follow: • The prototype will be outfitted with soft paddings for upholstery purposes. • The prototype will be used within nominal usage regulations. Thereby movements will be conducted upon the soft padding, putting different vertical and side forces upon the soft padding. • The test movements/ forces will be conducted for 60 min continuously by three individual persons (according to HAR-LSG-REQ-FUNC-0005). • After accomplishment of test procedure, the grade of detachment of the soft padding will be qualitatively evaluated and measured accordingly. • Qualitative evaluation of sitting comfort related to the soft padding (A: low, medium, high) • The usability test will be video recorded.

Figure 9. Intermediary manufactured and assembled chair (type one), including addition stabilization struts on the side of the chair [Source: LSG].

2.4.1 Static strength testing

The prototype demonstrator was tested by loading the chair with weights (10 kg steps) until a total weight of 100 kg was reached (see Figure 10).

The static strength testing showed that the chair could easily withstand the load of the weights applied, and therefore could support 100 kg.

Figure 10. Static load test of the demonstrator, where the chair held 100 kg [Source: LSG].

2.4.2 Usability, ergonomics, and handling testing

During the usability, ergonomics and handling test, the personnel had to use the chair for 60 min during normal office work. After the testing phase, a dedicated questionnaire was completed.

Three different people with different heights and weights sat on the prototype, performing nominal work procedures (e.g. writing/ office work, performing light work) (see Figure 11).

General usability, stiffness, and comfort of sitting received high and medium ratings, resulting into an overall positive evaluation within these categories. However, the criterium "movability" received low ratings (with one medium rating), resulting into a less positive overall evaluation.

For this test, a total of four tasks had to be performed in order to test the general handling capability of the demonstrator such as lifting-up, moving it from location 1 to location 2, manoeuvring the prototype through a door (hatch), and carrying it upstairs (about 25 steps). All four tasks were performed without major difficulties. Lastly, the soft padding, subsequently added to the chair, proved to be crucial in masking the coldness of the metal surface and providing warmth, while the Velcro® fitting was strong enough to prevent the padding from detaching when the person stood up.

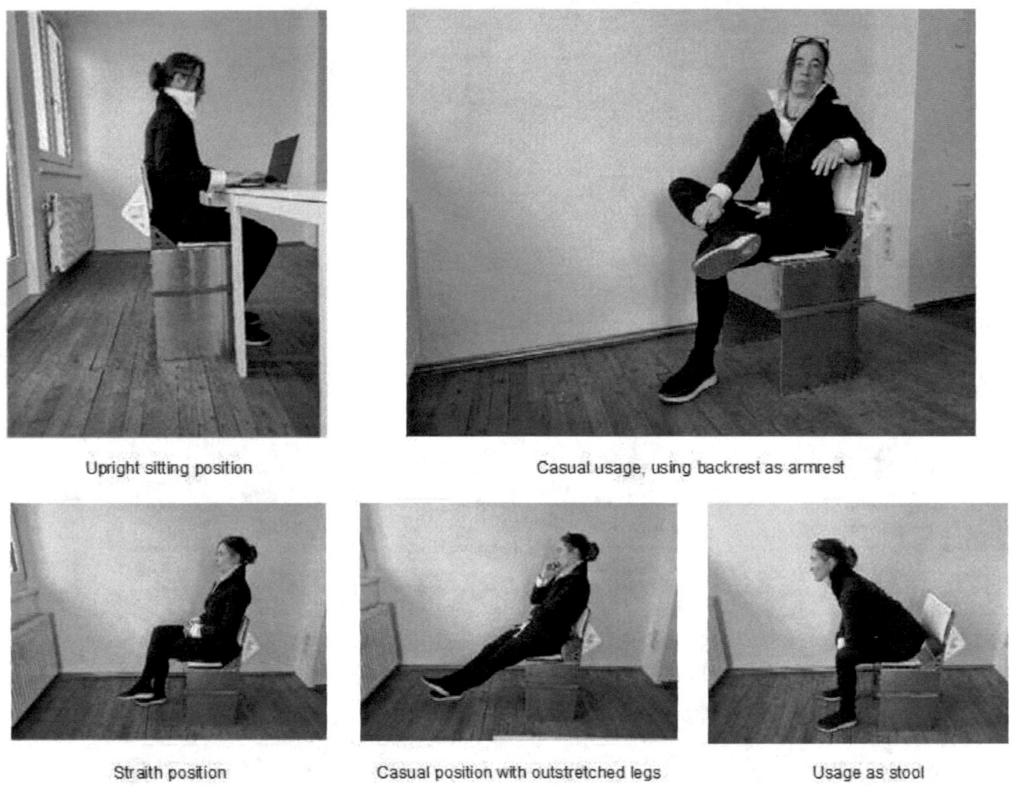

Figure 11. Different positions for the usage of the demonstrator; Holding procedure of the demonstrator (Transportability & Handling) [Source: LSG].

2.4.3 Assembly and disassembly testing

The demonstrator was assembled without major challenges. Overall assembly time for the demonstrator took approximately 1.5 h. This assembly time excluded production of the needed 3D-printed connectors. The disassembly time was much less and required only 30 min for the demonstrator. Both times include the time required for tooling and preparation (see Figure 12).

Figure 12. Different pictures of the assembly procedure. Overall assembly time was 1.5h; disassembly time was 0.5h [Source: LSG].

2.4.4 Overall test evaluation

In conclusion, the campaign has shown general positive overall results of the demonstrator capabilities, and the test campaign was performed without major challenges. The overall test results verified most of the before stated system requirements. However, after several hours of intensive use of the chair, signs of material fatigue within the 3D-printed connectors were visible in some places. These cracks did not lead to immediate failure of the overall construction, but allowed continued use.

The default structure of an FLD printed connector is a composition of a form defining shell of several layers filled with a volumetric geometry of adjustable density. An iterative process has shown that type and strength of stresses within one part can vary dramatically. This can be countered by varying shell thickness and infill density, in order to reach solid areas where needed.

In addition, the present demonstrator deliberately refrained from drilling additional screw holes to optimize the connections between the elements, which led to some signs of material fatigue during the several hours of testing. However, it can be assumed that further improvements to the connecting elements and optimization of the screw hole positions can achieve greater and lasting stability.

Also, it should be noted that all tests were performed under a 1G-environment, which adds to the application loads of the chair. A moon gravitational environment will further ease the handling condition, and the structural stress of the overall chair assembly will be significantly reduced.

Summarizing, the test campaign proved the chair's high degree of utility and high potential for recycled furniture. The referenced demonstrator is, thus, a good starting point for creating architectural elements for a future lunar or Martian base.

2.5 Implication for space scenario

The demonstrator has shown that the reuse of aluminium parts (close-out panels) for assembly using 3D-printed connector elements can be used to produce statically stressed components (the chosen example demonstrator was the "Chair type one"). The choice of the chair as a particularly demanding demonstration object has also shown that this construction method can even withstand dynamic loads to a certain degree.

In a reduced gravitational environment, the structural stability requirements for the connectors, and the overall geometric system, will be less severe. At 0.166g (Moon) and 0.38g (Mars) the 3D-printed connectors can potentially be built with less material, and still produce sufficient structural integrity and stiffness for an adequate usage inside the habitat.

Optimal connector design: dedicated 3D-printed connectors will play a major role for combining different elements together. These connectors can not only lint blind panels to create furniture items or secondary structural elements (e.g. divider walls), but also couple and combine components and structural parts in order to create tools and consumable items. An optimal connector design is essential for a successful creation of an up-cycled new item. Therefore, dedicated individual design layouts need to be predeveloped upfront and later serve as a blueprint for the onsite 3D-printing process. Important here is that the necessary support structure is designed in a way to avoid unnecessary printing waste.

By selecting and arranging suitable panel sizes, the ergonomic requirements of the user can be adequately met and further improved by combining them with other elements such as CTB (Cargo Transfer Bags) dividers, so that long-term use is feasible.

As the chair design only functioned as a demonstrator for recycling blind panels in future habitats on the Moon and Mars, the possibility of creating other structural elements has immense potential. Additional items (see Figure 13) like honeycomb panels, broken fans, carbon fiber reinforced polymers (CFRP), glass-fiber reinforced plastics (GFRP), housing cases, aluminium rods, different textiles, and cables/wires could be used to create further architectural elements such as beds, room dividers, ladders, cupboards, and tables. Furthermore, tools such as buckets, EVA (Extravehicular Activity) tools or intravehicular consumable items like plates, knifes or cups could be made partially from hand tool modified reused parts in combination with cast or 3D-printed elements produced on demand.

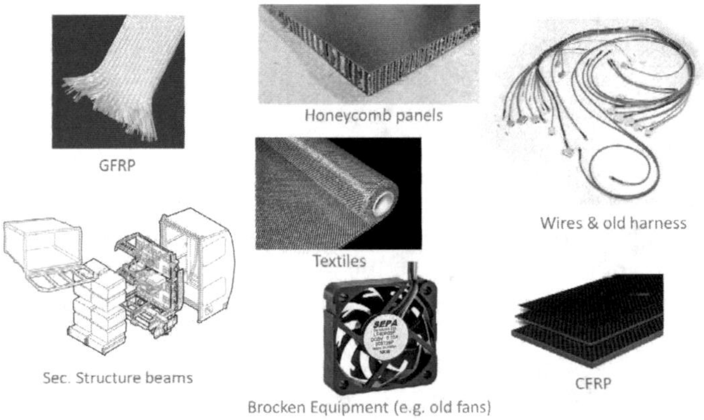

Figure 13. Example set of possible addition source material for recycling and up-cycling activities within future Moon or Martian habitats [Source: web].

Considering the overall system aspects, two major requirements and guidelines are essential for a successful recycling and up-cycling environment in future extra-terrestrial habitats.

IOP Conf. Series: Materials Science and Engineering **1328** (2025) 012018 doi:10.1088/1757-899X/1328/1/012018

Firstly, a recycling paradigm needs to be implemented at the beginning of design and development. Materials shall be selected, and components designed, as far as possible to be reusable in a second stage of their lifecycle.

The items targeted for recycling could be further optimised by including future design features into the original design, such as mounting/assembly holes, dedicated openings or interlocking mechanisms, as well as detachable connecting elements, for easier separation.

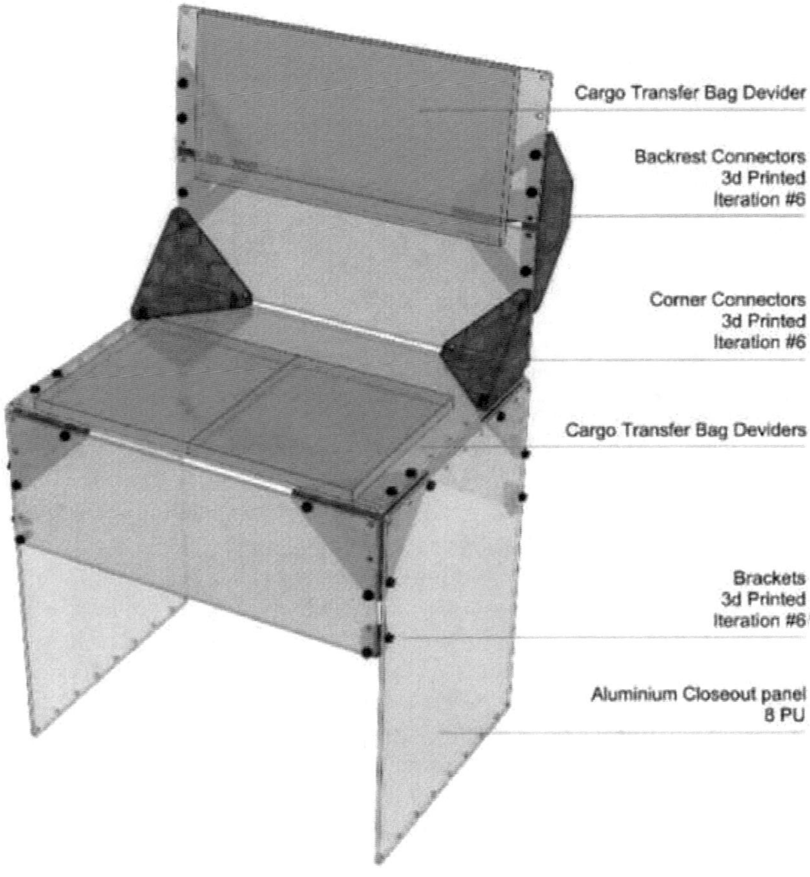

Figure 14. Final design of the chair demonstrator [Source: LSG].

The challenge is to incorporate these design features while meeting the primary lifecycle requirements of each item, and not compromising its function.

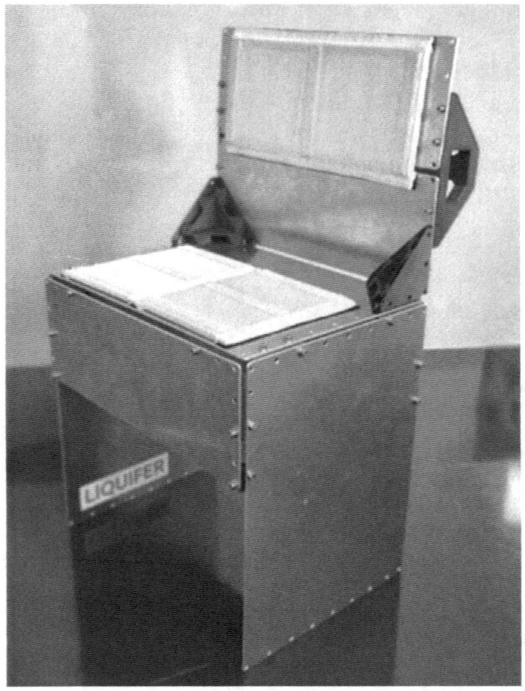

Figure 15. Final version of the chair demonstrator [Source: LSG].

Secondly, a local transformation infrastructure needs to be in place in order to allow the astronauts to assemble the new items. For assembling and reconfiguring the blind panels for the chair demonstrator, no specific technologies need to be developed. Nevertheless, an adequate set of handheld tools (e.g. screwdrivers, Allen® keys, socket wrenches, drills, metal saws, files), work bench (incl. assembly area), and an intermediate storage location need to be foreseen.

Figures 14 and 15 display the final design of the demonstrator after the final iteration procedure. This final design improved the demonstrator by providing additional drill holes where needed to utilise the full potential of the connector elements. This decision was based on considerations regarding reusability of items, as well as the resulting conclusion that contingency drillholes in the original design will facilitate reconfiguration, and that a future lunar infrastructure shall include the recommended needed tools and facilities to allow the modification and storage of hardware parts for further assembly.

3. Food package recycling into 3D-printable filament

3.1 Initial idea and concept trade-offs

The idea at the base of the demonstrator was to recycle the food packaging generated during a manned mission through a complete recycling process in order to obtain a low-density polyethylene (LDPE) 3D-printable filament. The plastics recycling for 3D-printing purposes in space defines a well-known field of interest (the Refabricator® on board the ISS is a clear proof of that), referring to basic materials such as PLA and ABS (Acrylonitrile butadiene styrene), but also to some high-performances thermoplastic such as PEEK (Polyetheretherketone) and PEKK (Polyetherketoneketone) [4]. The LDPE recycling is, however, a practice much less investigated and tested in the literature and, therefore, represents a challenging and newsworthy approach in view of a sustainable and Earth-independent human presence in extraterrestrial habitats.

3.2 Used material and recycling concept

Commercial low-density polyethylene Ziplock® bags of various sizes (see Figure 16) have been selected as suitable representatives of the food packaging materials commonly used in a number of space applications.

The LDPE Ziplock® bags that have to be recycled were firstly collected and cleaned, then they underwent a shredding process in order to be reduced to fine granulates. After this, the granulates were dried to eliminate moisture and dust, which influence the mechanical quality of the final product negatively. Finally, the recycled raw material was used as feedstock for the filament extruder, which turned it into a neat and tight filament, ready to be 3D-printed.

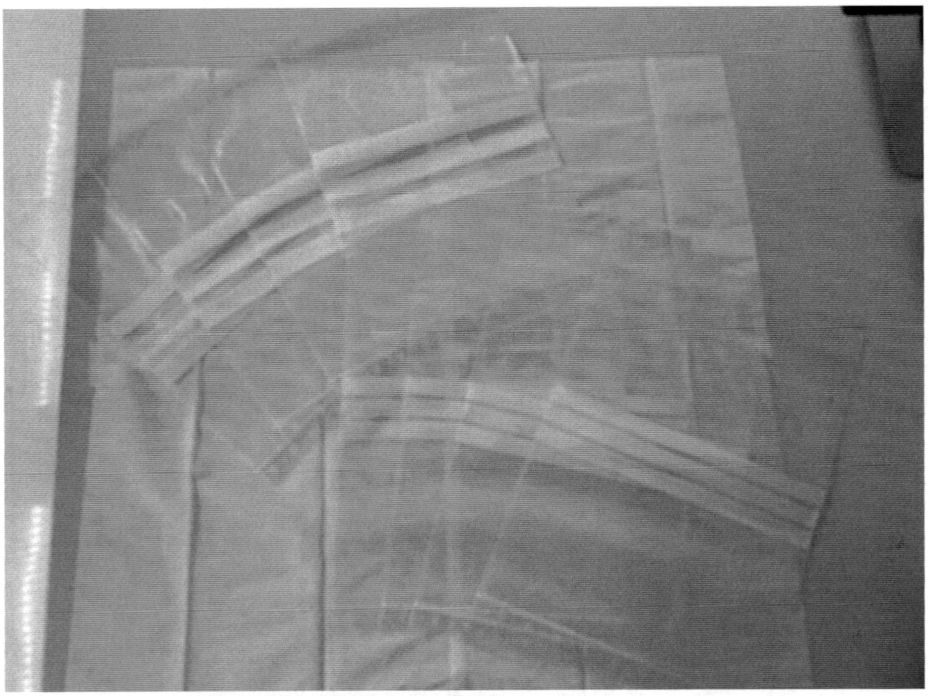

Figure 16. Commercial Ziplock® bags [Source: OHB].

3.3 Demonstrator design and recycling process

To perform the entire recycling process, the 3devo recycling machines SHR3D IT Shredder®, AIRID Dryer® and Precision450® filament maker [5] have been used. The recycling process started with the cleaning of the Ziplock® bags. This cleaning process was performed by washing the Ziplock® bags under rinsing water and using 2% Mucasol® solution. Before the real shredding process, a manual cutting step was performed (see Figure 17 a) given the specific geometry and shape of the Ziplock® bags and in order to simplify the following steps. Afterwards, the cleaned bags were manually chopped into smaller pieces and underwent a double-stage shredding process through the shredder (see Figure 17 b). In the frame of ESA research, three different mechanical shredding machines were evaluated and the SHR3D IT Shredder was the preferred choice over the others due to its good dimensional results, little contamination, and convenience of usage.

After the shredding, the drying process took place using the AIRID Polymer Dryer under a properly set fume hood (see Figure 17 c), so that undesired gas emission could have been controlled. The dryer was used to eliminate any moisture complication that may occur within a polymeric material. Moreover, the machine is equipped with a stirring rotator which guarantees evenly dried materials across all surface areas. On the basis of the LDPE peculiar characteristics, the drying temperature, the duration, and the input material morphology (i.e. granules or flakes) and quantity were selected to correctly start and perform the drying process. According to literature, in the case of polyethylene-based feedstock it was not mandatory to perform a drying process, being this material not so prone to absorb moisture and humidity [5]. However, this step was performed in order to give the re-LDPE the proper consistency for extrusion. The overall drying process lasted two hours and the set temperature was 65 °C.

Finally, the obtained dried material was used as feeding material for the PRECISION 450 Filament Maker for the filament extrusion process (see Figure 17 d). This machine, also able to process PLA, PEEK, PETG (Polyethylene Terephthalate Glycol), PEKK, can manage temperatures up to 450 °C. It is equipped with the 3devo's Precision Screw, which offers faster extrusion speeds and steady filament flow, and as a result more precise filament diameter [5]. On top of this, an automatic neat spooling system is also embedded in the machine, which, together with the *DevoVision App®*, which allows for a live data acquisition and monitoring, increase the efficiency of the overall process. During the filament extrusion process, the software receives real-time data of constant measurements displaying filament thickness, extruder RPM (Rotations Per Minute), puller speed, current filament length and mass, and other settings/information, allowing for a user-friendly fine-tuning of the process. Via its unique vertical extrusion set-up, automated fan cooling, sophisticated optical sensor and a dynamic puller wheels system, the filament maker is able to provide precise roundness, accurately guide the filament towards the spooling process and achieve a diameter precision tolerance of ± 0.20 mm.

Essentially, three parameters were adjusted in order to reach the desired filament outcome:

- screw RPM;
- fan cooling speed and orientation;
- heaters temperature profile.

Table 1. LDPE extrusion baseline setting used during the trials.

re-LDPE Extrusion Setting					
Heaters Temperature Profile				RPM	Fan Cooling
H n°4	H n°3	H n°2	H n°1	5.5-6.2	20-30%
185 °C	185 °C	185 °C	190 °C		

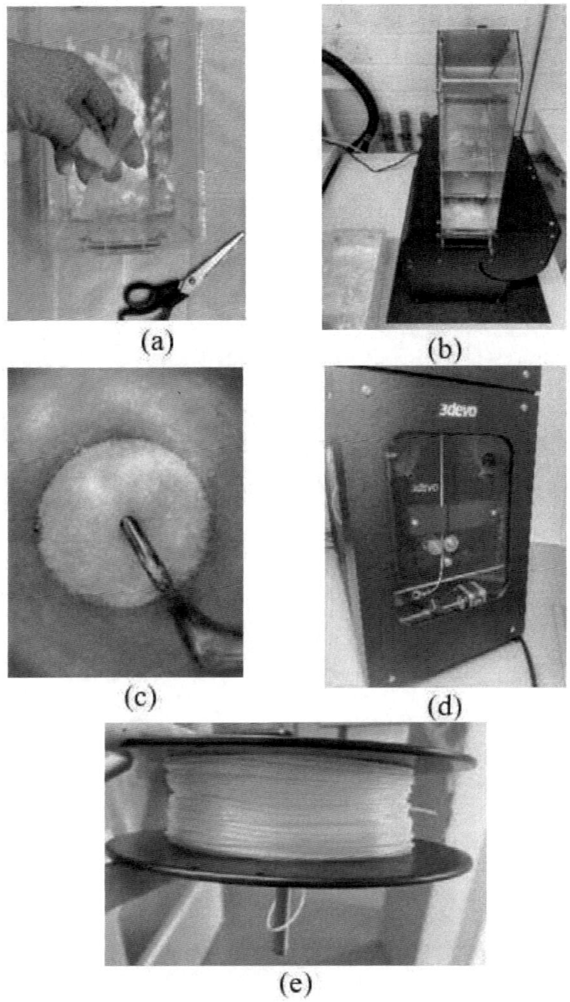

Figure 17. recycling of LDPE: (a) manual cutting phase, (b) shredding phase, (c) drying phase, (d) extrusion phase, (e) re-LDPE spool [Source: OHB].

Depending on the extruding conditions, the screw RPM and the fan cooling speed/orientation can play a more or less significant role, while the most important factor for a good quality filament extrusion is the heating system temperature profile. The temperature profile, indeed, consists of three different zones: feeding, transitioning and metering zones; where the material is respectively fed into the extruder, melted down, and finally where the obtained viscous fluid is pressurized and pushed out. Within these three zones, there are four heaters which together control the temperature profile.

The final extrusion setting for the re-LDPE was the one reported in Table 1.
Important to be mentioned is that at the end of every recycling test, a deep purging procedure of the filament maker must be performed. Cleaning the filament maker consists of extruding special cleaning materials such as HDPE and *DevoClean Mid Temp EZ* [5].

As previously mentioned, the process parameters were live-monitored during the extrusion with the *DevoVision App®*. In particular, the filament thickness was monitored with refer to the pre-set upper and lower limits (1.60 mm ± 0.20 mm, 1.75 mm ± 0.20 mm and 2.85 mm ± 0.20

IOP Conf. Series: Materials Science and Engineering **1328** (2025) 012018 doi:10.1088/1757-899X/1328/1/012018

mm). Then, when the desired thickness had been achieved, thanks to the combined action of the optical sensor and of the puller wheels, it was tightly spooled (see Figure 17 e) taking advantage of the spooling wizard program embedded in the filament maker. This was crucial since a poor spooling would have meant the impossibility to 3D-print the filament, therefore spoiling the overall recycling process.

3.4 Demonstrator testing phase

A total of 23 functional requirements were verified via review of design, analysis, inspection and testing.

The following requirements were addressed during the test phase of the demonstrator #2.

Table 2: Demonstrator #2 requirements.

Requirement ID	Functional requirement	Test Procedure
HAR-OHB-REQ-FUNC-0012	The filament shall have a diameter of 1.60 mm ± 0.20 mm (or better) along its overall length.	The test is carried out taking advantage of the *DevoVision* App software embedded in the filament maker, and by post-processing the data with Excel.
HAR-OHB-REQ-FUNC-0024	The filament shall not be either entangled or show signs of discolouration, contamination, degradation, ovalization and shall be in the aforementioned range of thickness with a consistent diameter roundness along its entire length to be 3D-printed. Moreover, the filament shall be neat and tight during the spooling process, but no overstretched and it shall not be spoiled by granulates of different densities, burnt particles and encapsulated bubbles to be 3D-printed. Finally, enough filament must be extruded (around 20 m at least) to be 3D-printed.ASTM samples shall be printed with the recycled filament suitable to be tested via ASTM D790 3-point flexure tests or different standard.	This verification is performed by simply gathering all the information derived from the visual quality inspection of the filament and on the base of the outcome of the test related to HAR-OHB-REQ-FUNC-0012.

The quality inspection campaign confirmed that all the recycled spools showed a neat and tight filament with negligible entanglement. Furthermore, the roundness consistency of the diameter along the filament spooled length was in the prescribed range of tolerance and it also confirmed the weight consistency which can be derived by the overall roundness uniformity and absence of degradation of the filament (i.e., absence of changes in the material properties).

Multiple filaments having a diameter of 1.60 mm ± 0.20 mm, 1.75 mm ± 0.2 mm and 2.85 mm ± 0.2 mm along their overall length have been extruded. The filament thickness has been live-recorded using the *DevoVision App*® and then transported into Excel for data post-processing as showed in Figure 18.

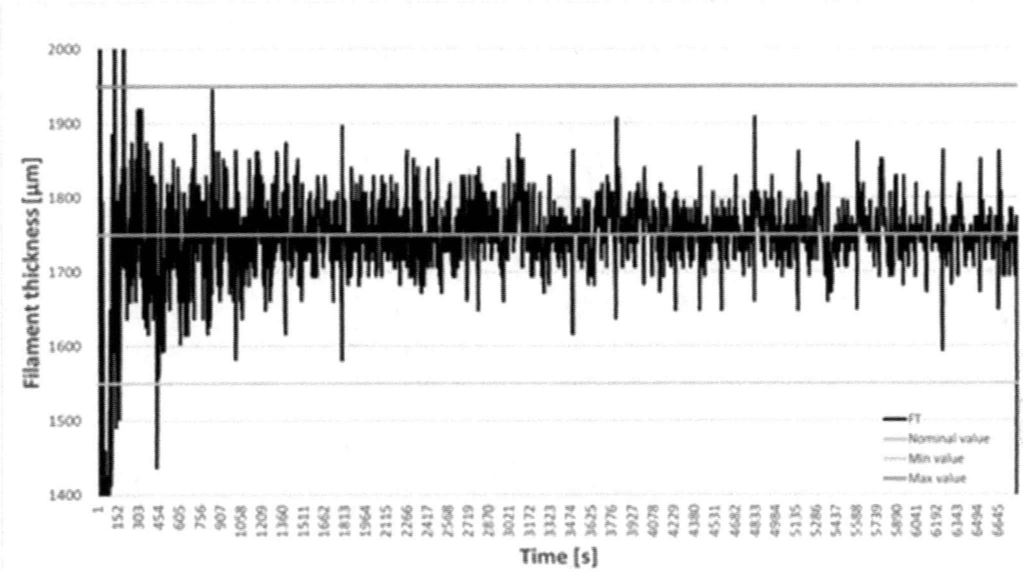

Figure 18. Filament thickness variation along its spooled length: case 1.75 mm ± 0.2 mm [Source: OHB].

After a visual inspection (see Figure 19), the extruded re-PE filaments appeared homogenous in properties without any visible sign of granules of different densities. No burnt particles were detectable as well. The use of a common lab microscope with a magnification factor of 10 also allowed to inspect the filaments more in detail in search of encapsulated air bubbles, discoloration, and contamination (dust and/or dirt).

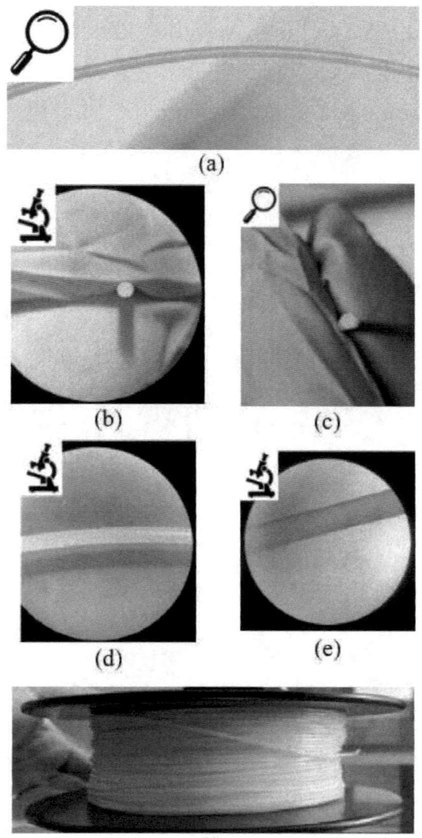

Figure 19. Recycled polyethylene filament inspection: (a) no burnt particles detected, (b) & (c) no ovalization spotted, (d) & (e) no air bubbles encapsuled, discoloration, contamination or degradation observed, (f) neat and tight filament spooled [Source: OHB].

The examination did not highlight any signs of material degradation, but a slightly rough filament surface was observed. This is expected, considering the extrusion process is complex and that the material is obtained from recycled hardware. Moreover, since the optical sensor embedded in the filament extruder machine is mono-dimensional, potential ovalization could not be detected by the app, therefore a dedicated test has been set-up to address this topic. By cutting the filaments in several random locations and observing through the microscope the cross section, no ovalization has been observed and the filament looked very rounded in all cases.

The various re-PE filaments were shipped to ESA laboratory at Harwell and tested with respect to:

- glass transition temperature via dynamic mechanical analysis (DMA) and differential scanning calorimetry (DSC);
- degradation onset temperature via thermogravimetric analysis (TGA);
- oxidation induction time via DSC.

IOP Conf. Series: Materials Science and Engineering **1328** (2025) 012018 doi:10.1088/1757-899X/1328/1/012018

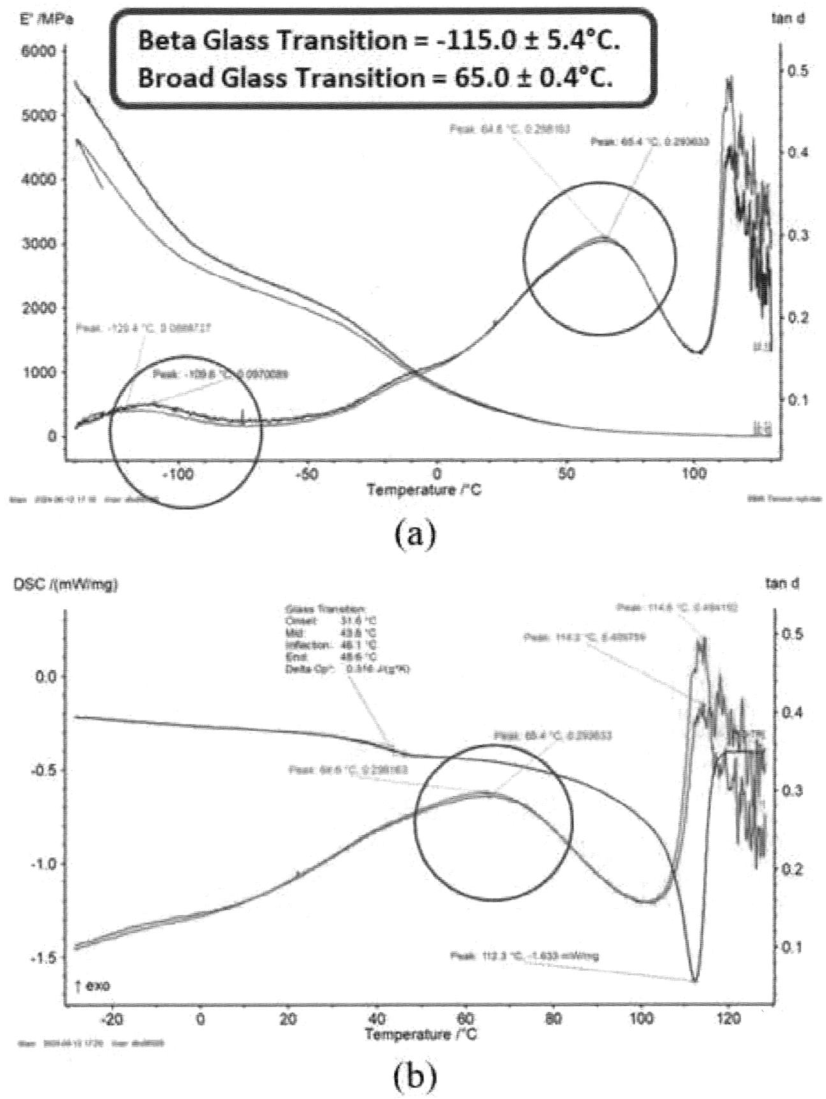

Figure 20. Recycled LDPE glass transition temperature via dynamic mechanical analysis (a) and differential scanning calorimetry (b) [Source: OHB].

Concerning the glass transition temperature, two DMA tests were run (see Figure 20 a) to confirm literature reference trends [6], [7] and [8].

These showed that if it is true that there is a small peak at -115.0 °C which implies a β-transition, a very broad peak is also visible which correlates with a transition shown in the DSC data (see Figure 20 b). Broadness of the peak could be due to a large range of chain lengths in the sample, each becoming more mobile due to the increased free volume approaching the melt. It was therefore concluded that β-glass transition (Tβ) occurred at -115.0 °C, while a broad glass transition (Tg) happened at 65.0 °C.

Regarding the degradation onset temperature, this was evaluated via TGA taking the intersection between the tangent lines of maximum and minimum gradient in a specified range.

The minimum of this range was found out to be 338.8 °C, therefore this value has been taken as the start of the degradation (see Figure 21).

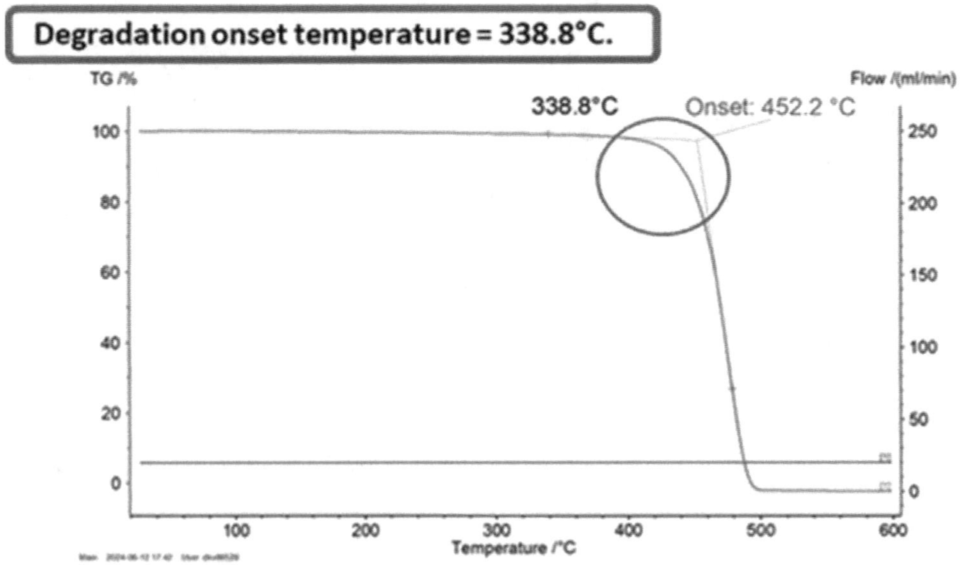

Figure 21. Recycled LDPE degradation onset temperature [Source: OHB].

The oxidation induction time study was conducted, via DSC, firstly at 205 °C since this was the temperature value at which the reference literature value was taken [10].

However, the oxidation turned out to be almost instantaneous, so the oxidation induction temperature and melting temperature were evaluated.

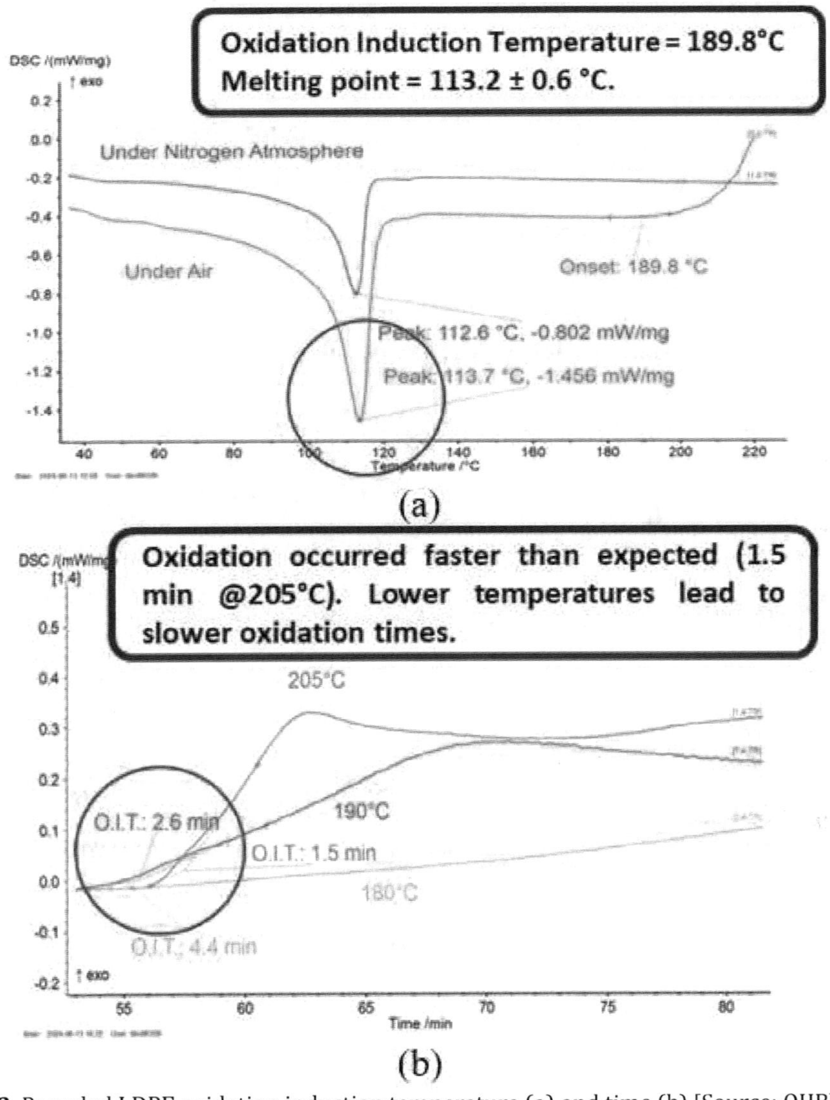

Figure 22. Recycled LDPE oxidation induction temperature (a) and time (b) [Source: OHB].

Melting temperature was found to be 113.7 °C, while oxidation induction temperature was found to be 189.8 °C (see Figure 22 a). The oxidation induction time test was therefore repeated at 180 °C and 190 °C such as to still comply with ISO 11357-6.

First tests showed oxidation induction temperature as onset of exothermic peak whilst under air, while under nitrogen no peak was observed. Consequently, the oxidation induction time at 205 °C was measured as 1.5 min, while at 180 °C as 4.4 min, resulting in both cases in a largely lower value than expected from literature (see Figure 22 b). One reason could be the lack of antioxidants usually observed in recycled material.

The conclusions that can be drawn after the thermal test campaign at ESA Harwell lab were the following:

- considering that in literature the glass transition temperature foreseen for PE ranges from

-130 °C to -100 °C, the test was considered passed since the obtained result met the success criteria with respect to the mean value;

- the evaluated degradation onset temperature was in line with the reference value [9] and respected the prescribed success criteria. Therefore, this test was considered passed;
- the measured oxidation induction time was far below the reference value and not respecting the prescribed success criteria. This test has been therefore considered failed. However, in literature it is well known that the PE quality (i.e., pristine PE or recycled PE) can influence the oxidation induction time and cause the reaching of temperatures up to 205 °C, close to the ones obtained during the test at Harwell. Therefore, it is reasonable to assess that the reference value was not appropriate to benchmark and judge the obtained results.

A recap of the results of the tests performed to thermally characterize the re-LDPE extruded filament, the selected literature reference values and the prescribed success criteria has been reported in Table 2.

Table 2. Results of the re-PE filament thermal testing performed at ESA Harwell lab [Source: OHB].

Test	Reference Value	Success Criteria	Result	Pass/Fail
Glass transition temperature	From -130°C to -100°C According to [3], [4], [5].	±5%*	-115°C	**Pass**
Degradation onset temperature	325°C According to [6].	±5%*	338.8°C	**Pass**
Oxidation induction time	17.19 min @ 205°C According to [7].	±5%*	1.5 min	**Fail**

3.5 Implications for space scenario

The lunar environmental conditions influence the design of any mission aimed for the surface of the Moon/Mars in various ways and to different degrees.

In the case of the recycling of LDPE, the shredder, the dryer, the extruder and the printer used in the filament making process will all have to be modified. The biggest challenge for the design is the thermal environment as all parts of the extruder and the printer will be affected. The very low temperatures will compromise any standard lubrication of the linear units and other moving parts (particularly in combination with microgravity). In space, all materials also become more brittle, especially plastics, which reduces the structural strength. Furthermore, the large temperature differences during day and night, as well as between the sun-exposed and sun-hidden sides, will introduce stress to the structure due to thermal cycling. This also affects the accuracy of the printer and the diameter of the filament that might come out of the nozzle.

Another problem would be the lack of significant gravitational acceleration, which results in a lack of convection. Hence, all electronics would require dedicated temperature control systems to avoid overheating, but also to prevent too low temperatures when the extruder or the printer will not be in operation. Furthermore, in a non-Earth environment, a conveyor belt or a system of pulling wheels should also be added to the extruder machine design, to cope with the lack of gravity driving force necessary to guide the filament from the nozzle to the spooling system. For an easier redesign of the machines without compromising safety, a Generative Design (GD) approach could be implemented in order to establish which parts can be removed, substituted or reused in other configurations having given prescribed mass and volume constraints.

On the other hand, electrical devices also are predominantly affected by radiation and need to be designed accordingly. A solution could be the radiation hardening of the electrical devices in order to prevent shorts and damage to the sensitive electronics. Furthermore, lunar or Martian dust could create several disruptions in the proper functioning of electronics (e.g., PCB), mechanisms and moving parts, which therefore need to be protected from contamination. For the case under discussion, this means that the whole shredder, dryer, extruder and printer should be properly sealed.

Even though an exact possibility of meteoroid impacts is hard to predict, and the probability of large impacts is small, a minimal protection against micrometeoroids should also be implemented.

Nonetheless, not only the design of the printer will be affected by the extraterrestrial environments, but also the feedstock and the operations.

The vacuum would actually be beneficial to the feedstock quality as it prevents air bubbles, but would also have a non-uniform temperature in the print chamber as consequence.

Additionally, the current layer could not be cooled with an airflow, but this might not be necessary anymore. The low gravity could cause the first few layers to warp-up on the outside. It would also be unavoidable to have dust inside the print chamber, which could reduce the print quality.

4. Aluminium scrap metals casted for tool fabrication

4.1 Initial idea and concept trade-offs

The idea for the final demonstrator was aluminium parts recycling. The material is available from hardware (defunct spacecraft, landers or rovers) remaining after various missions on Mars and, in particular, on the Moon.

The potential to recycle aluminium parts shows a large variety of resulting hardware. Such options as habitat structures, full-sized demonstrator of rover spare parts was considered, as well as a casted multi-tool to assist astronauts in construction and repairs. The demonstrator selected for this particular activity and based on the available resources, is a downscaled prototype of a multi-tool that could be potentially used for the construction of the habitat, fastening or screwing bolts and nuts etc... (see Figure 23).

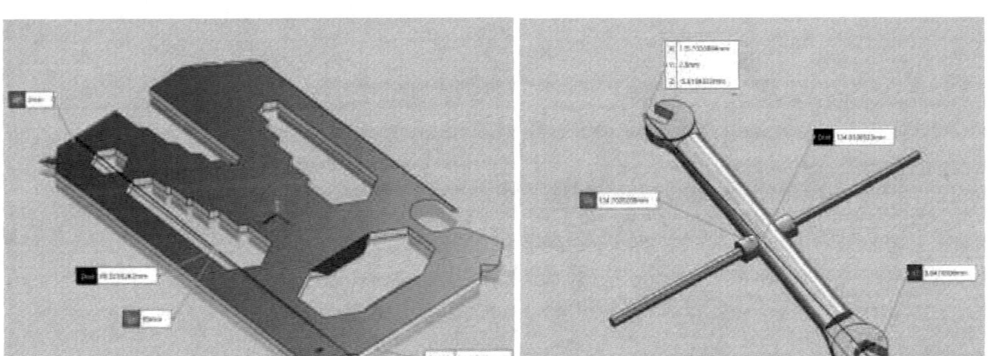

Figure 23. Considered prototypes for the multi-tool [Source: Azimut].

IOP Conf. Series: Materials Science and Engineering **1328** (2025) 012018 doi:10.1088/1757-899X/1328/1/012018

4.2 Used material and recycling concept

Aluminium was selected as a primary material for recycling due to such reasons as: high availability on space vehicles (especially on the Moon), machinability, nominal compatibility with load bearing applications, relatively low melting point compared to steel or titanium and high recyclability rates (up to 90% in terrestrial applications).

Additionally, recycling of aluminium via casting is well known and established on Earth, which means that adoption of such method, is more feasible for extraterrestrial applications.

Finally, since the aluminium is completely melted during the recycling process, the full thermomechanical history of the raw feedstock will be erased, and new pristine grains will be formed during solidification of that melted mass.

Based on the extensive research of past missions, it was decided to produce the casted tool demonstrator out of recycled aluminium rods composed of alloy 7075.

The 7075 alloy used for this demonstrator is a wrought aluminium alloy, which obtains its optimal mechanical properties through strain hardening or metal working. For this class of alloys, the as-cast state represents the least advantageous metallurgical state, for the mechanical properties. Therefore, other manufacturing routes should be investigated for recycling of this class of alloys. This includes potential grain refining or metal working processes.

4.3 Demonstrator design and recycling process

The finalised design is shown in Figure 24 below.

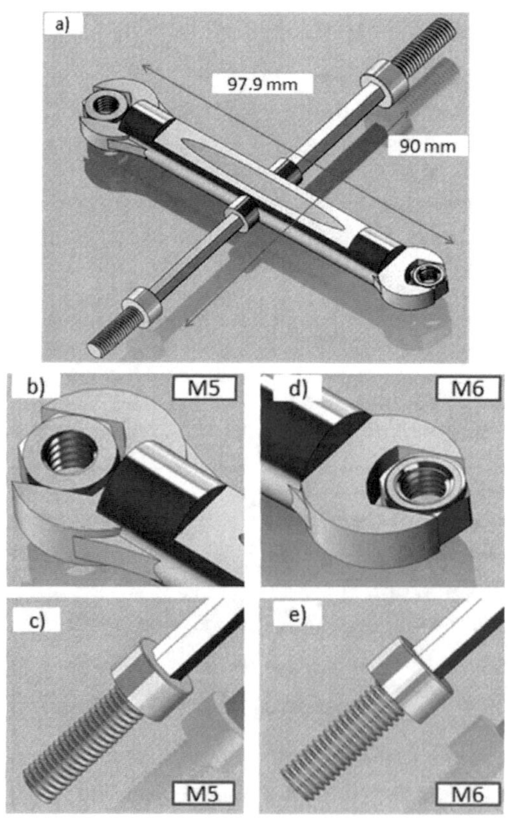

Figure 24. Multi-tool finalised design (a-e) to be casted with recycled aluminium [Source: Azimut].

IOP Conf. Series: Materials Science and Engineering **1328** (2025) 012018 doi:10.1088/1757-899X/1328/1/012018

The tool includes the following features: two wrenches suitable for M5 and M6 hex head bolts and Allen® keys compatible with M5 and M6 socket head bolts. Additionally, the tool possesses an opening in the centre that can be used as an attachment point to prevent accidental release.

The tool's design is envisioned with handles offering a secure grip, as well as a balanced mass distribution, which reduces the risk of strain or injury. Furthermore, the design aims for a compact footprint to be easily carried and stored within the limited space of the habitat. In addition, the tool is designed with a focus on sustainability and product circularity, allowing re-recycling of the part once the fatigue limit has been reached.

As a result of manufacturing trials, six distinct specimens were manufactured with different degree of success. Amongst all samples, two parts with the best quality achieved were selected for subsequent testing. These samples are shown in Figure 27.

The multi-tool demonstrator is manufactured via sand casting method. Firstly, a pattern is created via 3D-printing in the shape of the desired tool and then pressed within the sand. The pattern is then removed from the sand, leaving behind a cavity. These preparation steps are shown in Figure 25. The cavity is then filled with molten aluminium. Once the aluminium cools down and solidifies, the sand is removed to reveal the finished tool (see Figure 26).

Throughout the casting process such parameters as the pouring temperature, pouring speed, moisture level in the sand, quality of the 3D-printed insert, purity of the feedstock material etc... were assessed and partially controlled when possible.

Figure 25. Preparation of the casting container (a-c) [Source: Azimut].

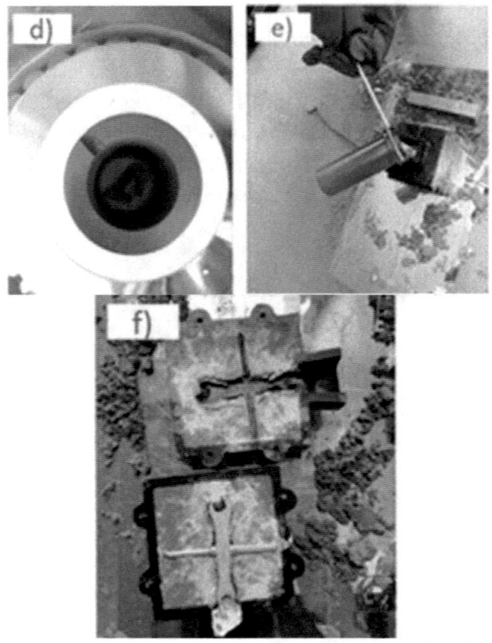

Figure 26. Melting of aluminium, pouring and extraction of the casted tool (d-f) [Source: Azimut].

As a result of manufacturing trials, six distinct specimens were manufactured with different degree of success. Amongst all samples, two parts with the best quality achieved were selected for subsequent testing. These samples are shown in Figure 27.

One of the goals of this activity was to minimize post processing of the tools in order to avoid release of particles/dust that inevitably results from machining or other types of post processing.

This goal was partially achieved with Trial 2 that required minimal post-processing relative to Trial 1. Namely, only residual melt left after the addition of extra air release openings had to be removed before the fit check for all the tools could be successfully performed with the corresponding fasteners.

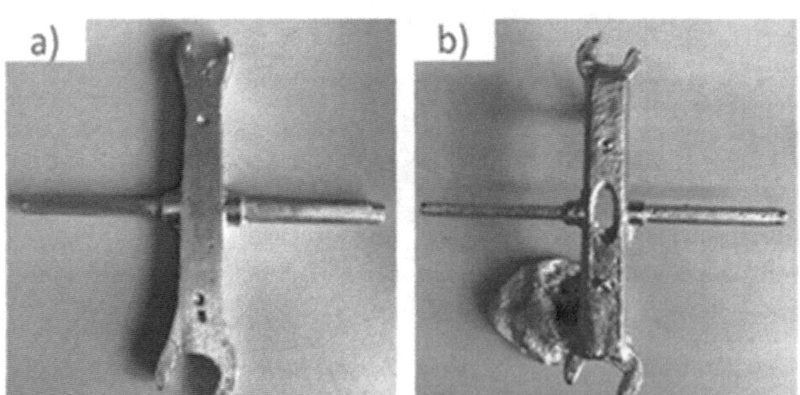

Figure 27. Manufactured specimen selected for testing: trial 1 (a) and trial 2 (b) [Source: Azimut].

IOP Conf. Series: Materials Science and Engineering **1328** (2025) 012018 doi:10.1088/1757-899X/1328/1/012018

4.4 Demonstrator testing phase

Following the manufacturing trials, casted specimens were inspected with respect to dimensional accuracy, surface roughness, functionality and other aspects. In total, two selected multi-tools were studied by various methods against 22 functional requirements.

The following requirements were addressed during the test phase of the demonstrator #3.

Table 3: Demonstrator #3 requirements.

Requirement ID	Functional requirement	Test Procedure
HAR-ASG-REQ-FUNC-0001	The hand-held casted tool shall be comfortable to hold and use by an operator for a working time of at least 1 h.	To be verified by a Human-in-the-loop (HITL) ergonomics study.
HAR-ASG-REQ-FUNC-0002	The hand-held casted tool shall be resistant to wear and operate at least 1.000 cycles without maintenance.	To be verified by cycle testing of multi-tool.
HAR-ASG-REQ-FUNC-0003	The casted hand-held tool shall allow dexterity of hand movement, enabling its grip and manipulation by medium hand sizes with a hand circumference of 21.6 to 22.9 cm.	To be verified by a Human-in-the-loop (HITL) ergonomics study.
HAR-ASG-REQ-FUNC-0004	The casted hand-held tool shall be designed compact and small to be held by a medium-sized human hand with a hand circumference of 21.6 to 22.9 cm.	To be verified by a Human-in-the-loop (HITL) ergonomics study.
HAR-ASG-REQ-FUNC-0005	The casted hand-held tool shall be light-weight and easy to carry.	To be verified by a Human-in-the-loop (HITL) ergonomics study.
HAR-ASG-REQ-FUNC-0008	The casted hand-held tool shall be recyclable for at least two re-recycle treatment.	To be verified by suitability for testing.
HAR-ASG-REQ-FUNC-0012	The casted hand-held tool shall be capable of tightening and loosening different types of fasteners, including M5 and M6 external HEX screws, flange head screws, slotted head screws, dome head screws and M5 and M6 nuts.	To be verified by verification of design including a Human-in-the-loop (HITL) ergonomics study and mechanical cycling tests with the mechanical fasteners mentioned.
HAR-ASG-REQ-FUNC-0013	The casted hand-held tool shall withstand a torque of at least 10 Nm.	To be verified by mechanical testing of the multitool.
HAR-ASG-REQ-FUNC-0016	Torques required to operate or control the casted handheld tool shall be less than the strength values for the 5th percentile female, defined as 60% of the calculated 5th percentile male capability.	To be verified by a Human-in-the-loop (HITL) ergonomics study.
HAR-ASG-REQ-FUNC-0019	The casted hand-held tool shall have a bulk density of at least 97%.	To be verified by suitability for testing.
HAR-ASG-REQ-FUNC-0022	The casted hand-held tool shall have a reliability with a failure rate of 1 out of 1,000 fastened bolts (λ=1/1,000) or better.	To be verified by suitability for testing.

Amongst different tests, the most interesting and insightful of the tools functionality is the cycling test where fasteners, depending on the tool, were repeatedly tightened and loosened to the specified torque.

Figure 28. Cycling test of M5 wrench (Trial 1) [Source: Azimut].

The cycling test was setup in the way displayed in Figure 28 above and executed repeatedly until the particular tool failed and the test could no longer be continued.

During this particular test, different tools showed distinct failure modes. As Figures 29 and 30 illustrate, the Allen® keys of the component obtained during Trial 1 and 2 were stripped after a few cycles and could not be used anymore. Meanwhile, M6 wrench cracked at the base of one of the jaws. This happened due to a combination of increased tolerances and high torque values. M5 wrench in Trial 2 could not be used at all, as the gap between the wrench and the corresponding fastener was oversized.

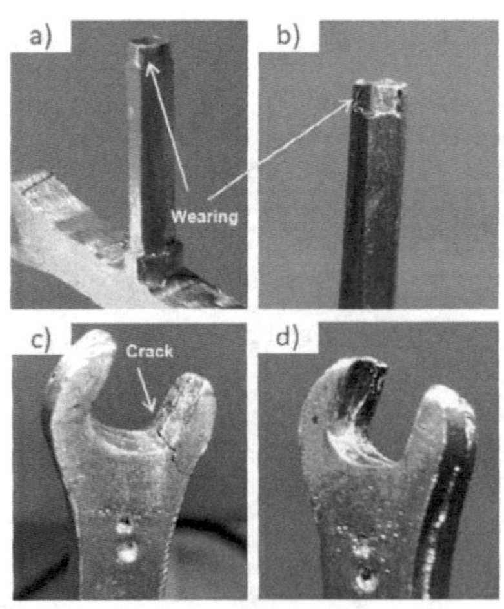

Figure 29. Damage on the Trial 1 specimen (a-d) [Source: Azimut].

There is a number of factors that lead to premature failure of the tools. Firstly, the mechanical properties of aluminium 7075 are decreased significantly during recycling. Although this effect can be mitigated with suitable pre and post-processing such as using grain refiner, heat treatment and other techniques, these activities were not performed since beyond the scope of the present project.

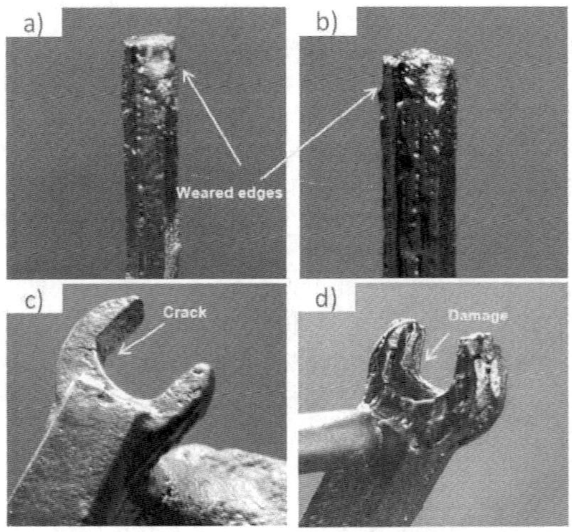

Figure 30. Damage on the Trial 2 specimen (a-d) [Source: Azimut].

Another outcome of the testing campaign is the identified density of the components that amounted to approximately 88% which is much lower than expected. Decreased density is expected given a basic sand-casting setup and limited resourced that did not allow a precise control over multiple variables that influence the quality of the manufactured components.

Additionally, significant variations of porosity with respect to manufacturing parameters were also observed.

4.5 Implications for space scenario

The adoption of the manufacturing workflow described in the previous sections for use in extra-terrestrial environment presents a number of challenges.

Firstly, the hardware and consumables needed to cast the tool have to be brought initially from Earth, which will require a significant mass and power budget.

Secondly, environmental factors are one of the most critical considerations when designing a multi-tool for Moon and Mars scenarios. Both Moon and Mars have harsh environments that can affect the performance and functionality of the tool. Extreme temperatures, dust, and low gravity are all important factors that shall be taken into account. On the other hand, vacuum casting (on the Moon) is likely to prove beneficial for the casting quality as this environment allows avoiding oxidation, reducing the number of contaminants and yields a better final quality of the manufactured parts.

The outcomes of the testing campaign suggest that the use of casted components with the simplified manufacturing process shall be limited to non-load bearing applications.

Indeed, aluminium 7075 is an inherently suboptimal alloy choice for casting as it is prone to embrittlement, cracking and requires post processing such as heat treatment in combination with aging. At the same time, utilization of the grain refiner is crucial for strength and hardness control of this material.

This approach shall be considered to improve mechanical properties of the casted parts, though this is only possible to a certain extent. Post-processing, though beneficial for mechanical properties of the manufactured component, comes at a cost of time and additional complexity which is an important consideration for extra-terrestrial applications. Finally, such an important aspect as the presence of contaminants on the feedstock coming from actual hardware is to be considered. Indeed, certain components made out of aluminium (for example, honeycomb panels) often include epoxy resin. Similarly, structural components are usually coated or treated, these inclusions might complicate pre-processing of the material prior to manufacturing.

The outcome derived from this demonstrator is the possibility of re-casting failed attempts, as well as re-casting a damage tool due to wear and fatigue failures. This ability to re-cast indicates the possibility of a fully circular process, relevant mainly for non-load bearing components, which is crucial for sustainable operations in space. The lessons learned from this demonstrator highlight the feasibility of creating a self-sustaining manufacturing cycle, reducing dependency on Earth-supplied materials over time, and enhancing the viability of long-term sustainable extra-terrestrial habitats.

5. Conclusions

During the HARMONISE project, extensive research of possible recycling opportunities for the set-up and support of extra-terrestrial habitats (lunar or martian) was conducted.

The first demonstrator has shown that discarded items such as blind panels in combination with CTB dividers, and 3D-printed connectors are suitable to create a usable chair. As the chair design was primary chosen as a demonstrator example, the spectrum of further designs and up-cycle solutions (e.g. furniture elements, secondary structural elements, tools) is multiple. The range of designable products can further be expanded by a general paradigm change in pre-flight design approaches. By implementing certain design adjustments into all habitat elements and consumables, the upcycling approach can be eased.

The campaign of the re-LDPE filament extrusion showed the capability of extruding and spooling up to 100 m of recycled LDPE with no signs of entanglement, discoloration, contamination, degradation or ovalization. The quality of the spools produced during the recycling process testified that this is indeed a suitable option for 3D-printing. As a way forward, the next goal could be to find either an additive compound that could improve the 3D-printability of the Ziplock® bag-derived re-LDPE or a suitable 3D-printer/3D-printer setting for ASTM samples production and testing.

The sand-casting trial performed within the project scope showed that it is a feasible manufacturing approach to produce the multi-tool in question in the Earth environment. In terms of performance, however, neither of the tools tested managed to reach the required number of cycles with the specified torque. This is true for both wrenches and Allen keys features. Additionally, it was found that wrenches are much more robust compared to Allen keys which failed after just a few tightening/loosening cycles. Finally, the density analysis indicated that the manufactured tools have a higher level of porosity than expected which further weakened the bulk material and lead to a premature failure. The main outcome of the trials was a set of recommendation such as to reinforce the wrenches and remove Allen keys from the tool to focus on hex-head rather than socket-heads bolts. Further steps include formulation of the de-rated torque values for the aluminium tool, improvement of robustness by adopting square head bolts that could increase the surface in contact and finally changing the design and geometry of the tool so it functions in the same way as a socket wrench. Moreover, it was understood that it is worthy considering other manufacturing processes for the aluminium 7075. Indeed, the alloy used for this demonstrator is a wrought aluminium one, which obtains its optimal mechanical properties through strain hardening or metal working. For this class of alloys, the as-cast state represents the least advantageous metallurgical state, for the mechanical properties. Therefore, other manufacturing routes should be investigated for recycling of this class of alloys. This includes potential grain refining or metal working processes.

Overall, as a take-home message, it was understood that the successful implementation of a recycling approach on the Moon or Mars involves a multifaceted approach encompassing the selection of appropriate materials, the adaptation of the technology to harsh off-Earth scenarios, design-for-recycling approaches, robust logistics, and stringent quality control and safety protocols. Beside the technical complexity and on a big picture, several multi-disciplinary challenges must be faced when dealing with the recycling of space artefacts, ranging from ethics to safety, from logistic (retrieval and storage) limitation to legal consideration, without forgetting the cost impact.

Acknowledgements

This work was funded by the European Space Agency (ESA) under Contract No. 4000139104/22/NL/AR.

The authors greatly acknowledge the Agency for the financial and technical support, and they wish to express their heartfelt gratitude to the entire team at Liquifer Systems Group, Azimut Space and OHB System AG for their restless dedication and invaluable technical and human contributions to the HARMONISE project success.

References

[1] Anna Daurskikh et all., Quality of am polymer parts produced using a printer without build volume constraint in space environment, ECSSMET2021
[2] Antonella Sgambati et all., URBAN: conceiving a lunar base using 3D printing technologies, IAC-18.D3.3.7x42605
[3] Dr. Manuel Ortega et all., Closing the loop in space 3D printing: effect of vacuum, recycling, and UV aging on high performance thermoplastics produced via filament extrusion additive manufacturing, Acta Astronautica AA-D-23-01616R2
[4] Linne D L et al NASA Glenn Research Center 2014 Waste management options for long-duration space missions: When to reject, reuse, or recycle *AIAA SciTech* AIAA 2014-0497
[5] LLDPE INJECTED PARTS – 3devo Shredding and Extrusion Walkthrough
[6] Fakirov S, Krasteva B 2000 On the glass transition temperature of polyethylene as revealed by microhardness measurements *J. of Macromolecular Science* Part B **39(2)** pp 297–301
[7] PE-LD: Polyethylene low density - NETZSCH Measurements
[8] Polymers' Glass Transition Temperature (Tg) in the Context of Plastic Injection Molding – Boyicnc
[9] Saikrishnan S et al 2020 Thermo-mechanical degradation of polypropylene (PP) and low-density polyethylene (LDPE) blends exposed to simulated recycling *Polymer Degradation and Stability* 182 109390
[10] Oxidation Induction Time Measurement by DSC – Hitachi High-Tech Science Corporation

IOP Conf. Series: Materials Science and Engineering **1328** (2025) 012019 doi:10.1088/1757-899X/1328/1/012019

Accelerated ageing of tapes for spacecraft structures: a comparison of rubber and silicone-based Pressure Sensitive Adhesives tapes (PSA)

Théo Henry[1], Bahar Kilitoglu[2], Małgorzata Hołyńska[1*]
[1] European Space Agency (ESA), European Space Research and Technology Centre (ESTEC), Keplerlaan 1, Postbus 299, 2200 AG Noordwijk, The Netherlands
[2]Airbus Defence and Space, 88039 Friedrichshafen, Germany
*E-mail: Malgorzata.holynska@esa.int

Abstract. Pressure Sensitive Adhesives (PSA) also commonly named as "tapes" are of common use, either by being applied definitively on spacecraft structures, or temporarily, for instance to ease the assembly and integration processes (e.g. fixing, masking, marking). Benefiting of their ease of manipulation and versatility, the tapes are used in many situations. However, PSAs can easily leave residues and contaminate surfaces directly or by cross-effects. In addition, in most of the cases, PSAs are suffering from a poor ageing resistance and can quickly evolve in time. To reduce these risks of stability, silicone-based tapes are considered in many situations. Unfortunately, silicone is a type of contamination source that is undesired because it is difficult to mitigate. In the scope of this work, commercial rubber-based and silicone-based tapes were compared at regular intervals during more than 130 days. Accelerated ageing experiments were also performed by varying the temperature parameter and the combination of temperature and humidity. The peel forces at 180 ° were measured, the samples inspected, and the contamination levels investigated. The adhesive layer resistance upon contact with solvent was also tested. Results show that silicone-based PSA adhesives require a higher force to be peeled-off, have a higher environmental stability than rubber-based PSAs and leave less residues on aluminium substrates after a certain period. On the other hand, rubber-based PSA may leave fewer residues during the first days of applications in comparison to silicone-based, besides the contamination layer is easier to identify and clean while its adhesive layer is more resistant.

1. Introduction

1.1. Generalities on pressure sensitive adhesives

According to the Pressure Sensitive Tape Council [1], a pressure sensitive adhesive tape (PSA) is defined as followed: "Pressure sensitive adhesive (PSA) tape can be defined as a continuous flexible strip of cloth, paper, metal or plastic coated on one or both sides with a permanently tacky adhesive at room temperature which will adhere to a variety of surfaces with light pressure (finger pressure) with no phase change (liquid to solid) and usually on a roll. PSAs can be blends of natural or synthetic rubber and resin, acrylic, silicone, or other polymer systems, with or without additives.". PSAs typically stick to a surface upon contact without any chemical reaction

[2]. The PSA can be single coated, double coated, reinforced, or unsupported (adhesive transfer tapes). For a single coated tape, it is composed of three parts, the adhesive, the carrier (also named backing) and the release liner. The rear side of the backing can also be coated, and a primer can be present at the interface between the backing and the adhesive. There are three main types of adhesives: the rubbers, the acrylics, and the silicones. The rubbers are usually used for indoor applications with low stress. They have a good removability and adhere well to low surface energy substrates. Because rubbers are not tacky by nature, resin additive are used to create the adhesive. Rubbers are often used for masking tapes, sealing tapes, duct tapes... Rubbers can be divided into natural and synthetic rubbers. They are known to oxidize upon time [3] with an induction time followed by an abrupt change of color and adhesive consistency (e.g. sticky, oily). For extended ageing, the rubbers tend to crosslink, become harder or more brittle up to, sometimes, the impossibility to remove the tape [3]. Nowadays a crosslink step is usually performed during their manufacturing process [2]. The acrylics are used both for indoor and outdoor applications for long-term bonding. Tackifier are often added to improve the quick-stick properties. They exhibit high intrinsic adhesive properties, have a low glass transition temperature, anti-ageing characteristics and are resistant to oxidation [4] and water. Finally, silicone adhesives are mainly used for critical application requirement high stability (UV, moisture, chemicals, temperature...). They adhere well to low surface energy surfaces (non-polar) to which many acrylics will not or poorly bond [4]. Note that some hybrid adhesives can also be developed for improved and specific performances [4]. The characteristics of the adhesive are highly linked to their rheological properties and therefore their chemistry, molecular weight, and crosslinking; with a strong coupling between the geometry and mechanical properties [2, 5-6]. Often, the exact formulation of the adhesive is proprietary of the supplier and tackifiers or other additives can be included to tailor the thermal and rheological properties of the adhesive. Generally, PSAs are developed for a dedicated application. The backing can be made of many materials depending on the application, most of the time with polymers (polyesters, polyethylene, polyvinyl, polytetrafluoroethylene), woven cloth or metallic foils (e.g. aluminum, copper).

When a PSA is applied on a surface, three characteristics can be distinguished. The adhesion which is the addition of all forces at the interface between the adhesive and the substrate, the cohesion, which can be seen as the inner strength of the adhesive, or its elastic behavior and the adhesive strength, which is the combination of adhesion and cohesion. They are function of many variables, being mainly:

- The time: full bond strength between the adhesive and the substrate can take several days. Often, a duration of 72 h is considered as a baseline to reach full bond strength;

- The adhesive: with the main factors being the type of adhesive, its thickness, hardness etc. A soft adhesive will flow easily on irregular surfaces but will have a limited resistance while a hard adhesive will have a lower tack but a higher stability;

- The pressure: the more pressure is applied, the better a tape bonds, due to a higher wet-out coverage;

- The environment: the conditions such as the atmosphere, the temperature, humidity, the light exposure;

- The substrate: the bonding will depend on the surface state (topology), its shape, its cleanliness, its surface energy etc.

According to Creton [2], the mechanism of the debonding of a PSA starts with a cavitation event, then the growth of the cavitations and possibly the formation of fibrils. The cavity growth is mainly governed by the elastic modulus of the adhesive and can be followed by a coalescence process; or in absence of coalescence of the cavities, the formation of fibrils. When fibrils are formed, the elongation of the fibrils and the detachment is related to the elongational properties of the adhesive. Besides, the ability of a PSA to stick to a surface depends to the capability to make bridging fibrils during the pulling. To obtain fibrils, the specimen needs to have a resistance to fracture by crack propagation.

The failure of adhesion can be two types: an adhesive failure which occurs at the interface between the adhesive and the substrate or a cohesive failure which occurs within the center of the adhesive. Qualitatively, the type of failure is tightly linked to the viscoelastic properties of the adhesive [6]. As a general trend, at low peeling rates, the deformation by flow of the adhesive governs the peel behavior and leads to a cohesive failure. On the opposite, at high peeling rates, there is little or no viscous deformation of the adhesive, and the peeling force is independent of the rate leading to an adhesive failure. In between, the failure mode is instable [7]. The peeling force and failure type are also eventually dependent on the molecular weight of the adhesive, the thicknesses of the tape and the backing and the angle of peeling. As an example, for the experimentations and simulations of Mohammed et al. [8] in the case of a cohesive failure, the peel force decreases when the peeling angle is increased, increases with the peel speed, increases with the adhesive thickness, is mainly dependent on the viscoelastic / elastic property of the adhesive, not the backing, tends to decrease when the elastic modulus of the adhesive increases. The minimum peel force is typically reported with a peeling angle around 135º [7, 9].

Despite the complexity of the chemistry and the physics behind PSAs, they are now in the everyday life with hundreds of references from DIY to super specialized applications in every industry. This diversity is explained by the flexibility offered by the PSAs and the infinite tailoring possibilities, on top of the flexibility offered by these products from an operational point of view. The space industry is however a peculiar case where the selection of PSAs becomes more and more stringent.

1.2. Pressure sensitive tapes in the space industry

When applied to space applications, two situations of PSA use can be distinguished and will have different requirements: the tapes applied on a surface for a permanent use: they are not supposed to be removed at any time after their application, and the tapes for a temporary use: they are applied for a limited period. In the first case, the tapes will have to comply with all the flight requirements of the system, including having acceptable outgassing characteristics and a good resistance to the seen environment. In the second case, the tapes are applied temporary, on the

IOP Conf. Series: Materials Science and Engineering **1328** (2025) 012019 doi:10.1088/1757-899X/1328/1/012019

fight hardware, the ground support equipment or even possibly on the room surrounding (e.g. floor); typically during the manufacturing of parts, the assembly and the integration. Therefore, their use must be compatible with the flight hardware during the whole manufacturing, assembly and integration of the spacecraft. As these PSA aimed to be removed and the flight hardware must remain as clean as possible, particular attention is paid to the risk of residues. Residues are typically a part of the adhesive layer which is transferred unintentionally from the PSA adhesive layer to the surface initially covered by the tape. At a further extend, the risk of transferring adhesive residues by manual operation should also be minimized to limit the risks of contamination. These tapes should also be resistant to solvents to limit the risks of cross-contamination if some are used, and the residues should also ideally be cleanable with soft solvent to have a curative solution to face anomalies. Ultimately, these undesired residues might degrade the optical properties of the elements directly by reducing the overall optical properties of an optical element, or indirectly by outgassing during test or in-operation and contaminate other surfaces. Unfortunately, these specific aspects are often not evaluated by tape manufacturers as space requirements are high and specifics (for some applications, the surface cannot be clean, and few nanometers of contamination is already beyond the requirements), and the demand in comparison to other markets is low (e.g. automotive). In addition, the use of temporary tapes for relatively long periods of time (several months) is uncommon in other application but not rare in space projects. For instance, when tapes are used for vacuum bagging or masking, they are typically removed after few days' maximum. These aspects might also become of importance as some project might include long-term storage (storage of several years of flight hardware).

To illustrate these aspects, the presence of residues, the interaction of the tapes with surfaces (stains, discoloration), the delamination of the tapes during their removal or the inability to clean the residues are issues encountered in different projects, contexts and with different tapes and surfaces. Unfortunately, these problems are not well documented in the scientific literature, probably because of the very specific scope, and the quantities of residues which are of interest for space applications are very low compared to other applications. It is also worth to mention that these topics are also of interest for NASA according to internship reports [10-11].

Considering this scope, two tapes were selected to investigate their behavior on aluminum substrates as a baseline. One silicone-based and one rubber-based adhesive tape were selected. This choice was motivated by the fact that silicone-based adhesive is known to have a good stability and resistance to ageing, and the rubber-based adhesive as a possible replacement of the silicone-based one. Indeed, silicone contamination has the drawback to be difficult to clean and have severe yellowing ageing when exposed to space-environment. The main objective of this work is to evaluate and compare their behavior when they are left on a surface several weeks in a cleanroom environment, and to qualitatively identify the main ageing variable affecting the tape viability by accelerated ageing. To evaluate their characteristics, this investigation relies on peeling experimentations and inspections with a specific focus on the residues left on the surface. Additional tests are performed to test the response of the tapes when facing different solvents.

IOP Conf. Series: Materials Science and Engineering **1328** (2025) 012019 doi:10.1088/1757-899X/1328/1/012019

2. Experimental

2.1. Materials

The tapes used for this study are manufactured by AirTech and known under the commercial names Flashbreaker 2 and Flashbreaker 2R. They will be named in this document FB2 (the "blue tape") and FB2R (the "green tape") for convenience. Their characteristics reported by the manufacturer are summarized in Table 1. They have been delivered in April 2023. An old Flashbreaker 2 roll from 2008 was also retrieved for comparison. This old tape roll has no storage traceability but is assumed to have been stored in laboratory cabinet in air. The blue tape is, according to the manufacturer, be used for: *"[...] multi-purpose functions in metal bond and composites, such as holding down vacuum bagging materials, thermocouple wires, masking-off, chem-milling protection. It offers further protection in chem-milling or acid bath etching. Flashbreaker® tapes remove easily after the cure.".* The green tape is according to the manufacturer, *"[...] holds down vacuum bagging materials and thermocouple wires on metal or composites parts and tooling. This tape is designed for use in sensitive areas where silicone adhesives are not permitted.".*

The substrates for peeling tests were squares of 50 x 50 mm with a thickness of 2 mm of 5083's aluminum series (contains ~ 4 % to 5 % of Mg). Prior to taping the samples, they were clean on both sides with hexane by rubbing a wet iso-wipe and applying pressure. It has been renewed two times. A last cleaning was performed with acetone. Note that a good cleaning was found essential to have reliable results. The samples were not reused, as preliminary experiments suggested that the history of the sample might affect the results. Hence more than 130 different plates were used for this investigation.

The tapes were applied manually in the middle of the samples on the smoothest face having an Ra roughness about 0.4 μm. PSAs were disposed always in the orientation of the grain's lamination. When placing the tapes, cautions were taken to avoid gaps or air bubbles and the tape was accompanied by a finger with a slight and gently repeatable pressure, unquantified.

For the contamination quantification clean CaF_2 crystals were used. The samples were applied manually by pressing the thumb on the crystal to avoid the tape to unstick due to the crystal frame.

2.2. Ageing set-ups

The non-accelerated ageing was performed by leaving the samples in the TEC-QEE laboratories for variables periods. The nominal conditions of the laboratory are 22 ± 3 °C and 55 ± 10 %RH. The laboratory illumination has an approximate flux of 0.15 $mW \cdot cm^{-2}$ procured by fluorescent-like tubes.

The temperature ageing was performed in an Heraeus T 6060 oven at 50 °C for 168 h (1 week). Prior to the ageing, the samples were left in a cleanroom for 7 days after having been taped to ensure the adhesion of the tape on the substrate reached stability.

IOP Conf. Series: Materials Science and Engineering **1328** (2025) 012019 doi:10.1088/1757-899X/1328/1/012019

The damp heat ageing was performed in a VCS3 7027-10 chamber from Vötsch Industrietechnik. The test was performed at 50 °C and 70 ± 3 %RH for 168 h according to the ECSS-Q-ST-70-16C. Note that this approach differs from the ASTM D3611 as the suggested ageing conditions were considered as potentially overstressing. According to the datasheet of the chamber the temperature homogeneity in space goes from ± 0.5 °C to ± 2.0 °C and a deviation temperature from ± 0.3 °C to ± 0.5 °C. Prior to the ageing, the samples were left in a cleanroom for 7 days after having been taped to ensure the adhesion of the tape on the substrate reached stability.

The light ageing was performed by a LED lamp composed of 40 LEDs of SMD type having a power of 3 W. A polycarbonate plate acting as lenses covers the front face. The LED are bright white with a cooler temperature of 6000-6500 K. They do not emit any UV and are composed of a sharp intense peak center around 450 nm, a broad peak between 500 nm and 700 nm. To avoid a local heating of the tapes, the ageing was performed in a ClimEvent C/270/70/15/M chamber from Weiss Technik set at 22 °C and 55% RH. According to the datasheet for climatic test the temperature homogeneity in space inside the chamber goes from ± 0.5 °C to ± 1.0 °C, a deviation temperature from ± 0.1 °C to ± 0.3 °C and the humidity constancy in time from ± 1 %RH to ± 3 %RH. The illumination was performed for 168 h. Prior to the ageing, the samples were left in a cleanroom for 7 days after having been taped to ensure the adhesion of the tape on the substrate reached stability. The irradiance was in average of 13 $mW{\cdot}cm^{-2}$.

2.3. Solvent ageing

The first approach consisted of deposition a small drop of a solvent onto the adhesive layer of the tape with a glass stick and wait until its complete evaporation prior to inspection.

The second approach aimed to fully immerse the tape into the solvent for 20 minutes. After removal, the tape is scratch with a spatula, and then with a wipe. The PSA are then inspected.

The third approach aimed to evaluate the solvent capability to clean the residues. Two approaches were considered on residues left on aluminum samples after damp heat ageing. The first approach is to deposit a small drop of solvent on the residue and leave it evaporates. The second approach was to soak a cotton-bud of the solvent and make 10 strokes from the inner towards the outer side of the samples.

These methods are comparative and have been followed with isopropyl alcohol (IPA), acetone, methyl ethyl ketone (MEK), hexane and toluene.

2.4. Measurements

Optical inspections were performed with a Keyence VHX-6000 numerical microscope with a VH-Z20UR/W/T objective.

Thermo-Gravimetric Analyses (TGA) were carried out using a Mettler-Toledo TGA/SDTA 851e/LF1600 from 25 °C to 1000 °C at 10 $°C{\cdot}min^{-1}$ under 50 $mL{\cdot}min^{-1}$ N_2 or air gas flow with alumina crucibles of 70 μL.

IOP Conf. Series: Materials Science and Engineering **1328** (2025) 012019 doi:10.1088/1757-899X/1328/1/012019

Modulated Differential Scanning Calorimetry (MDSC) was conducted on a DSC2500 device from TA Instruments from 20 °C to 300 °C at 3 °C min^{-1} and a modulation of 1 °C over a period of 1 minute with a N_2 gas flow of 50 mL·min^{-1}. Pierced "Tzero Aluminium Hermetic" pans were used for the MDSC with a typical sample mass ~ 5 mg.

Fourier Transform InfraRed (FTIR) spectroscopy measurements were performed on a Bruker Optics VERTEX70v vacuum FTIR device. For the tapes, the samples measurements were carried out in ATR (Attenuated Total Reflectance) mode with a germanium crystal, a KBr beam splitter and a resolution of 4 cm 1 for 64 scans. Measurements on CaF_2 samples were performed in transmission on the same FTIR instrument with the same parameters.

Raman spectra were acquired with a WITec's Raman microscope alpha300R using a 785 nm laser excitation. Measurements were performed in the confocal mode with an objective Zeiss LD EC Epiplan-Neofluar HD Dic 100x / 0.75 between 200 cm^{-1} and 3000 cm^{-1} with a resolution ranging from 2 cm^{-1} for lowest wavenumbers to 1 cm^{-1} for highest wavenumbers. The integration time was 0.5 s for 100 repetitions a power levels of ~ 5 mW.

Peeling experiments were conducted on a LR5K Plus from Lloyd Instruments equipped with a load cell of 50 N, grade 0.5 %. The peelings were performed at 180 ° and at a speed of 5 mm·s^{-1} (as per ASTM D3330). The force was recorded all over the peeling until the tape detached. The averaged peeling force was determined as the mean of the recorded force between ~ 4 s and ~ 16 s of the peeling experiment.

The contamination quantification was performed by FTIR with respect to calibrated data, according to the ECSS-Q-ST-70-05C entitled "Detection of organic contamination of surfaces by infrared spectroscopy".

An Accurion EP4 imageing ellipsometer from Park System have been used with a Nanochromat_7x objective (analysis surface ~ 600 µm x 600 µm with 2.5 µm resolution) was used with a "RCE" method at 42 ° between 300 nm and 800 nm. The considered model was an aluminum substrate with a native oxide of Al2O3 and the contamination layer was modelled by a Cauchy models with no extinction coefficient to simplify the approach.

3. Results and discussions

3.1. Pristine characteristics

The general characteristics of the tapes are reminded in the Table 1 here below. Both tapes have the same polyester based carrier with the same thickness, but their adhesive layers differ by their nature and their thickness. The FB2 is silicone-based and have a thickness of 33 µm while the FB2R is rubber-based and have a thickness of 50 µm.

Table 1. Summary of the characteristics of the flashbreaker 2 and 2R PSA.

Characteristics	FB2	FB2R
Colour	Blue	Green
Width	1 inch	1 inch
Carrier	Polyester (50 μm)	Polyester (50 μm)
Adhesive	Silicone (33 μm)	Rubber (50 μm)
T_{max}	204 °C	204 °C
TGA; 1 %wt loss	310 °C	190 °C
TGA; 5 %wt loss	340 °C	395 °C
DSC; T_{amb}->300°C	-	Irreversible exothermic peak at [130-180°C]
Pigments (Raman) **The metallic atom might differ*	Copper phthalocyanine (CuPc)	Chlorinated copper phthalocyanine

The TGA of the FB2R and FB2 tapes were performed under air and N_2. The results are shown in Figure 1. Both tapes have a main degradation around 400 °C, starting around ~ 350 °C and ~ 390 °C for the FB2R and FB2 respectively. The main difference happens from ~ 130 °C where the mass starts to decrease before the degradation event. This suggest FB2R has a lower thermal stability than FB2. At 350 °C the FB2R lost ~ 7 % of its mass against ~ 2 % for the FB2 tape. This event happens before the recommended maximum temperature referred in the technical datasheet of the product. Under air, a new degradation process is present between typically 450 °C and 550 °C. The Table 1 reports some characteristics values. From DSC measurements (not shown) and apart from the polyester carrier features (T_g ~ 75 °C and T_{melt} ~ 260 °C), no particular feature is observed in the measured range for FB2 while FB2R shows an irreversible exothermic peak between 130 C and 180 °C. This is related to the adhesive layer and is likely to relate with the observed mass loss but TGA, for instance due to a crosslinking event. Therefore, the FB2R has a lower thermal stability than the FB2 tape. FTIR measurements (not shown) confirmed the nature of the carrier, and the adhesives advanced by the manufacturer as mentioned in the Table 1. From Raman analyses, a major part of the spectra obtained would fit with the tapes pigments and were attributed to phthalocyanine and chlorinated phthalocyanine for FB2 and FB2R respectively, the complex metal being unknown (Cu, Zn ...).

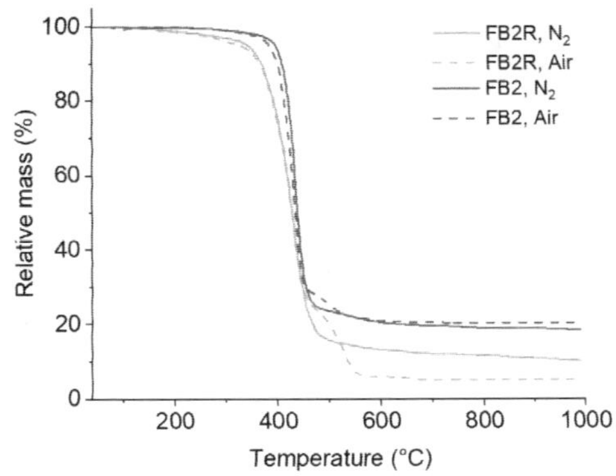

Figure 1. TGA measurements of the rubber-based (green lines) and silicone-based (blue lines) adhesive tapes in N_2 atmosphere (full lines) and air atmosphere (dash lines). Credits: ESA

3.2. Aging of tapes

3.2.1. Non-accelerated ageing and old tape

FB2 and FB2R tapes were stored in cleanroom conditions for several durations. The resulting peeling forces are shown in Figure 2. The FB2 tape has a starting peel force (t_0) higher than the FB2R suggesting a higher tack. This is coherent with the experience while operating the tapes. With time, the peeling force quickly evolve in a log10 or exponential trend during the first days. For FB2, it stabilizes after approximately one week around ~ 0.6 N·mm-1. The FB2R behaves differently with a peeling force increasing in log shape at least until ~ 100 days and reach ~ 0.55

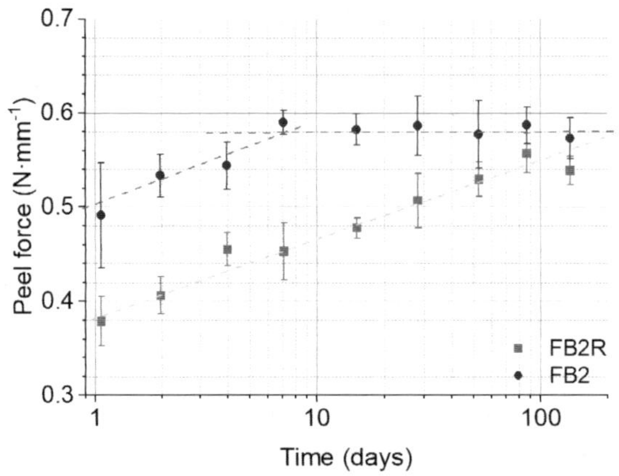

Figure 2. Evolution of the peeling force for non-accelerated ageing of FB2 and FB2R. The error bars correspond to the standard deviation of five samples. The dash lines are displayed as guidelines. Credits: ESA

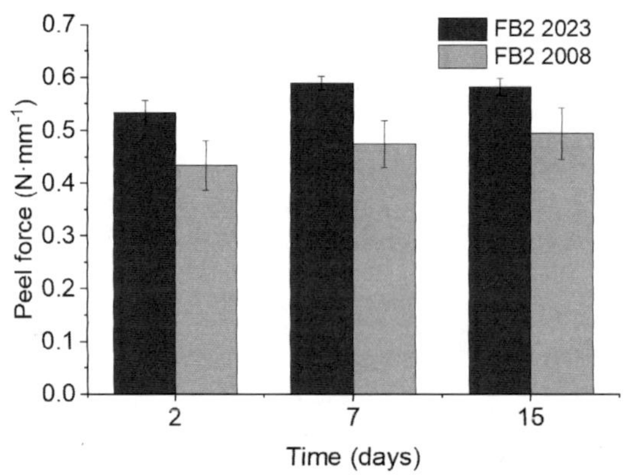

Figure 3. Averaged peel force for non-expired (2023) and expired (2008) FB2 tapes after 15 days of application. The error bars correspond to the standard deviation of five samples. Credits: ESA

N·mm^{-1}. Hence, the silicone-based adhesive sticks stronger and faster than the FB2R tapes according to the applied procedure.

When these data are compared to an old tape from 2008 after 2 days, 7 days and 15 days after applications for the FB2 tape (Figure 3), a drop about 20 % of the peeling force is present for the old tape compared to the fresh one. The peeling force along the experiment is also less homogeneous, which is represented by larger root mean square (error bars in Figure 3). This suggest that the tapes tend to be altered with time during storage over long periods. This could also be seen on the adhesive layer by inspection with local parts of the adhesive layer shrinking when unrolling the tape.

3.2.2. Accelerated ageing

The Figure 4 shows the peeling forces measured after the accelerated ageing experiments (1 week of stabilization under cleanroom conditions and 1 week of ageing) in comparison to the non-accelerated values (after 2 weeks). Based on these values, the averages and the standard deviations are also reported in Figure 5.

For the FB2 tapes, no obvious ageing exists, even though a slight increase of the peeling force is observed, particularly after the damp heat test. It however remains within the standard deviation. For the FB2R tapes, while the performed white light exposure do not affect the adhesion of these tapes, the heat and damp heat exposure show a clear increase of the peeling from approximately 0.48 $N{\cdot}mm^{-1}$ to 0.60 $N{\cdot}mm^{-1}$. The heat is therefore a parameter which tends to age the rubber based adhesive layer. Overall, the tape with rubber based adhesive layer is less stable with environmental factors compared to the tape with silicon based adhesive layer. It is notably affected by heat and damp heat ageing. The tape with the silicone-based adhesive layer might be unsensitive to the ageing, except eventually by the damp heat.

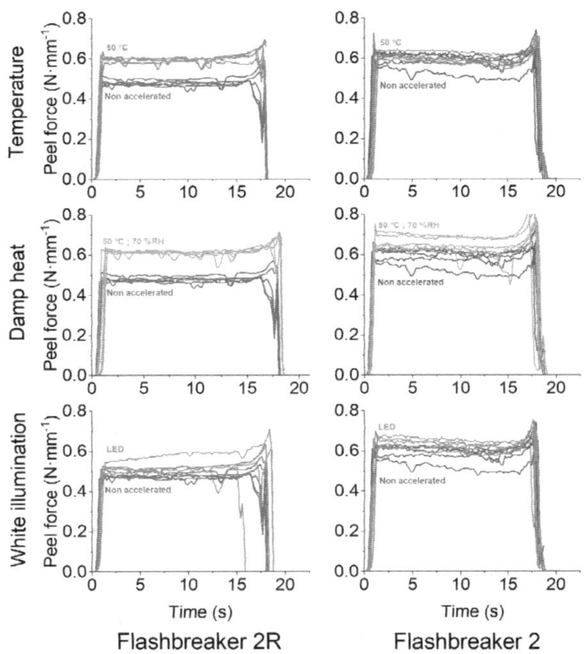

Figure 4. Results of the peeling experiments comparing the tapes after the accelerated ageing procedures and after 15 days of non-accelerated ageing of FB2R (left) and FB2 (right) tapes. Credits: ESA

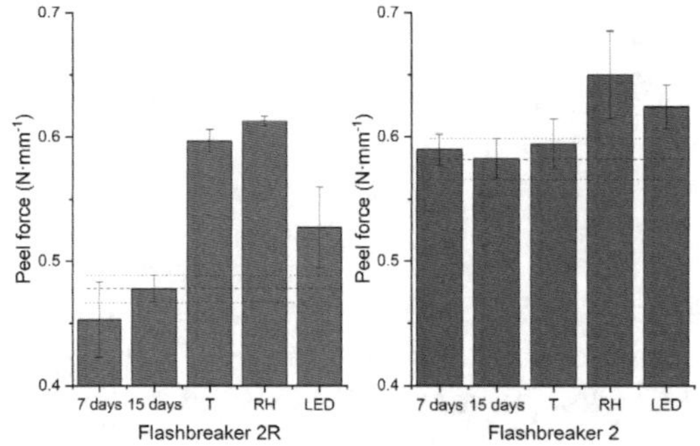

Figure 5. Averaged peel force after non-accelerated ageing (7 days and 15 days) compared with accelerated ageing procedures (T for temperature ageing, RH for damp heat ageing, and LED for white illumination ageing). The error bars correspond to the standard deviation of five samples. Credits: ESA

3.3. Contamination

After the peeling experiments, the aluminium substrates were imaged under partial light permitting to enhance the visualization of the contamination residues from the tapes. Results are summarized in Figure 6 with a sample representative of the trends for each batch of conditions. Note that the square pattern on the samples is due to the stitching overlaps and is an artefact produce by the image assembly reconstruction. According to these results, the residues layers get more and more visible with time after the FB2R tape applications. The residue layer is notably affected by heat ageing and hugely impacted by the damp heat effect, while the LED ageing does not seem to affect much the residues deposition on the substrate. Regarding FB2 tapes, no obvious residue is observed, except slight features after damp heat ageing. Note that is does not mean there is not residue, as it might be invisible by naked eyes. These results are coherent with the peeling experiments observations. They confirm the critical effect of the damp heat ageing on the rubber-based adhesive tape. The contamination deposition was estimated by the method proposed in the ECSS-Q-ST-70-05C by the direct method after applying the tapes on CaF2 substrates by applying strong pressure with a thumb. The results after 0, 1, 7, 15 days, heat and damp heat ageing are shown in Figure 7. The results shows that the contamination deposit increases with time and by heat and damp heat ageing for the FB2R. For FB2 tapes, the contamination residue on CaF_2 increases during the two first days and is then stabilized without being affected by accelerated ageing. These results are qualitatively coherent with peeling experiments and visual inspections. However, it cannot be compared quantitatively to peel test as the substrates and the application methods of the tapes are different. Chemically and based on FTIR spectra, no clear difference was observed between the residues and the bulk adhesive (not shown) for both tapes. This suggest that if chemical changes happen, they are only minor or superficial. Interestingly, after an instant deposition of the tape and peeling, the left residues were higher for the silicone based adhesive layered tape compared to the rubber based adhesive layered tape. This is also coherent with the high peeling force observed at t0 for the FB2 tapes. From an experimental point of view, this suggest that if an operator is prompt to tape and instantaneously remove the tape from a surface (e.g. wrong positioning), the risk of leaving residues is higher with the FB2 tape than the FB2R. Note that the quantitative values should only be considered in relative as the deposits were not very homogeneous due to the application method and the peeling manual.

IOP Conf. Series: Materials Science and Engineering **1328** (2025) 012019 doi:10.1088/1757-899X/1328/1/012019

Table 2. Typical thicknesses of the residues left after peeling FB2 and FB2R tapes from the aluminium substrates determined by ellipsometry.

Characteristics	FB2	FB2R
Real ageing	~ 5-10 nm	~ 5-10 nm
Temperature; 50 °C	~ 10 nm	~ 20 nm
Damp heat; 50 °C & 70 %RH	~ 25 nm	~ 65 nm

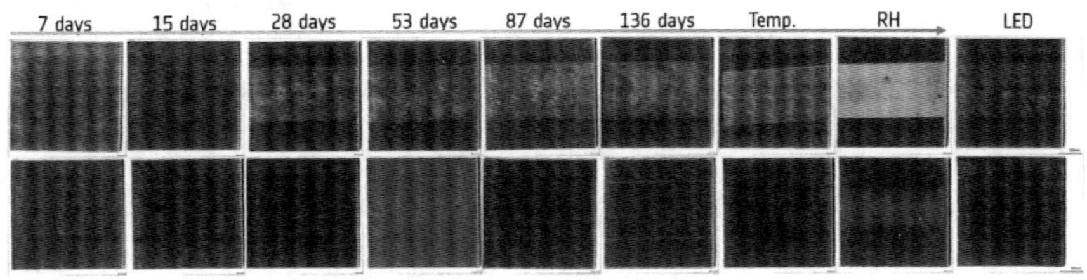

Figure 6. Inspection by partial light of aluminium samples after peeling off the FB2R (top row) and FB2 (bottom row) tapes. The square pattern is due to the stitching image composition. Durations expressed in days refer to non-accelerated ageing, Temp. to the temperature ageing, RH to the damp heat ageing and LED to the white illumination ageing. Credits: ESA

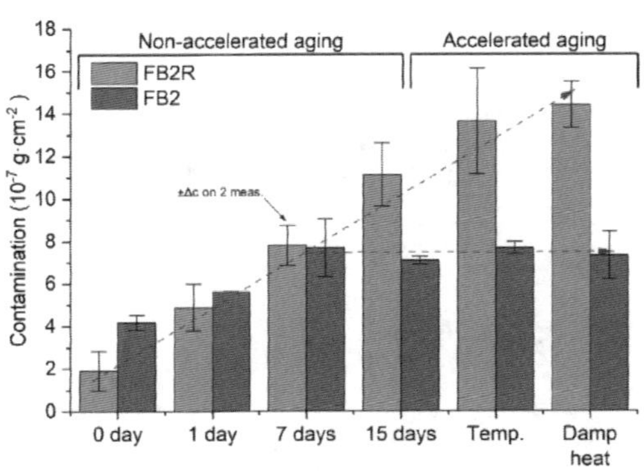

Figure 7. Contamination from residues deposited on CaF_2 windows determined by FTIR after different no-accelerated ageing durations and after temperature and damp heat accelerated ageing. Dash lines are displayed as guidelines. Credits: ESA

To have a better idea of the quantity of the left residues and aluminum samples, ellipsometry measurements were performed on the samples. A summary is reported in Table 2. According to these results, it seems that even though no residues can be seen for the FB2 tape, by naked eyes, a small layer of contamination is left on the surface, about 5-10 nm. It barely increases after heat treatment while the effect of damp heat is more obvious. This is also coherent with visual inspection shown above. For FB2R, the deposit is more important, particularly after ageing. For both tapes, under real ageing up to 100 days, no trend or residue build-up with time was observed.

It is assumed it is due to experimental conditions which are not optimal or precise enough to observe it.
The effect of the FB2R residues on the substrates can be observed in Figure 8 as seen by naked eyes. It can easily be identified after the damp heat ageing, seen without many difficulties after the heat ageing, and very difficult to see without accelerated ageing. Note that the bluish color of the residue after ageing is assumed to mainly coming from constructive interferences.

3.4. Solvent ageing

The Figure 9 reports the inspection of the adhesive layers of the tapes after having deposited a drop of the solvent and left it evaporates. On the FB2, all the solvents are damageing the adhesive layer, but in a different manner. IPA, acetone and MEK lead to similar features with a surface having wrinkles. Hexane and toluene induce deeper damages in depth of the adhesive layer. The damaged surfaces have a wavy pattern.
One hypothesis to explain these topological changes could be that IPA acetone and MEK slightly dissolve and fluidify the adhesive layer. When the solvent evaporates, the adhesive at the border of the evaporated solvent tends to get more cohesive and to condense on localized sites leading to a wrinkling pattern. For the hexane and the toluene, the adhesive layer might be dissolved completely. On contrary, the FB2R seems a lot more resistant to solvents. For IPA and acetone, only the contour of the evaporated drop can be distinguished, with difficulties. For MEK, the damages are more pronounced, and the location of the drop gets more evident. The effects of hexane and toluene are even more important for the FB2R adhesive layer with a notable change of the topology. As a global picture, the FB2 adhesive layer is more sensitive to solvents compared to FB2R. The biggest damages are both observed for hexane and toluene. The IPA and acetone do not to affect the FB2R adhesive layer with the considered experimentation procedure.

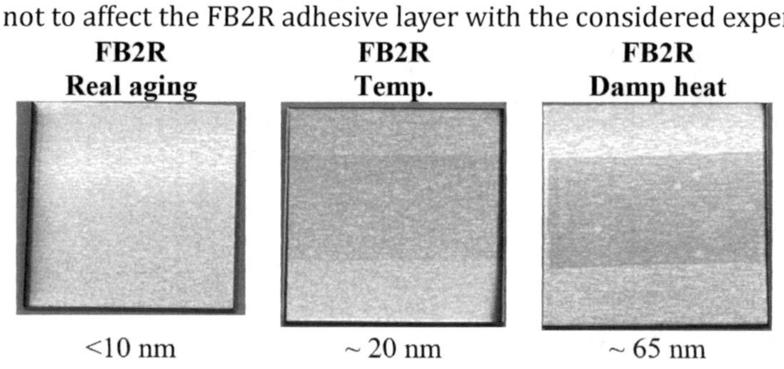

Figure 8. Observation of the residues of FB2R tape by naked eyes after 15 days of non-accelerated ageing (left), after temperature ageing (middle) and after damp heat ageing (right). Credits: ESA

The rubbing procedure was processed as explained in the section 2.3. results are displayed in Figure 10. On this figure the upper part of the tapes (above the yellow line) was rubbed with a wipe, and the lower part with a spatula. For the FB2, all solvent leave stains on the wipes after rubbing. In general, the adhesive remains attached to the backing. The damages were more pronounced in the order IPA < acetone < MEK < hexane ≈ toluene. However, when a spatula was used, IPA permitted to completely remove the adhesive layer while the acetone did not. The spatula experiment confirms that hexane and toluene have a higher effect of the FB2 adhesive. For the FB2R, only hexane and toluene leave stains on the wipes after rubber (MEK as well but at a very small level). IPA do not affect the tape. Acetone seems to lower the adhesion with the back because the adhesive layer did not sticked or physically changed and remain elastic but was easily detached from the backing. With MEK the removal of the backing was less obvious than for acetone. However, when using hexane or toluene, the FB2R adhesive completely dissolved and created a wet paste. Overall and according to these tests, it was generally observed that the FB2 adhesive had a better adhesion to the backing in comparison to FB2R but is more sensitive to "soft" solvents compared to FB2R. For both tapes, hexane and toluene have a drastic effect on the adhesive layer, particularly for FB2R. The solvent efficiency to clean the residues was also tested by depositing a drop and letting it evaporate and rubbing the surface with a wet cotton stick toward the exterior of the sample. The results are shown for the samples after being aged by damp heat and peeled in Figure 11. Unfortunately, as the residue of the FB2 is thin and almost invisible not conclusions can be drawn. For the FB2R. IPA wet very well on the contamination / substrate but the contamination is barely removed. Acetone is more efficient than IPA but does not completely remove the contaminant. MEK and hexane seems to be the most efficient with a clearer border between the cleaned and contaminated areas than with toluene (which could be related to wetting/evaporation properties). Toluene seems as efficient or a bit less than MEK and hexane. Hence, while the FB2R adhesive seems more resistant to solvents, its residues might still be affected by strong solvents. IPA is not efficient, and acetone only removed a partial part this the applied cleaning process.

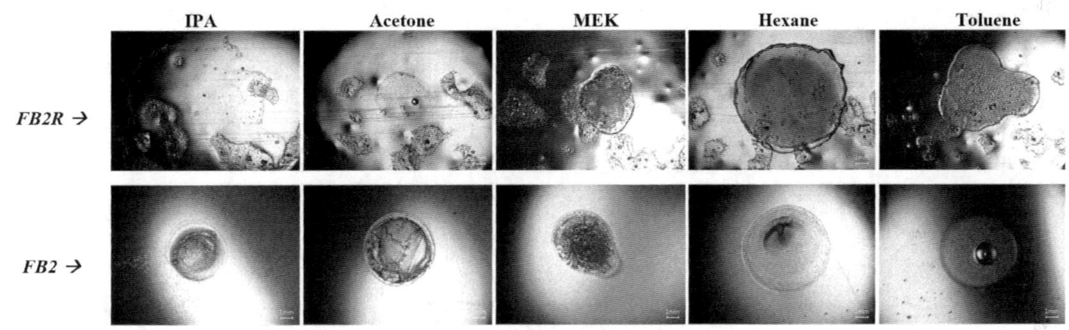

Figure 9. Micrographs of the damages induced by the drop evaporation of various solvents on the adhesive layers of the FB2R and FB2 tapes. The observations were done under coaxial illumination. Note that the adhesive topology of the FB2 tape is more homogeneous than the FB2R, which originally shows some craters and hills like embedded particles. Credits: ESA

4. Conclusions and recommendations

The FB2 and FB2R, respectively having silicone-based and rubber-based adhesive layers with a polyester backing were investigated. The nature of the tapes was confirmed. In addition, it was found out that the tapes contain organometallic complex dyes based on phthalocyanine molecule (potential metallic atom Cu, Zn ...). The inspection of the tapes suggests an adhesive layer with few inhomogeneous spots for FB2R while FB2 is more homogeneous (with an alteration for the old FB2). In term of thermal stability, the FB2R is less stable than FB2, with a thermal degradation starting around ~ 130 °C for FB2R against ~ 350 °C for FB2. At ~ 130 °C, for the FB2R, an exothermic peak is present (DSC) which might be related to a crosslinking phenomenon. It was also noticed that after accelerated ageing, the FB2R tape released some smells, which were absent for FB2 tapes which also goes in the directly of a lower thermal stability.

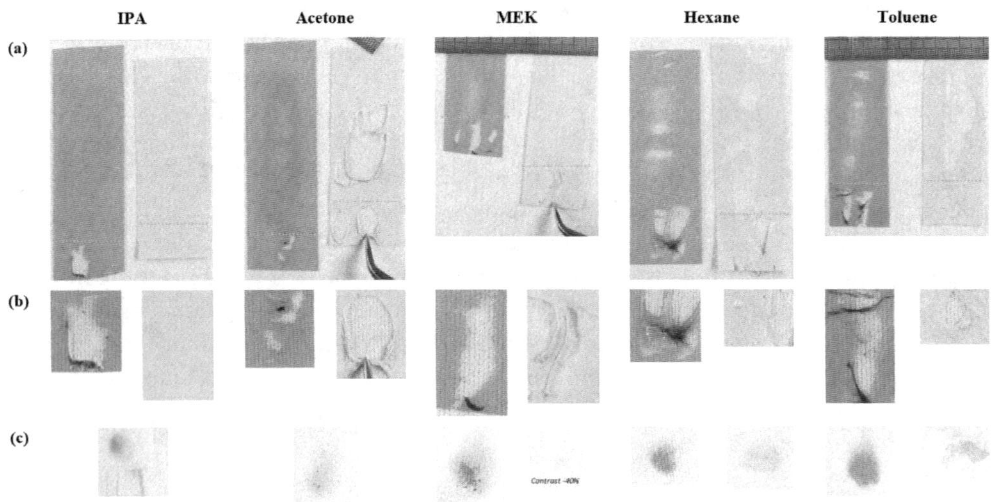

Figure 10. Overview of the damages induced by the rubbing procedure on the adhesive sides of the FB2 (blue) and FB2R (green) tapes. (a) Overview of the two methods considered, rubbed with the wipes above the yellow line and with a spatula below the yellow line; (b) magnification when rubbing with the spatula; (c) ISO wipe after having rubbed it against the adhesive side of the tapes. Credits: ESA

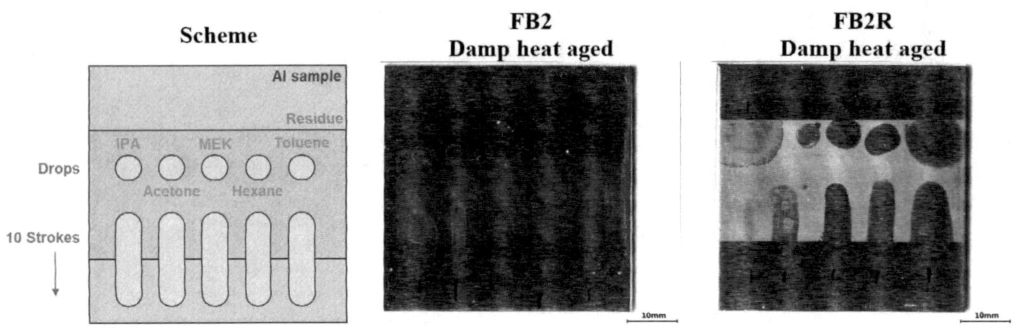

Figure 11. Contamination residues cleanability of the FB2 and FB2R tapes according to the third procedure introduced in the section 2.3. The scheme (left) maps the used solvents and applied processes. The FB2 (middle) and FB2R (right) residues cleanability is inspected by partial light. Credits: ESA

To follow the ageing of the tapes, peeling and contamination experiments were carried out. Under non-accelerated conditions, the old FB2 tape have a lower adhesion than the fresh FB2. The FB2 requires a higher force to be peeled compared to the FB2R. After approximately 7 days, the peeling force of FB2 stabilizes while it continues to increase for FB2R. When the samples are aged, FB2 (silicone-based adhesive) remains insensitive to accelerated ageing (or eventually a few to damp

IOP Conf. Series: Materials Science and Engineering **1328** (2025) 012019 doi:10.1088/1757-899X/1328/1/012019

heat). On the contrary, FB2R (rubber-based adhesive) is notably affected by the temperature and the damp heat ageing with an increase of its peeling force. Looking to the residue deposit, it increases the first days and then stabilizes for FB2 after 7 days and the accelerated ageing does not affect the level of contamination. For FB2R, the contamination and residues are considerably increased by the accelerated ageing conditions. Visually, similar conclusion can be drawn based on the inspection of the samples. FB2 residue is almost invisible while FB2R residue can be revealed by the proper characterization. The reason of FB2R sensitiveness to ageing is unclear since no obvious chemical changes are seen based on the characterization of the FTIR residues. The main hypotheses are a thermal oxidation, a crosslinking change and/or by moisture through adsorption and/or plasticization. Nonetheless, as the damp heat ageing has the most critical effect, the humidity is certainly the variable affecting the most the ageing, possibly in synergy the heat (e.g. diffusion, oxidation etc.).
When looking to the adhesive layer solvent sensitivity, both tapes are very sensitive to hexane and toluene. In addition, on one hand, FB2R is more resistant to softer solvents such as IPA, Acetone or MEK compared to FB2. On the other hand, the adhesive layer of FB2 seems to have a better adhesion of the backing compared to FB2R.
From an application point of view, it has been shown that the FB2R residues could be, in most cases, observed by naked eyes. This is not the case of the FB2 residues. Moreover, at some extend, FB2R leave quantitatively more residues than FB2 but remain difficult to clean by soft solvents. Hence, the selection strategy can be questioned: is it worth to have a tape leaving few residues but difficult to spot, or a tape leaving more residues but easier to identify, and therefore, most likely to be easier to clean? Ideally and with respect to the surrounding environment and constrains, the approach should be based on a compromise between the contamination sensitiveness and nature, its identification and its cleanability. If a PSA leaves more residues but it can be identified and cleaned, it can be worth than taking the risk of an invisible contaminant which cannot be seen and removed.
As a final word, tapes properties are a very complex, multidisciplinary topic, dependent on many factors such as tapes' characteristics, substrates' characteristics, the application and environmental conditions. Hence, it is highly recommended to test and qualify any new use case or modification of processes with PSAs as results might not be transferrable from one to another application case, substrate or product. The behavior of a tape on a substrate should be considered as a case-by-case work and in absence of evidence, it should be considered difficult to anticipate the affinity of a tape vs substrate (nature, roughness etc.) and the left residue. It is also believed that the effect of the tape application and the cleanliness of the surface should not be underestimated.

5. Disclaimers

This work shows an example of investigation with trendlines. It does not aim to qualify or precisely quantify properties nor promote or demote any product.
This research was supported by ESA/ESTEC (research fellowship). The views expressed herein can in no way be taken to reflect the official opinion of the European Space Agency (ESA) and are not intended to endorse particular technologies, companies, or products.

IOP Conf. Series: Materials Science and Engineering **1328** (2025) 012019 doi:10.1088/1757-899X/1328/1/012019

References

[1] https://pstc.org/why-tape/psa-tape-glossary/, consulted 08/10/2024
[2] Creton C 2003 *MRS bull.* **28** 434
[3] Smith M A *et al* 1984 *J. Am. Inst. for Conserv.* **23** 101
[4] Mapari S *et al* 2021 *Polym. Bull.* **78** 4075
[5] Sun S *et al* 2013 *Int. J. of Adh. and Adhes.* **41** 98
[6] Gibert F X *et al* 1999 *J. Adh. Sci. and Techn.* **13** 1029
[7] Aubrey D W et al 1969 *J. Appl. Polym. Sci.* **13** 2193
[8] Mohammed I K et al 2016 *J. Non-Newt. Fluid Mech.* **233** 85
[9] Chivers R A 2001 *Int. J. Adh. and Adhes.* **21** 381
[10] McFadden S J 2019 *Spacecraft Adhesive Residue On-Composite Analysis* (No. KSC-E-DAA-TN67208)
[11] Smiley A 2019 *The Analysis of Adhesive Residue Failure* (No. KSC-E-DAA-TN67212)

IOP Conf. Series: Materials Science and Engineering **1328** (2025) 012020 doi:10.1088/1757-899X/1328/1/012020

Euro Material Ageing – A European experiment on International Space Station for materials sciences research and technology development

S Perraud*, E Laurent, L Marelli, N Kerboub, D Faye and I Savin De Larclause

CNES, Toulouse, France

*E-mail: sophie.perraud@cnes.fr

Abstract. Low Earth orbit (LEO) space environment is very harsh for materials used in satellites. Before using in-flight, materials have to be validated on ground but the on-ground facilities cannot combine simultaneously all the space environment parameters so synergetic effects are not seen. Therefore, Euro Material Ageing (EMA) is a significant opportunity to expose materials samples in real space conditions to improve our understanding of their degradations and to collect environmental data thanks to environmental sensors to refine models and analysis.

EMA – a collaboration between CNES, CNRS and ESA – is made of two experiments:

- SESAME (Space Environment Study of Ageing of MatErials) for studying materials behavior in space environment and measuring in-orbit parameters such as atomic oxygen flux, contamination, radiation, micrometeoroids and debris population for a better knowledge of the LEO space environment and its effects.
- IR-Coaster (principal investigator of IR-Coaster is at LISA/CNRS) for characterizing the behavior of organic compounds when exposed to solar UV radiation by in situ infrared spectroscopy.

The synergy between the two experiments will be highlighted in this paper as well as the assembly, integration and testing performed on 2024.
EMA has been launched to the International Space Station (ISS) on November 2024 and will be exposed between 6 and 18 months (nominal duration 12 months) outside the International Space Station (ISS).

1. Introduction

When spacecraft progresses in Low Earth Orbit (LEO), it is exposed to severe environmental conditions including extremes temperatures, large thermal cycling, radiations (ultraviolet (UV), charged particle radiations, and x-ray), atomic oxygen (AO), micrometeoroids and debris.
This environmental exposure can degrade materials properties as thermo-optical and optical properties, physical properties, material integrity (erosion), etc...These damages are likely to reduce spacecraft's performance and durability.

For new materials, it is important to validate their behavior in a definite environment and to know their end-of-life properties before implementing them on satellites. The material validation is often assessed against one parameter of the environment after another or against the most

IOP Conf. Series: Materials Science and Engineering **1328** (2025) 012020 doi:10.1088/1757-899X/1328/1/012020

prominent one because on-ground technical facility cannot allow to combine simultaneously all the constraints of the environment.

That is why there is a significant science benefit to submit satellites materials samples to real in orbit space conditions to observe and analyze potential synergistic effects between the different environmental parameters.

To increase the knowledge of materials ageing in LEO environment, many materials experiments were installed on external platforms on the ISS until now. The Materials International Space Station Experiments (MISSE) from NASA have been regularly exposed since 2001 [1] and especially PEACE on MISSE 2 [2-3] dedicated to polymeric and composite materials and their Atomic Oxygen erosion. The MISSE series have tested some 4,000 material samples and specimens — from lubricants and paints to fabrics, container seals and solar cell technologies — to demonstrate their durability. Today, the 15th MISSE mission was exposed, by NASA. JAXA also developed different missions to study space debris, material-ageing mechanisms and observe surface reactions and degradations for example MPAC (Micro-Particles Capturer) and SEED (Space Environment Exposure Device) exposed on the Russian Service Module (SM) in 2001 and Material Degradation Monitor 2 (MDM2) in 2015-2016 [4-5]. In 2022, Chinese sent an experiment to expose 407 pieces of materials outside of their Tiangong Space Station (in Low Earth Orbit between 340 to 450 km) which will be returned to Earth by a spacecraft in 2024 for handover the researchers ground [6].

A French experiment was launched on-board ISS/Columbus named MEDET (Material Exposure and Degradation Experiment) from February 2008 to September 2009 [7]. Different types of materials were exposed like polymeric films, quartz, thermal control coatings and cover glass adhesives, and analyzed during the flight with a UV-Visible and Visible –near infrared (IR) spectrometer. Many post flight analyses from the different samples have been done like optical microscopy inspection, mass, thickness and thermo-optical measurements, transmission and reflectance Fourier Transform infrared (FT-IR), FT-IR-ATR (Attenuated Total Reflectance) spectrometry, surface roughness...

CNES newly developed a European experiment on ISS for materials space research: Euro Material Ageing (EMA) with the collaboration of ESA. EMA is composed of the two payloads, SESAME and IR-Coaster, linked together with a mechanical bracket and an electrical harness with four connectors, as shown on Figure 1.

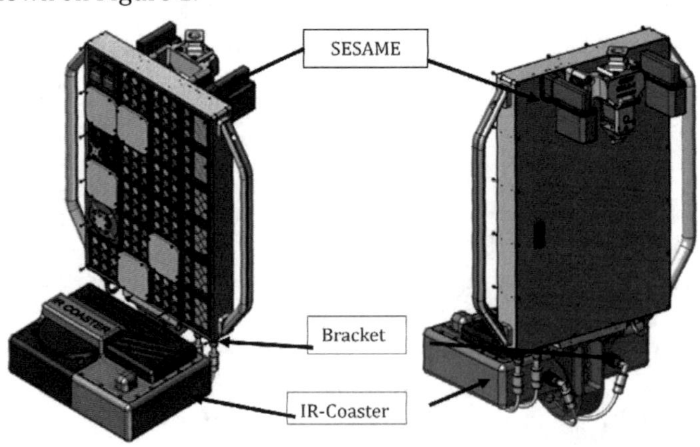

Figure 1. EMA payload ©CNES.

IOP Conf. Series: Materials Science and Engineering **1328** (2025) 012020 doi:10.1088/1757-899X/1328/1/012020

SESAME is developed for studying materials and coatings behavior in space environment and measuring in-orbit parameters such as atomic oxygen flux, contamination, radiation, micrometeoroids and debris for a better knowledge of the LEO space environment and its effects on materials.

IR-Coaster aims at characterizing the stability or evolution of biological and organic compounds when exposed to solar UV/VUV radiation.

This paper presents EMA facility focused on SESAME: design, integration, qualification and scientific objectives.

2. Description of the External Environment on International Space Station (ISS)

ISS operates at an altitude of 400 km with an orbital inclination of 51,6°.

EMA will be exposed on slot 4 (Figure 2) of the European Bartolomeo module, developed by Airbus Defence and Space in partnership with ESA, mounted on ESA's Columbus Module on ISS. The environment has been defined by the module supplier and EMA has to comply with the associated requirements. EMA will be exposed outside ISS on the RAM face, the duration of the mission is scheduled for 6 to 18 months, with a nominal duration of 12 months.

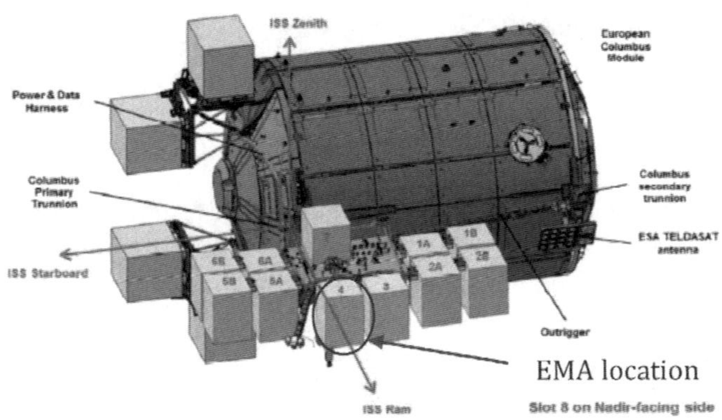

Figure 2. Bartolomeo module on Columbus on ISS. ©ADS

On-orbit minimum pressure environment is 3.6 x 10^{-8} Pa.

Atomic oxygen flux considered is 5.0 × 10^{21} atoms/cm^2/year. Micrometeoroids and orbital debris environment has to be taken into account for design and resistance purposes.

The solar electromagnetic energy flux near the Earth at one Astronomical Unit is about 1371 ± 5 W/m^2. The irradiance spectrum approximates a black body curve, T = 5762K which peaks in the visible frequency range. The integrated irradiance of the near ultraviolet portion of the electromagnetic flux at wavelengths 0,18 µm < λ < 0,4 µm is approximatively 118 W/m^2, ie about 8,6% of the solar constant. The far ultraviolet portion at λ < 0,18 µm contributes about 0,023 W/m^2.

On-orbit solar electromagnetic radiations are detailed in Table 1:

Table 1. Solar electromagnetic radiations

Type	Wavelength (nm)	Level (W/m²)
UV	100-150	7.5E-3
EUV	10-100	2E-3
X-rays	1-10	5E-5
Flare X-rays	0.1-1	1E-4

The specified qualification temperature range for EMA is -40°C to +60°C, but thanks to its mass and thermal control system, it is expected temperature excursions to be limited to the range -10°C to +30°C during operations. Samples closely follow the behavior of the main structure, with a temperature difference with respect to it varying between +2°C and +10°C (depending on exposure and thermo-optical properties of the given sample).

2.1. *Main Mechanical environments*

Launch environment

EMA will be launched in 2 parts, SESAME and IR-Coaster are packaged separately and will be assembled on board. This configuration avoids to have a too big container and limits the mechanical constraints on EMA integrated. Payload packaging is a foam Plastazote (LD24FR) box, with 15 mm thickness. Random vibration level is 9g RMS during 1 min.

Due to foam damping, beyond 100Hz vibration load level strongly decreases. Consequently, this packaging allows to have fragile elements in the payload.

Mission environment

There are 2 main mechanical environments during the mission:

- astronaut kick or tool impact: payload shall sustain 556N on an area of 4 inch by 4 inch
- robotic arm impact: payload shall sustain an impact of 1J in some areas where clearances between the payload and some other structure.

2.2. *Human spaceflight requirements*

To ensure safety, payload design and operations in human spaceflight are governed by many requirements on materials, margin philosophy, parts containments, etc.., mainly described in the SSP 51721 "ISS Safety Requirements Document". EMA design and operations take into account all these requirements and they were validated by ESA safety team during the process build around 3 safety reviews.

3. Description of the payload and assembly on ISS

For the first EMA flight opportunity, 141 passive materials samples and 5 types of sensors will be exposed on SESAME in ISS Ram direction, offering unrestricted view to space.

Sensors (active and passive) will provide space environmental data such as AO erosion data, contamination, radiation measurement and, micrometeoroids and debris evaluation. To ensure the safety of the crew, as some samples / sensors are fragile, some sample holders are protected

IOP Conf. Series: Materials Science and Engineering **1328** (2025) 012020 doi:10.1088/1757-899X/1328/1/012020

by a Makrolon transparent cover, enabling the integrity of the sample to be checked before opening.

IR-Coaster also embeds environment sensors:

- 5 thermo-luminescent dosimeters are installed in different parts on the instrument (inside and outside).
- 6 UV photodiodes are disposed on the structure to measure UV dose received along the mission.

In addition, 10 temperature sensors (Pt 1000) are distributed inside IR-Coaster to have a temperature mapping and an absolute clock will date all the data.

All these environmental data and housekeeping will be used by SESAME during exploitation phasing to correlate material ageing and environment seen by the samples.

After the flight to ISS, EMA will be handled and prepared inside the station by ISS crew under pressurized atmosphere. The following steps will be performed: removal of the main aluminum cover (see Figure 3), visual inspection (particularly fragile samples) then removal of the transparent protective covers. Finally, the crew will assemble the two experiments SESAME and IR-Coaster.

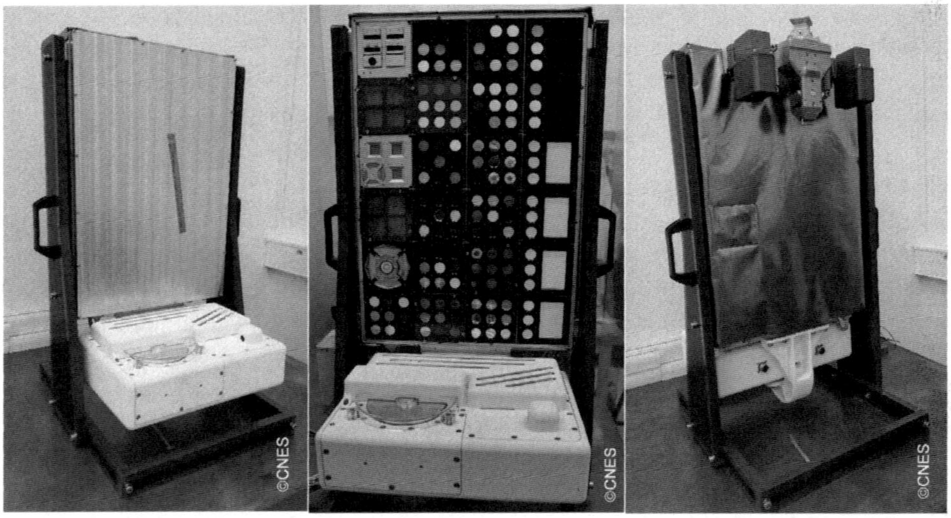

a) b) c)

Figure 3. Pictures of EMA payload without the handles –(black handles are for GSE) in different configuration, a) front face with the main cover ,b) front face without the main cover and c) rear face. ©CNES

EMA assembled will be transferred to the Nanorack airlock to be depressurized and then installed outside ISS, at Bartolomeo location slot 4 with the robotic arm. After checking of the good mating by Bartolomeo and reaching the operational temperature (-30°C), EMA will be powered on.

The process of returning to the ISS will be reversed: power off, return to the Nanorack airlock with the robotic arm, repressurization, recovering by the crew, dismounting of SESAME and IR-Coaster payloads in the manned pressurized compartment, reinstallation of the main aluminum cover (Figure 3 a)). SESAME will be then packed in a hermetically sealed depressurized bag and returned on Earth in the same foam kit.

The size and the mass repartition is in Table 2.

Table 2. Size and mass of EMA

Experiment	Dimensions (mm)	weight (kg)
SESAME	750 x 420 x 200	19
IR-Coaster	400 x 280 x 140	10
EMA	870 x 430 x 470	30

EMA telemetries are limited. During the mission, the only parameters downloaded are the current to have access to the power consumption, temperature acquired at the backside of a sample holder, gold connector mating status and health status of the payload.

Data registered by EMA, scientific data (from active sensors and IR-Coaster) and housekeeping are stored on SESAME memories (triplicated). These data will be downloaded on ground after EMA retrieval.

During the mission, at least two pictures of SESAME will be taken outside ISS on its location, one at the installation and one before removal to check the overall payload integrity.

In the following part of the article, the focus will be made on SESAME experiment.

CNES develops SESAME to understand materials and coatings behavior in space environment and to record some environmental parameters on sensors: AO fluence, radiation dose and contamination. This exposure on ISS is a real opportunity to complete existing data and still enhance our knowledge of the LEO space environment and its effects on materials.

The goal of SESAME is to expose to LEO environment for a dedicated period of time two types of samples (see Figure 4 and Figure 5):

- Passive materials samples requiring no power supply (only exposed to the space environment):
  - Materials: composite materials, paints, films, optical materials, adhesives
  - Passive sensors: silica aerogels for micrometeoroids and debris capture, optical fibers for radiation measurement, adhesives for contamination, polyimide materials for AO fluence determination, tapes for contamination
- Active samples requiring power supply, environmental detectors:
  - Quartz Cristal Microbalance (QCM) for contamination measurement
  - RESISTACK for AO measurement

The properties of interest of the different materials (thickness, mass, thermo-optical properties, adherence, surface analysis) will be studied on ground before and after flight.

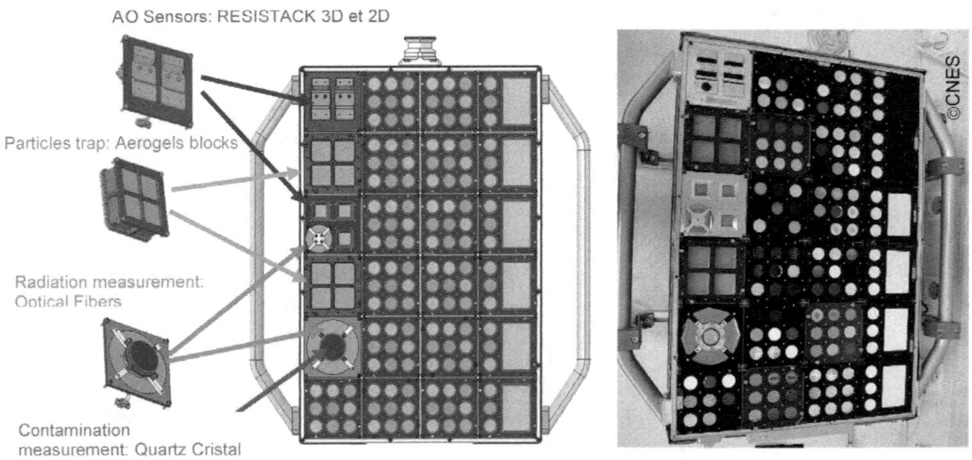

a) b)

Figure 4. SESAME exposed face a) drawing b) photography. ©CNES

4. Material samples

SESAME could take away 141 samples.
Table 3 presents the 2 sizes of passive materials samples:

Table 3. Size of SESAME samples

Type	Size (mm)	Number of samples
Small square or round	25,4 ± 0,1	72 CNES samples + 63 ESA Samples
Large rectangle	77,2 ± 0,1 48,7 ± 0,1	6 CNES samples

The sample dimensions (25,4 mm) are standard for on ground analysis, especially for thermo-optical and mass measurement after environmental stresses. Large rectangle samples are specific for surface electrical measurement and adherence test.

The samples can have 3 different thicknesses, 1 mm, 2 mm or 3 mm.

CNES provided 78 samples (including 9 from JAXA) and ESA 63.

Rationale to choose CNES material samples has been to select materials which are:

- already used on LEO missions with insufficient data (on specific properties)
- new: validated but with lessons learnt on flight missions
- comparable to other previous flight experiments or to compare to ground test data.

The materials families on SESAME are listed in Table 4 :

Table 4. Different families of materials

Family of materials	Goals
Polymer Thermoplastics (films and composites) and thermosettings	Degradation / Contamination
Adhesives	Degradation / Functional properties / Contamination
Thermal coatings	Degradation / Functional properties / Contamination
Optics	Degradation / Functional properties / Contamination
Ceramics	Degradation / Contamination
Contamination trap	Contamination
Contaminated samples and witnesses	Degradation / Functional properties / Contamination

Regarding especially contamination traps, double-sided tapes provided by JAXA were installed at different locations on EMA/SESAME structure, close to the samples and on the edges in order to capture induced particles from erosion of materials under atomic oxygen for instance. They were qualified in terms of outgassing (according to ASTM E595 standard) and it was demonstrated that their adhesive properties were preserved especially under vacuum and atomic oxygen exposure during on-ground tests [8]. After retrieval, the captured particles will be characterized and their chemical natures compared to those of another traps such as aerogels. Additionally, aluminosilicates adsorbents and optical witnesses will also be exposed to collect potential molecular contaminants coming from materials outgassing (see Figure 5).

Figure 5. Locations of double-sided tapes (in red), molecular adsorbents (in green), optical witnesses (in yellow). ©CNES

5. Material samples

5.1. Aerogels

Two modules of 4 aerogels blocks (Figure 6) are integrated to capture micrometeoroids and debris particles (Figure 7). This type of aerogel blocks already flown on MEDET experiment in 2008-2009 [9-10], and [11]. For SESAME experiments, aerogels blocks of 37x37x30 mm were synthesized.

The density of these aerogels is ultra-low (around 0.1 g/cm^3) then it allows to capture intact microparticles by allowing deceleration without modification. The aerogels will be analyzed after on-ground return to study the microparticles trapped, their depth of penetration and assess their speed. Post-analyses should allow to study composition of the particles in order to differentiate debris type and origin. They have been manufactured by Prime-Verre.

To help monitor space debris and micrometeoroids/ fragments, it is planned to analyze all EMA surfaces to list impacts on different materials as already done on MEDET [12]. The aim is to be able to set particles size, sorting and to compare it with environmental models of meteoroids and fragments.

Figure 6. One Aerogel block before integration. ©CNES

Figure 7. Four integrated Aerogel blocks with Makrolon protective covers. ©CNES

5.2. Optical fibers passive measuring device

Two optical fibers spools co-developed by the French company EXAIL and the Hubert Curien Laboratory - Jean Monet University from Saint-Etienne (France) - are integrated to measure the radiation received by the materials (Figure 8 and Figure 9). These specialty fibers are already in use in the ISS (LUMINA experiment) since 2021, and provided detailed mapping of the ionizing radiation deposited inside the ISS across the orbit [13-14].

Figure 8. Optical fiber before integration. ©CNES

Under radiation, optical fibers are generally subjected to different radiation effects. The most prominent one is the Radiation Induced Attenuation (RIA), which is caused by localized defects created in the silica matrix of the fiber, during the irradiation. Under certain circumstances, the related attenuation increase can be used in the fiber to extract information about the radiation environment. In this case, the selected optical fibers have been specifically designed to be more sensitive to radiation, in a controlled way. The degradation is calibrated and well known in order to be able to extract information on the deposited dose, directly from attenuation measurement. Whereas in the case of LUMINA experiment, the dose reading is in real-time, in the case of SESAME, a post-mortem reading is used to determine the averaged dose received during the mission.

A disk of Kapton coated with MAPATOX K covered the optical fibers to protect them from AO (Figure 9).

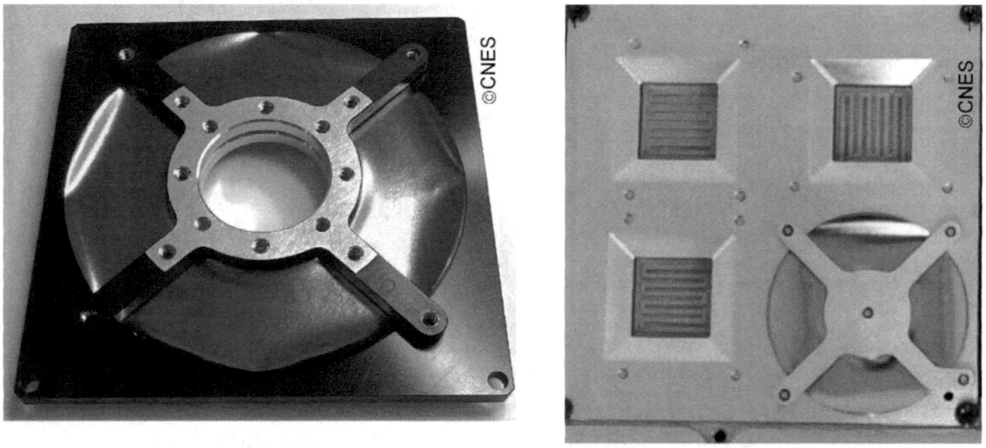

a) b)

Figure 9. Optical fiber integrated and protected against AO a) around the QCM location, b) near RESISTACK. ©CNES

6. Active sensors

6.1. A quartz crystal microbalance (QCM)

Quartz Crystal Microbalances are commonly used under vacuum for real-time molecular contamination monitoring on ground and in orbit. They are highly sensitive measuring devices based on the piezoelectric properties of quartz and can detect extremely low concentration levels of condensable gas.

Microbalances for space use are often composed of the following 3 main parts:

- A metallic case with a mounting flange,
- A base with an oscillator/mixer hybrid chip designed to operate the device with very low power consumption,
- A mechanical heart comprising 2 matched piezoelectric quartz crystals, one exposed to the environment and the other one not exposed and a temperature sensor in between.

The measurement principle relies on the oscillation frequency difference between the exposed sense quartz which collects contaminants and a reference quartz inside the device. A linear mass to frequency relationship defined by the well-known Sauerbrey equation allows to calculate the amount of mass deposited and the rate of deposition according to the sensitivity of the instrument. QCM can operate over a wide range of temperatures (e.g.: from -196°C to +125°C for standard temperature).

For EMA/SESAME, a cryogenic quartz crystal microbalance has been chosen and equipped with an internal heater that will be used to boil off the condensed material to avoid saturation on the crystal and allow accurate mass measurements all along the exposure period.

In operation, temperature and frequency will be measured. 3 operating modes have been defined (OFF mode, measurement mode and decontamination mode).

Figure 10 presents views of the chosen QCM model (QCM Research MK18 CQCM with a quartz crystal frequency of 15 MHz) before and after integration.

a) b)

Figure 10. QCM a) before integration b) QCM integrated in the center surrounded by the longest optical fiber. ©CNES

6.2. *RESISTACK*

RESISTACK is a real-time monitoring AO sensor. It is an original ONERA design whose development has been supported by CNES [15]. It is a stack, multi-layer detector made of metallized Kapton foils. The monitoring of equivalent resistance of the device allowed for detecting resistance step-increase that can be correlated to AO calibrated fluence. References [15] and [16] give the principle of Resistack sensing and validation made on ground.

5 sensors are mounted on EMA SESAME (Figure 11):

- 3 sensors on 2D module for direct AO measurement. 3 specimens are exposed to correlate the results obtained.
- 2 sensors on 3D module with a specific geometrical setup to assess AO reflexion. Principle of use of this setup has already been proven on ground [156]. For SESAME need, the setup has been adapted. A comparison between on-ground results, forecast of AO fluence after reflexion and results from EMA SESAME will allow to refine modeling method.

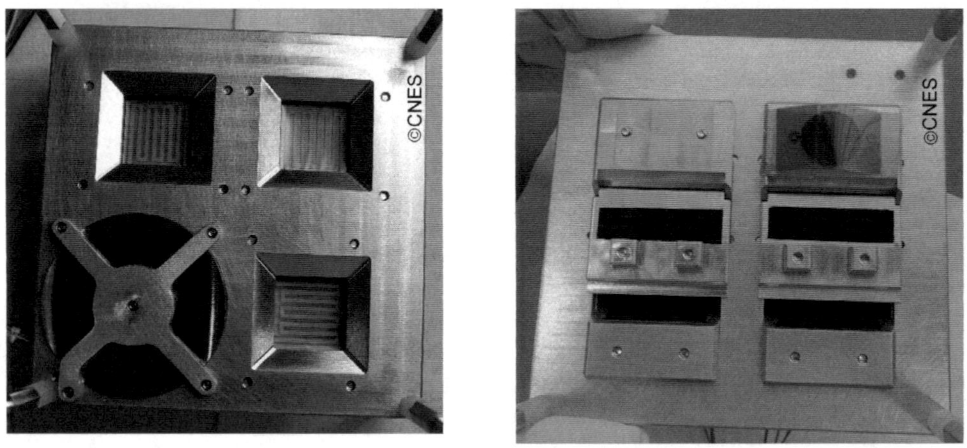

a) b)

Figure 11. RESISTACK a) 2D module integrated b) 3D module integrated. ©CNES

7. Integration Plan

Samples and sensors management all along the project phases have been planned, managed and tracked with a dedicated plan (Figure 12). It sets the rules for:

- Procuring samples and sensors, with specific requirements relative to samples dimensions, geometry, absence of burrs, sharp edges, technical criteria for materials selection, REACH conformity, associated documents. The cleanliness level of samples is defined as a well as possible.
- Receipting the different samples models (flight, spare and reference) i.e. incoming inspections and characterizations,
- Defining their initial state and selecting flight models,
- Storing them before integration in a controlled environment
- Integrating them in SESAME sample holders
- Integrating the sample holders in SESAME structure

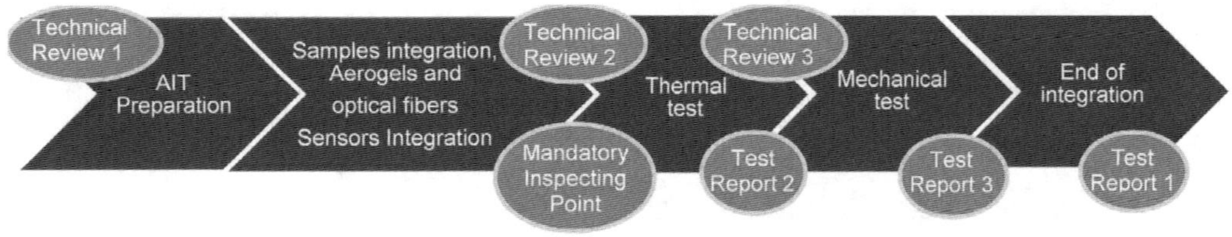

Figure 12. AIT flowchart. ©CNES

The cleanliness control has been managed through a dedicated plan. For example, ISO 8 level was ensured from reception to integration in EMA SESAME structure. Spare and reference samples will be kept in the same conditions until return from flight. For sensors, once integrated in their sample holder structure, ISO 8 environment was also ensured.

During testing activities, materials samples have been checked regularly with mandatory inspection points.

For incoming inspections, a specific setup has been used to take pictures of all samples allowing as much as possible observation and reproducibility conditions. Samples have been checked against dimensions and general aspect to select flight, spare and reference models.

For visual inspections, the following steps have been followed:

- Unaided inspection, general aspects of SESAME and samples holders
- Specific samples inspections, sample holder by sample holder
- Cleanliness specific issues (illumination with UV and/or white light at grazing angle)

For SESAME sample holders and samples inspection, the visual criteria have been checked at each mandatory inspection point. They are for example, physical integrity, no evolution of existing defaults or aspect criteria (color, homogeneity), no appearance of new defaults (marks, scratches). Each suspicious element has been noted, photographed and analyzed relative to SESAME goals.

8. Qualification

8.1. Mechanical environment

To qualify SESAME to mechanical environment, a vibration test was performed on SESAME packaged in the foam box, in launch configuration. The kit was installed on the shaker and hard maintained by a dedicated equipment, as presented on Figure 13 a). Two triaxis accelerometers were installed inside the kit, one on the gold connector, and one on the bracket and two triaxis and three singles axis were fixed on the tool and on the shaker.

Random vibration qualification levels are detailed in table and graph of the Figure 13 b).

After vibration, a detailed inspection was performed, in ISO 5 environment. No damage was observed on the SESAME samples. A functional test was also made: all the results were nominal. Qualification to vibration environment was declared successful.
Other mechanical environments were covered by analysis.

IOP Conf. Series: Materials Science and Engineering **1328** (2025) 012020 doi:10.1088/1757-899X/1328/1/012020

Frequency	Level
20 Hz	0.057 g²/Hz
100 Hz	0.057 g²/Hz
200 Hz	0.071 g²/Hz
500 Hz	0.071 g²/Hz
2000 Hz	0.018 g²/Hz
Overall	9.0 grms
Duration	60 s / axis

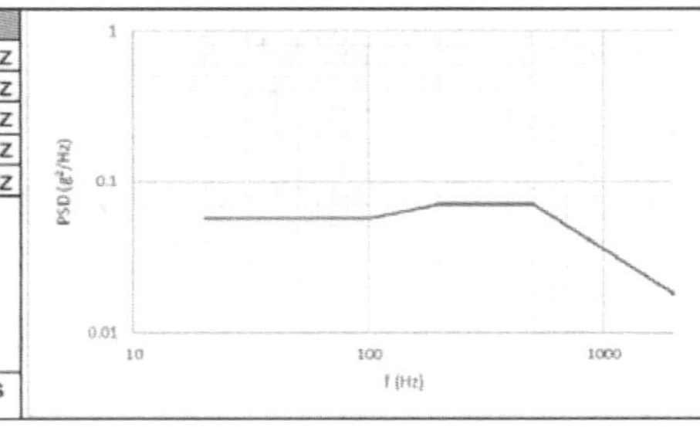

a) b)

Figure 13. a) SESAME on shaker for vibration test (MECANO ID equipment) and b) level applied for qualification random vibration. ©CNES

8.2. Thermal environment

EMA underwent two separate Thermal Vacuum Cycling (TVAC) campaigns: a thermal cycling test campaign for qualification and a thermal characterization test campaign.

Thermal cycling was performed in MECANO ID's thermal vacuum chamber, in early 2024. In order to fit inside the vacuum chamber, SESAME and IR-Coaster had to be mechanically separated from the bracket (which was not used during the test). Moreover, SESAME was laid down on one side, the handle's I/F acting as a mechanical I/F with the chamber's temperature-regulated plate (see Figure 14).

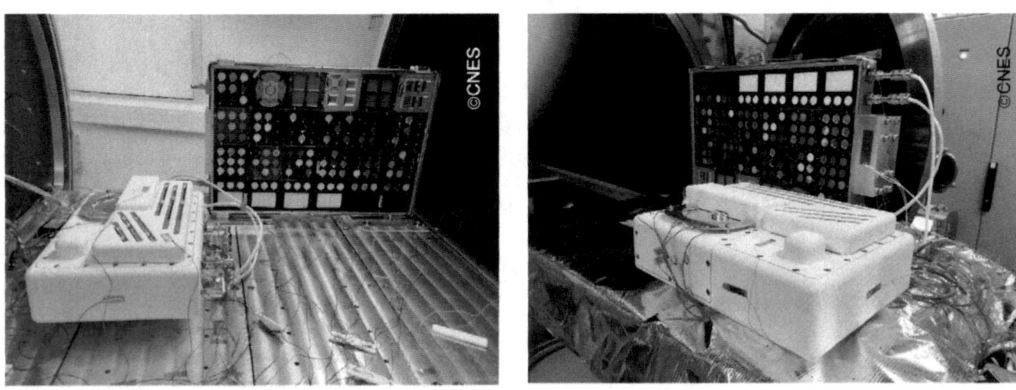

Figure 14. Thermal cycling (left) and thermal characterization (right) test configuration. ©CNES

Thermocouples distributed on the experiment's structures ensured surveillance throughout the whole test. Figure 15 shows the as-run temperature profile. EMA was cycled between cold NOP (non-operating) temperatures (-40°C/+60°C), then a series of functional tests through a series of short temperature plateaus running from -30°C to +40°C.

IOP Conf. Series: Materials Science and Engineering **1328** (2025) 012020 doi:10.1088/1757-899X/1328/1/012020

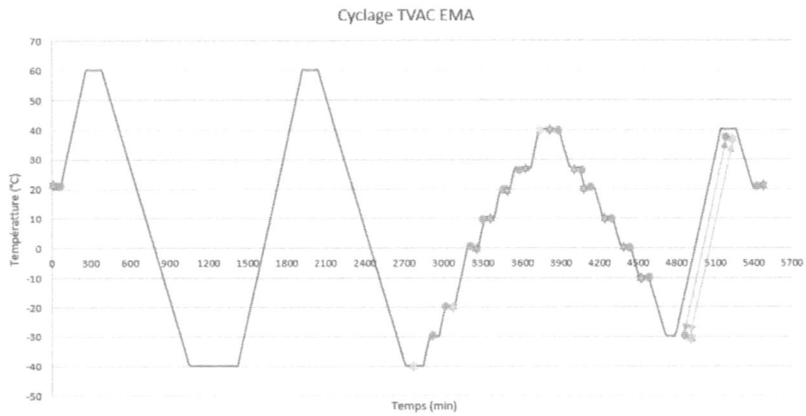

SESAME		
	statut	acquisition f µbalance + resistack+HK
	ON	x

IRCoaster			
	statut	EBT	EBT réduit
	ON	x	
	ON		x

	EMA
	activation ligne heater 120V
	x

Figure 15. Qualification thermal cycling as-run temperature profile. ©CNES

A similar mechanical and thermal configuration was used for the thermal characterization test, run in CNES's 2M³ thermal vacuum chamber (Figure 16).

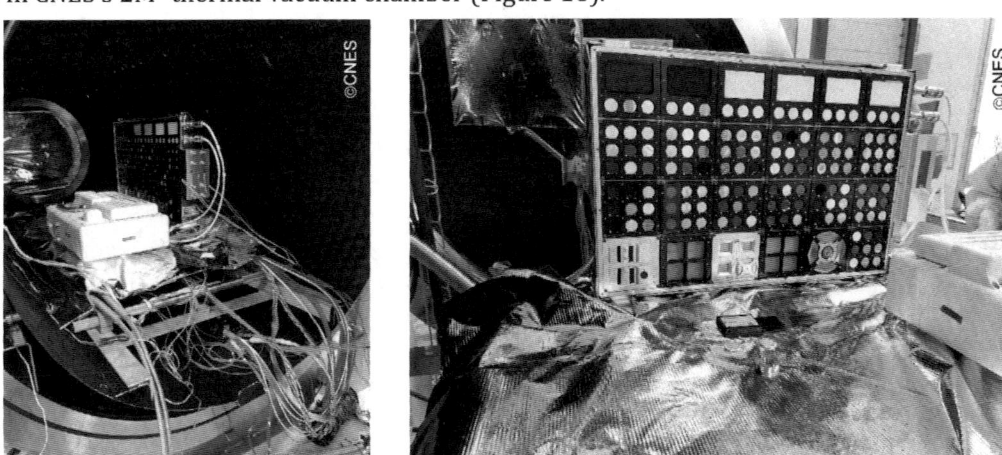

Figure 16. EMA inside CNES thermal chamber. ©CNES

We ran this second TVAC campaign to gather better knowledge of the thermal behaviour of the experiment and to correlate the existing numerical model. Figure 17 shows the temperature and pressure profile plot from the measured data. Numerical model correlation is currently ongoing.

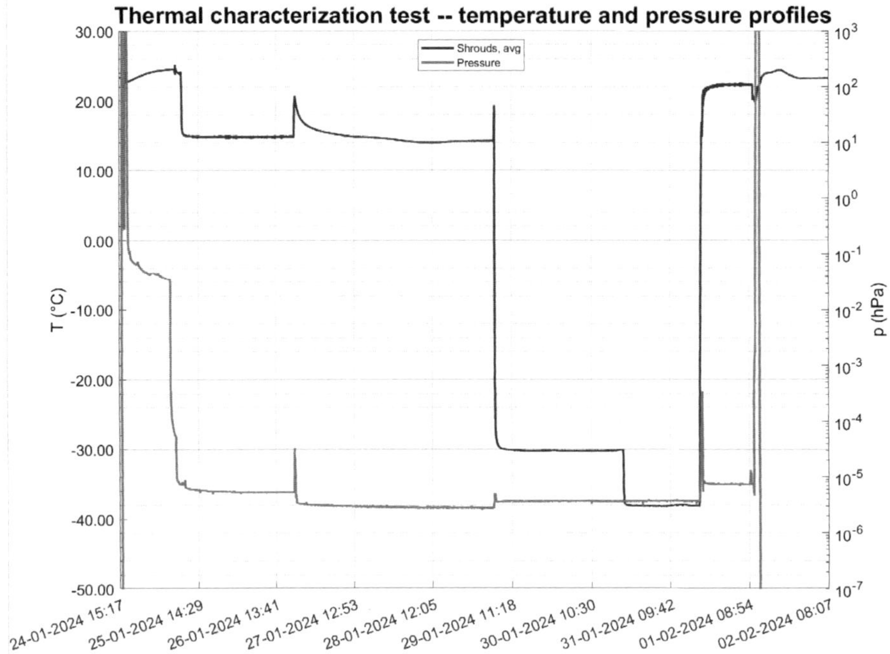

Figure 17. Thermal characterization temperature and pressure profiles. ©CNES

8.3. Electromagnetic compatibility test

A test campaign was performed to verify the EMA compliance to electromagnetic requirements defined by Bartolomeo, in terms of emission and susceptibility. Each test was made in radiated and conducted modes. Radiated tests were conducted in a semi anechoic room. As contamination was not controlled in the anechoic room, EMA has been covered with a transparent box with nitrogen flushing as shown on Figure 18. All electromagnetic tests were performed in flight configuration, i.e. SESAME and IR-Coaster assembled. For each configuration, the worst EMA operational mode was run, and after each test, a functional test was made to ensure the non-degradation of the payload. All tests were successful, except a variation of the behavior of the QCM, probably not linked to electromagnetic test but to the use at atmospheric pressure. A nominal behavior is expected during the mission.

IOP Conf. Series: Materials Science and Engineering **1328** (2025) 012020 doi:10.1088/1757-899X/1328/1/012020

Figure 18. EMA during susceptibility test installed in anechoic room for EMI/EMC testing. ©CNES

8.4. Contamination monitoring

Cleanliness and contamination control requirements applicable to EMA/SESAME for installation on ISS were specified by Bartolomeo. Those linked to the potential risk of in-flight contamination of the station and experiments on board nearby by EMA concern:

- a general surface cleanliness level: EMA exposed and accessible surfaces shall conform as a minimum to the Visibly Clean-Sensitive (VC–S) level as per ECSS-Q-ST-70-50C
- a maximum size of 100 µm for particles
- a limitation of the outgassing flux for chemical species or through vents
- a limitation of the deposition rate of condensable species

Note: The last two requirements were checked through the integrated ISS contamination analysis carried out by NASA based on the Declared Material List of materials and in-flight temperatures seen by the experiment all along the mission.

Other requirements refer to handling and packaging, in particular the need to protect all surfaces sensitive to contamination such as optics and sensors (aerogels and microbalance).

Thus, a cleanliness and contamination control plan has been established to fulfill these requirements during Assembly and Integration activities in ISO 8 cleanroom at CNES, prime contractor and other partners premises. Preventive actions were described and corrective actions anticipated. Cleanliness constraints were taken into account very early in the project phases and were managed at the same level for the different kinds of sample models (witness, spare and flight).

Several key points for inspection were carried out at critical stages, typically at reception, after integration in the sample holders, before and after environmental tests (see Figure 19)... until delivery of the flight experiment with even more stringent requirement whenever applicable (e.g. Visibly Clean Highly Sensitive + UV).

Couples of molecular and particulate cumulative or stage witness plates were also installed for contamination monitoring, allowing to identify the most contaminating phase and then a need for appropriate cleaning (vacuuming or wet cleaning) depending on the characteristics for each surface.

Before final shipping, high resolution pictures of each sample holders, MLI covering and other equipment with different lighting sources were taken during a last in-depth examination for a strong traceability of the final configuration.

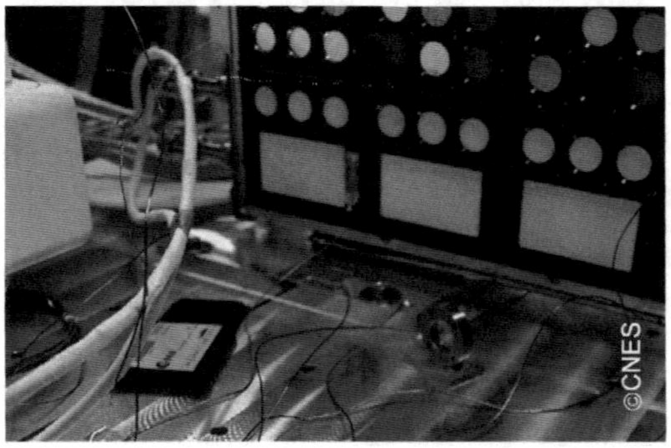

Figure 19. Molecular and particulate witness plates close to SESAME during TVAC test. ©CNES

9. Conclusion

EMA is a European experiment initiated by CNES and developed by COMAT in order to expose materials under LEO environment. This development was an important work in collaboration between numerous different teams, Laboratories and Space agencies involved for samples and sensors supplies and with different companies. This paper describes the design of the experiment to answer to the need and to the flight constraints (mechanical, thermal, environmental, human spaceflight, radiation...). SESAME integration and the qualification tests are detailed.

EMA flew to ISS aboard SpaceX CRS-31 on November 2024, then was installed outside ISS and its mission began mid-December 2024.

The ground scientific expertise after exposure in space environment during several months will provide a wealth of information about the environment, debris and materials ageing in LEO environment around ISS expected for the end of 2026.

10. Acknowledgements

EMA is a payload under CNES leadership developed by COMAT for SESAME and integrated tests and LISA laboratory for IR-Coaster.

Authors would like to acknowledge the collaborative effort of ESA (NL), and ADS (Gr).

Many people have contributed to the experiment, and in particular the dedicated efforts of Agnieszka Suliga and her colleagues from ESTEC/ESA (NL), Sebastien Vincent-Bonnieu from ESA (NL), Sophie Duzellier and David Leveque from ONERA (Fr), Nicolas Balcon from CNES (Fr), Martin Roche, Arnaud Meyer and Sylvain Girard from the University of St Etienne (Fr) Laurent Duffours and Pascale Colombel from PRIME Verre (Fr), Riyo Yamanaka, Yugo Kimoto, Aki Goto, Kazuki Yukumatsu and Yuko Kubo from JAXA (Jp) and Advanced Engineering Services Co., Ltd for JAXA samples preparation, and the contribution of ADS (Fr) an TAS (Fr).

Our thoughts go out to Aurelien Hot, our project manager who unfortunately did not see this project through to completion.

IOP Conf. Series: Materials Science and Engineering **1328** (2025) 012020 doi:10.1088/1757-899X/1328/1/012020

References

[1] De Groh K and Banks BA 2021 MISSE-Flight facility polymers and composites experiment 1-4 (PCE 1-4) *No. NASA/TM-20205008863*

[2] De Groh K, Banks BA, McCarthy C, Berger LA, Roberts LM and Rucker R 2006 Analyses of the MISSE PEACE polymers international space station environmental exposure experiment *Proceedings of the 10 th ISMSE, 8 th ICPMSE.*

[3] De Groh K, Banks BA, McCarthy CE, Rucker RN, Roberts LM and Berger LA 2006 MISSE PEACE polymers atomic oxygen erosion results MISSE 2006 *Post-Retrieval Conference (No. NASA/TM-2006-214482)*

[4] Goto A, Umeda K, Yukumatsu K and al 2021 Property changes in materials due to atomic oxygen in the low Earth orbit *CEAS Space Journal* **13** 415–432

[5] Neish MJ, Kitazawa Y, Noguchi T, Inoue T, Imagawa K, Goka T and Ochi Y 2005 Passive measurement of dust particles on the ISS using MPAC: Experiment summary, particles fluxes and chemical analysis *Proceedings of the 4th European Conference on Space Debris* 221-226

[6] Gao M, Zhao G and Gu Y 2022 Recent progress in space science and applications of China's space station in 2020–2022 *Chin. J. Space Sci* **42** 503-510

[7] Rejsek-Riba V, Soonckindt S, Duzellier S, Remaury S, Durin C, Desmarres JM, Tighe A, Van Eesbeek M, Lobascio C, Nebiolo M 2013 Post-flight analysis of materials exposed on the spectrometer sub-unit of MEDET (18 months on-board ISS) *In Protection of materials and structures from the space environment. Astrophysics and Space Science Proceedings* **32** 41-55

[8] Yamanaka R and Faye D, On-ground and in-orbit experiments: new insight to reveal Atomic Oxygen Induced Contamination, 2025 *IOP Conf. Series: Mater. Sci. Eng.*, to be published

[9] Tighe AP, Iwanosky B, Van Eesbeek M, Duzellier S, Dinguirard M, Falguere D, Pons C, Inguimbert V, Rejsek V, Durin C, Mandeville JC, Gabriel S, Goulty D and Roberts G 2009 Overview of results from the materials exposure and degradation experiment (MEDET) after 18 months in orbit on the ISS *Proceedings of the 11th ISMSE*

[10] Tighe A, Gabriel S, Goulty D, Dinguirard M, Durin C, Chambers A, Gabriel S, Goulty D and Roberts G 2001 Materials exposure and degradation experiment (MEDET) 2001 *Conference and Exhibit on International Space Station Utilization* p. 5070

[11] Durin C and Mandeville JC 2009 SODAD: First results and post-flight analysis Proceedings of 11th ISMSE

[12] Durin C, Remaury S, Mandeville JC, Rejsek-Riba V, Duzellier S, Duffours L, Colombel P, Woignier T, Labat L 2013 CNES Experiments on MEDET: Lessons Learned *Protection of Materials and Structures From the Space Environment Astrophysics and Space Science Proceedings* **32** 217-234

[13] Roche M, Balcon N, Clement F, Cheiney P, Morana A, Di Francesca D, Malapert JC, Kerboub N, Oro Marot L, Ricci D, Mekki J, Canton R, Marin E, Melin G, Robin T, De la Fuente Molina G, Girard S 2024 Solar particle event detection with the LUMINA optical fiber dosimeter aboard the International Space Station *IEEE Transactions on Nuclear Science* **71(8)** 1580-1588

[14] Di Francesca D, Balcon N, Cheiney P, Chesta E, Clement F, Girard S, Mekki J, Melin G, Morana A, Roche M, and Ricci D 2023 Low radiation dose calibration and theoretical model of an optical fiber dosimeter for the international space station *Applied optics* **62** E43-E50

[15] Duzellier S, Soonckindt S, Chardon JP, Pons C, Laurent E, Remaury S, Minton T, Murray V 2018 Validation of atomic oxygen active detector RESISTACK *14th International Symposium on Materials in Space Environment (ISMSE)*

[16] Leveque D, Duzellier S, Vilaranda Fernandes M, Chanteperdrix G, Nguyen Van Sang D, Perrin-Bailly V, Jouanne P and Laurent E 2023 Interaction mechanisms between atomic oxygen and materials: investigation on reflected beam *IOP Conference Series: Materials Science Engineering* **1287** 012010

Synergistic Effects of Dual Source Irradiation with Protons and Electrons

Erik M. Klein[1]*, Patric Seefeldt[1] and Thomas Renger[1]

[1]German Aerospace Center (DLR), Institute of Space Systems, Mechanics & Thermal Systems, Bremen, Germany

E-mail: *erik.klein@dlr.de

Abstract. Interplanetary space is dominated by radiation from protons, electrons, ions, and photons, making it a key environmental factor. While material qualification has focused on individual radiation sources, synergistic testing remains largely unexplored. This study builds on prior research into proton/electron radiation and VUV interactions. Though interactions between energetic particles and artificial light sources have been examined, the effects of dual irradiation with protons and electrons remain a significant knowledge gap. Additionally, low-energy particles, despite their prevalence in the solar system, require further study.

The primary objectives of this research encompass demonstrating the feasibility of dual source irradiation with protons and electrons and concurrently augmenting the understanding of the interactions and processes unfolding when materials undergo simultaneous exposure to these particles.

To achieve these objectives, the study leverages the capabilities of the Complex Irradiation Facility (CIF) to conduct simultaneous irradiation of materials with protons and electrons. Building upon former research that subjected solar sails to proton irradiation [1], a methodical comparative analysis between dual source and single source irradiation is undertaken, drawing upon well-documented outcomes to provide a robust basis for the study.

The findings of this study not only present best practices for establishing dual source irradiation setups but also identify operational constraints, explore synergistic effects within the CIF and on materials. The investigation delves into the processes inherent in these interactions, conducting in-situ examinations to contribute insights to the scientific discourse.

Keywords: electrons, protons, synergistic effects, irradiation tests, material testing

1 Introduction

In the vast expanse of interplanetary space, the omnipresence of radiation, consisting of protons, electrons, ions, and electromagnetic radiation (ranging from VUV to FIR), stands as a crucial environmental factor, see figure 1 for reference. While substantial attention has been dedicated to material qualification and engineering concerning individual radiation sources, the exploration of synergistic testing remains a relatively uncharted domain within the radiation testing community. This study builds upon previous investigations into proton/electron radiation and VUV light interactions [2]. Although the interactions of energetic particles with artificial light sources have been explored, the specific effects of dual irradiation involving protons and electrons, hereafter referred to as dual source irradiation, persist as a significant

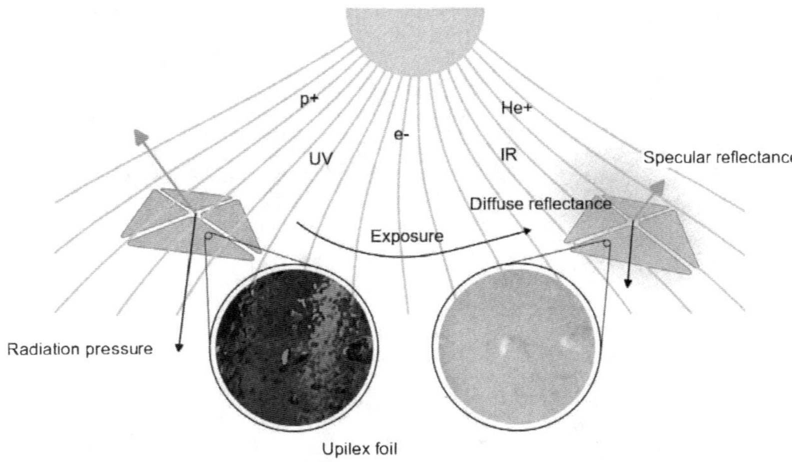

Figure 1: Shown is simple sketch depicting the composition of the radiation environment in space and the influence these elements can have on functional surfaces like space sail membranes.
Image created by the author.

knowledge gap, even within established standards like ECSS [3, 4]. Moreover, despite their prevalence among charged particles in the solar system, low-energy particles continue to be an understudied facet, highlighting the necessity for comprehensive examinations aligned with recognized standards.

This publication documents the effort to produce synergistic radiation consisting of protons and electrons made by the German Aerospace Center (DLR) Bremen. A preliminary irradiation has been conducted showing the novel capability to irradiate samples with protons and electrons simultaneously.

During the following three subsections, the capabilities of the Complex Irradiation Facility will be listed, preceding tests will be shown and eventually synergistic testing is introduced.

1.1 Complex Irradiation Facility

The CIF is operated by the Institute of Space Systems (IRS) of the DLR Bremen [5]. It is a sophisticated instrument designed to study the effects of various types of radiation on materials. The facility consists of a vacuum chamber where samples are placed, and four tubes for connecting different radiation sources to the irradiation chamber. These sources include a proton/electron dual beam system, solar simulator, deuterium lamp, and a Vacuum Ultra Violet (VUV) source [6].

The CIF offers corpuscular irradiation capabilities with its proton/electron dual beam system, which allows users to irradiate samples simultaneously or separately with protons and electrons. This versatility enables researchers to explore the interactions of different particle species with materials, leading to a deeper understanding of material properties and behaviour under various radiation conditions. The feature of simultaneous irradiation is being discussed in this publication.

In addition to corpuscular irradiation, the CIF also supports non-corpuscular irradiation through its VUV source, deuterium lamp and solar simulator. The VUV source is a calibrated source that can produce high-quality spectra for spectroscopic analysis [6]. Users can adjust parameters such as gas flow rate and emission current to customize the spectral output.

The facility's technical design includes features that enable precise control over radiation exposure, such as magnetic manipulators for sample handling and positioning. Additionally, the CIF features a quadrupole mass spectrometer installed at the irradiation chamber, allowing for in-situ analysis of out-gassing composition changes caused by radiation exposure.

The CIF is designed to accommodate various types of samples, including thin films and bulk materials. By providing both corpuscular and non-corpuscular irradiation capabilities, the CIF expands the scope of research in fields such as materials science.

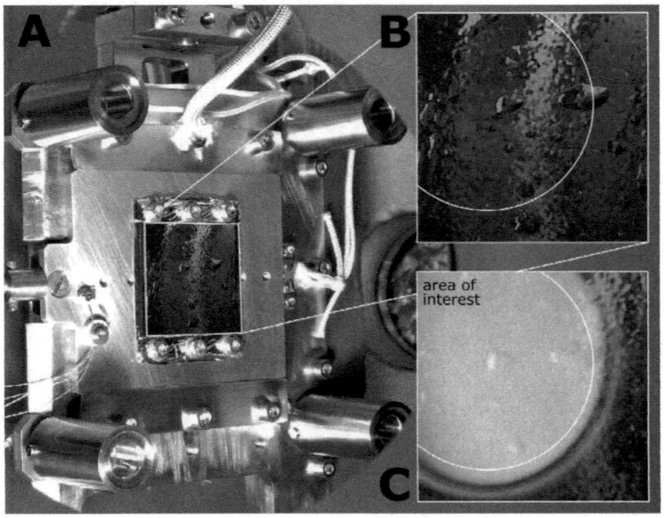

Figure 2: A) Sample S1 prior to radiation inside of the sample station of the CIF. Camera used was a DSLR Canon EOS 800D. B) and C) both being transformed into top view perspective, before and after irradiation, respectively. For brightness evaluation, only the pixels inside the area of interest, as marked above, were used. Irradiation has been conducted at -176 °C using 2.5 keV protons.
Image created by the author.

1.2 Preceding Tests

Preceding work investigated the influence of proton fluence [7], magnitude of proton flux [8], and a target material temperature [1] on the phenomenon of hydrogen blistering, which is the formation of hydrogen blisters that can cause irreversible change to the surface of metals. In the context of proton radiation, the implanting protons form hydrogen by recombination with free electrons of the metal lattice.

Eventually blistering occurs as hydrogen agglomerates into H_2-clusters. Their appearance is defined by blisters filled with H_2 bulging the metallic surface outwards. These bubbles than act like a diffuser for reflected light. The effect it has on the optical properties can be seen in figure 2 or in the context of environmental exposure in figure 1. It shows the sample setup inside the CIF including before and after images of a proton irradiation conducted at -176 °C. More details can be found in [1].

1.3 Synergistic Radiation

Solar wind is neutral, hence protons and electrons are bombarding satellites simultaneously [9, 10], see figure 1. Looking at aluminium for example, we know that there are electrons readily available in the form of electron gas, that exist in most metal lattices. When these kind of materials are irradiated with protons, the available electrons form together with them hydrogen.

But what is with materials such as insulators, where electrons are not free. Does the simultaneous irradiation of protons and electrons have an influence here? These kind of questions have to be asked when looking at synergistic radiation consisting of two or more radiation sources.

Previously this question has been worked on in the CIF for the interaction of the VUV-source and corpuscular radiation [2].

2 Setup of the Complex Irradiation Facility & Preliminary Tests

To setup the CIF for dual irradiation is a complex process. In figure 3 the basic components of the CIF can be seen. It is comprised of bending magnets for steering the particles onto the beamline, several steering magnets for fine tuning and a beam sweep in x- and y-direction to homogenise the irradiated area.

Due to the magnets being in each others immediate proximity they influence each other. Especially the two bending magnets do have a heavy influence on each other. The calibration process of the beam includes several iterations between proton and electron beam calibration, as changing one also inflicts

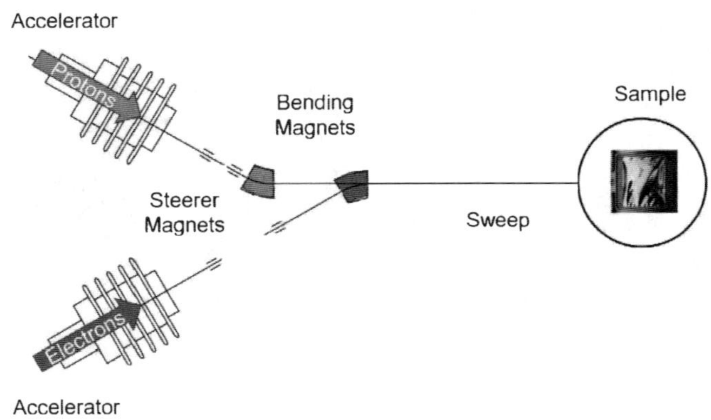

Figure 3: Sketch of the setup of the CIF. Shown are, from left to right, proton/electron sources, steering magnets, bending magnets, beam sweep and irradiation chamber including sample. The horizontal shown path of the particles, from bending magnets to sample is the so-called beamline.
Image created by the author.

change onto the other particle beam. Eventually the beam can be seen on the sample area, where a phosphor screen was placed that illuminates under ionizing irradiation. This is shown in figure 4a.

As can be seen in figure 4a the beam shapes have been different for protons and electrons. Due to the vast weight difference of protons and electrons, each of them have different velocities after going through the same acceleration voltage. Further more, although having the same kinetic energy, each act differently under the same electric or magnetic field, e.g. for the example the magnetic field of the Earth influencing the movement of charged particles by the LORENTZ-force. In order to compensate the latter, especially for electrons, electromagnets are used, that cause the shape of the electron beam. Protons are less susceptible to this deviating force.

The beam sweep has not been used in order not to reduce the beam flux on the sample area. Furthermore, being the irradiated areas as they are, there are three differentiated areas. One where only protons, one where only electrons and one where both implant into the target area, thus effects to all different irradiation types could be observed. Additionally, for proton and electron beams with the same energy, the beam sweep deviates the particles differently, since electrons are faster, while protons are heavier.

The final beam parameters can be seen in table 1. Energy chosen was 4 keV, since at this energy most of the protons are expected to get stuck in the aluminium layer on top of the Upilex substrate [1], additionally, the average particle energy of solar wind protons is of this magnitude [11, 12]. For electrons the same energy was chosen. For the same energy, electrons are expected to have slightly higher penetration depth [13], hence a majority can get implanted into the Upilex and deposit charge there. Flux and Fluence have been chosen such that hydrogen blistering would appear.

Table 1: Irradiation properties of protons and electrons.

Parameter	Electrons	Protons
Energy	$4\ keV$	$4\ keV$
Beam area	$\sim 6.4\ cm^2$	$\sim 16.53\ cm^2$
Flux	$7.7E12\ e^-/cm^2s$	$1.2E12\ p^+/cm^2s$
Fluence	$5.788E17\ e^-/cm^2$	$9.02E16\ p^+/cm^2$

Having set up the CIF, first irradiation could be conducted. Since, previously, many tests have been done concerning hydrogen blistering as explained in the sections before, the logical next step was to test with dual irradiation. Therefore aluminium coated Upilex foil was mounted onto a sample holder and equipped with a temperature sensor. Throughout the following experiments, the temperature was close to room temperature (17 °C). See figure 4b for assembly of the PT100 element that measured temperature.

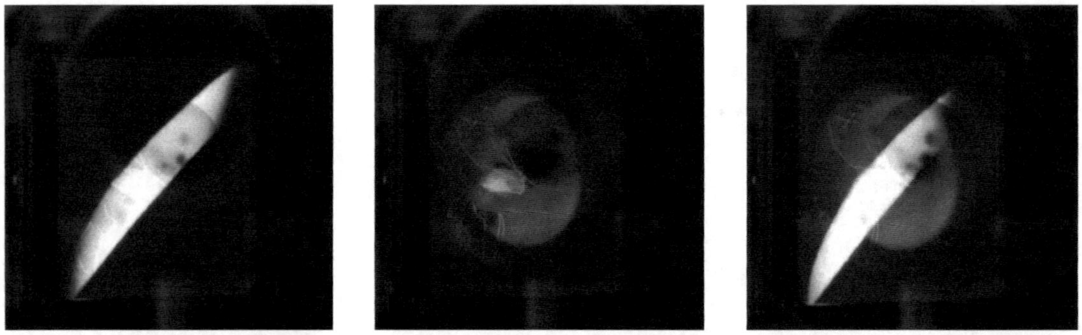

(a) Visualization of beam calibration. From left to right, only electron, only proton and dual irradiation is shown.

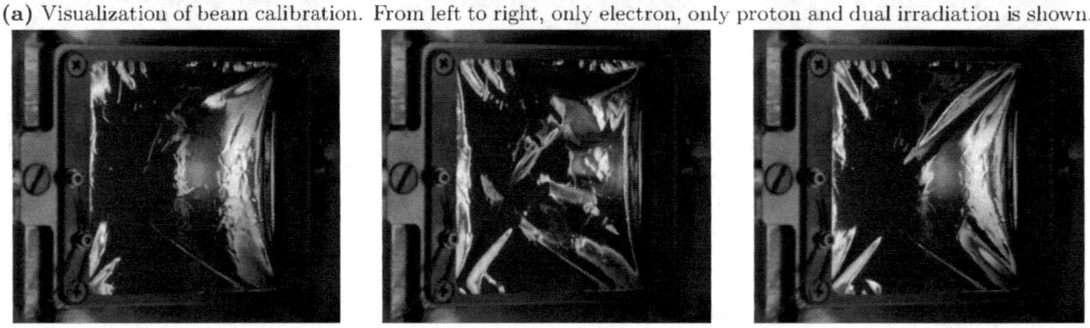

(b) Foil irradiated with electrons and protons with beam shown in figure 4a, right image. From left to right, before, during and after irradiation is shown.

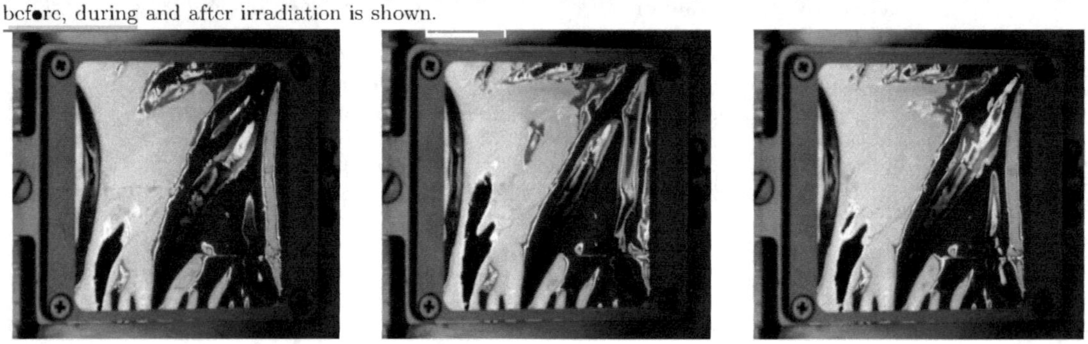

(c) Foil irradiated with protons only with beam shown in figure 4a, middle image. From left to right, before, during and after irradiation is shown.

Figure 4: Compilation of all made tests and calibration.
Images created by the author.

The measuring element has been clamped into the frame, hence did not measure the foil directly. The sample foil being clamped hinders the heat flux, thus possibly created heat due to the radiation can only slowly drain. But since no temperature increase due to proton radiation is to be expected, temperature of frame, e.g. clamp and sample holder, and foil will have been identical. The sample holder has been grounded, hence any charge deposited in conductive material could be drained.

Additionally a second sample has been irradiated with protons only, since the test setup differed slightly from past experiments, e.g. in the past, foils have been glued onto a copper plate instead of being clamped in a frame of the sample holder (compare figure 2 with figure 4b).

In figures 4b and 4c the comparison between dual and proton only irradiation is shown. Further, after irradiation, microscopic images have been made as seen in figure 5. The dual irradiation sample developed the typical appearance of metallic surface affected by hydrogen blistering that can be seen in figure 5a on the right side and in figure 5b as white curved band. The blistered area correlates with the proton only area of the dual irradiation sample. This is peculiar, since neither did they appear where the electron beam hit the samples surface, nor did the proton only sample show any signs of hydrogen blistering as seen in figure 5c. It suggests that the electron beams influence extends over its actual beam area.

3 Discussion & Comparison

Both, dual and protons-only samples, can be seen in figure 4, additionally the beam profile was given in order to possibly assign any changes to types of the beam area. Immediately after irradiation start the dual irradiation sample, started to wrinkle, where the proton samples did not. This could be due to the build up of charge in the Upilex layer due to the implanting electrons. The Upilex, being an insulator, resists the drainage of charge. That protons cause charge in the foil is seen in figure 4c, where also tiny wrinkles can be seen but to far less extend.

Eventually thermo-optical properties have been measured which can be seen in table 2. The measurements have been conducted using the handheld 410-Solar reflectometer by Surface Optics. The device uses an integrating sphere to determine total and specular reflectance. After extraction from the vacuum chamber, the foils showed no deformation, e.g. bigger wrinkles, that could have non-negligible influence on the measurements. The results don't show a clear influence of dual electron/proton irradiation. Variation of the total and specular reflectance are inside of the measurement accuracy of the used measurement device. The measurements of the dual irradiation sample have been made at areas where both particle beams hit the sample surface, not where the blistering could be seen. The standard deviations where determined doing several measurement over the samples area, hence e.g. the proton sample showed higher variation of specular reflectance over several measurements.

Table 2: Overview of investigated samples. Measurement accuracies are coming from several measurement conducted at different places over the irradiated area.

Sample	**Total reflectance** ρ	**Specular reflectance** ρ_s
Unirradiated	0.923	0.919
Synergistic (proton + electron)	0.907 ± 0.006	0.893 ± 0.005
Protons	0.909 ± 0.007	0.863 ± 0.079

4 Conclusion & Outlook

It has been successfully shown, that the CIF can irradiate samples with protons and electrons simultaneously. First preliminary tests indicate interesting future results and open new possible testing and qualification scenarios, e.g. instead of successive proton and electron irradiation, which also halves irradiation time and possibly testing costs.

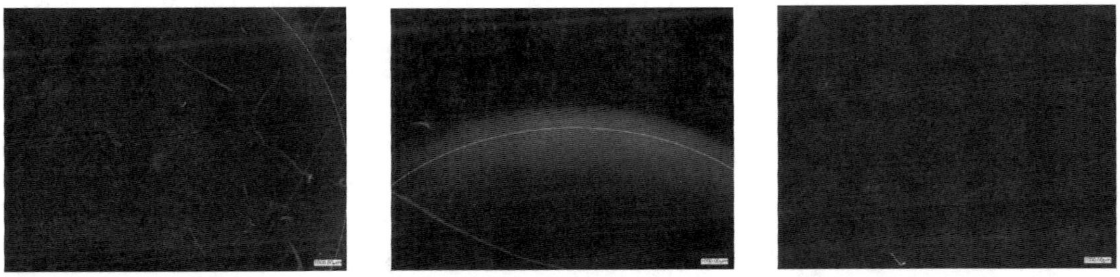

(a) Dual irradiation sample (50x magnification)

(b) Dual irradiation sample (200x magnification)

(c) Proton irradiation sample (50x magnification)

Figure 5: Images made using a microscope of the dual irradiation sample with magnifications of 50x and 200x and one of the proton only sample with 50x magnification. The blistered area is hinted via yellow lines.
Images created by the author.

Although total and specular reflectance differ only slightly, different behaviour for dual and proton samples have been observed during irradiation.

In a first test, energies have been chosen to be identical. In future test, firstly the electron energy will be matched, such that penetration depth will be identical, hence decreasing electron energy. Looking at the scenario of solar wind onto insulating substrate with an conductive coating, it is of interest what phenomena can be observed when both particle fluxes implant into the non-conducting substrate, here the Upilex, hence increase penetration depth by increasing the energy compared to the previous scenarios.

Eventually the diffusion parameter shall be observed by varying the temperature, e.g. by this changing the hydrogen blistering rate, since temperature influences the moveability of hydrogen molecule inside the metal cluster. Questions asked for example are, will hydrogen blisters form inside Upilex although no presence of the electron gas, such as in aluminium, supports the creations of hydrogen molecules.

Looking past the scenario of solar sails, other materials are of interest as well. Future materials for dual irradiation testing could be FEP, ceramics, CFRP and coatings such as SiO2, Al2O3 and ITO among others.

Conflict of Interest Statement
The authors declare that the research was conducted in the absence of any commercial or financial relationships that could be construed as a potential conflict of interest.

Author Contributions
All authors listed have made a substantial, direct, and intellectual contribution to the work and approved it for publication.

Acknowledgments
No external funding, collaborations, or third-party materials were involved in the creation of this article.

Data Availability Statement
For original data, inquiries can be directed to the corresponding author.

References
[1] Sznajder M, Seefeldt P, Spröwitz T, Renger T, Kang J H, Bryant R and Wilkie W 2020 *Advances in Space Research* ISSN 02731177

[2] Klein E M, Seefeldt P, Sznajder M and Renger T 2023 *IOP Conference Series: Materials Science and Engineering* **1287** 012013 ISSN 1757-8981

[3] 2020 ECSS-E-ST-10-04C Rev.1 – Space environment (15 June 2020) Tech. rep. ECSS

[4] 2008 ECSS-Q-ST-70-06C– Particle and UV radiation testing for space materials (31 July 2008) Tech. rep. ECSS

[5] Renger T, Sznajder M, Witzke A and Geppert U R 2014 *Advances in Solar Sailing* (Springer) p 541557

[6] Sznajder M, Renger T, Witzke A, Geppert U and Thornagel R 2013 *Advances in Space Research* **52** 19932005 ISSN 02731177

[7] Sznajder M, Geppert U and Dudek M 2015 *Advances in Space Research* **56** 71–84 ISSN 02731177

[8] Sznajder M, Geppert U and Dudek M R 2018 *npj Materials Degradation* **2** 1–8 ISSN 2397-2106

[9] Meyer-Vernet N 2007 *Basics of the Solar Wind* Cambridge Atmospheric and Space Science Series (Cambridge: Cambridge University Press) ISBN 978-0-511-53576-5

[10] Tribble A C 2003 *The Space Environment: Implications for Spacecraft Design* 2nd ed (Princeton, N.J.: Princeton Univ. Press) ISBN 0-691-10299-6

[11] Klein E M, Sznajder M and Seefeldt P 2022 *Frontiers in Space Technologies* **3** 14 ISSN 2673-5075

[12] Sznajder M 2023 *Advances in Space Research* ISSN 02731177

[13] Seltzer S 1993 Stopping-Powers and Range Tables for Electrons, Protons, and Helium Ions, NIST Standard Reference Database 124 Tech. rep. National Institute of Standards and Technology

IOP Conf. Series: Materials Science and Engineering **1328** (2025) 012022 doi:10.1088/1757-899X/1328/1/012022

Overview of the European FIAMMA (Flammability facility for human space missions)

Antonia Simone[1*], Cathal Mooney[2*], Lucia Propato[1*], Lucia Grizzaffi[1*]
[1] Thales Alenia Space Italia, Turin 10143, Italy.
[2] European Space Agency (ESA), European Space Research and Technology Centre (ESTEC), Keplerlaan 1, Postbus 299, 2200 AG Noordwijk, Netherlands.
*E-mail: antonia.simone@thalesaleniaspace.com ; cathal.mooney@esa.int ; lucia.propato@thalesaleniaspace.com ; lucia.grizzaffi@thalesaleniaspace.com .

Abstract. Materials for human spaceflight must withstand harsh conditions and meet safety standards. The TASI facility, in collaboration with ESA, tests materials' flammability, focusing on Lunar Human Explorations with higher oxygen levels than the ISS. Compliance with NASA STD-6001B and ECSS-Q-ST-70-21C is crucial. The FIAMMA facility conducts standard and R&D flammability tests in environments up to 50% oxygen and pressures from 0.7 to 2.0 bar, ensuring materials are safe for missions, spacecraft, and crew.

1. Introduction

Materials located in habitable areas of spacecraft, including the materials of the spacecraft, stowed equipment and experiments and all materials located in other areas, shall be evaluated for flammability.

Spacecraft atmospheres operate within a narrow band of conditions, specifically along the normoxic atmosphere control band, where the partial pressure of oxygen is maintained at Earth-normal levels (21 kPa) and total pressure is varied by adjusting the diluent composition. This allows for a focused approach to materials flammability testing, as the normoxic band constrains the test parameter space [1]. For manned missions and habitable environments additional flammability requirements are in place and materials must comply to these scenarios (Figure 1). For the Lunar Gateway project (e.g. the international habitable module – iHAB) materials must exist in conditions up to 24.1% oxygen with nominal operating functions at nominal atmospheric pressure of 101.3 kPa (14.7 psi) and with elevated oxygen levels of up to 30% oxygen for atmosphere pressures up to 70 kPa (10.2 psia).

Within these higher-level flammability requirements, system requirements (*e.g.* ECLS) exist also for flammability. For example, in enclosed bays containing an ignition source and no forced air flow, Gateway should provide a means of Fire Suppression within the Bay, or all exposed components located within the Bay shall be composed of materials that comply with NASA-STD-6016B materials flammability requirements at an atmospheric oxygen concentration that is 10% higher than the maximum O_2 concentration expected within the bay during nominal operations. If for example the maximum atmospheric O_2 concentration expected within the bay during nominal operations was 30%, all exposed materials within the bay would need to be certified as self-extinguishing at 40% O_2 in order to remove provisions for fire suppression from the bay. In

IOP Conf. Series: Materials Science and Engineering **1328** (2025) 012022 doi:10.1088/1757-899X/1328/1/012022

addition, ground launcher and transportation flammability requirements must be complied to through the launcher and payload Integration Process [2].

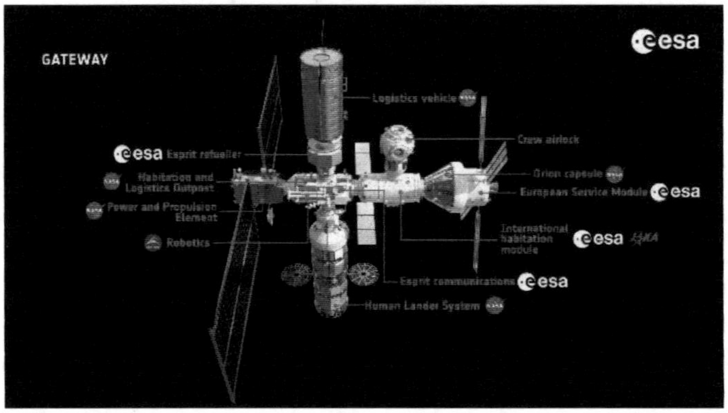

Figure 1: The Lunar Gateway- Credits: ESA. [3].

Currently, materials for space missions are tested following two standards, i.e. the ECSS-Q-ST-70-21C and the NASA STD-6001B, using an upward propagating flame (UPT test) on the material which is set vertically in the chamber. This is a pass/fail test to determine if the materials are self-extinguishing and do not transfer burning debris which could ignite adjacent materials. The test method is defined in ISO 14624-1 & ISO 14624-2 [4-5].

In the frame of a joint activity with ESA, Thales Alenia Space Italy – Turin conceived a high-volume flammability chamber; the FIAMMA- Flammability facility for Future HuMan Missions SpAceflight, with tuneable oxygen concentration and total pressure [6].

The FIAMMA chamber is adaptable to various flammability test methods, not only capable of performing the Upward Propagation Test (UPT) and the Maximum oxygen concentration (MOC) tests but is also capable of performing the Upward limiting oxygen index test (ULOI). The FIAMMA facility, apart from testing at materials level, is the only facility within Europe which can perform flammability testing at component level *e.g* as COTS (commercial-off-the-shelf), following the configurational test according to the ECSS-Q-ST-70-21 flammability standard. Additional tests are performed to test the response of the tapes when facing different solvents.

2. FIAMMA capabilities

The FIAMMA facility is a 1500L (litre) chamber, connected to a vacuum pump with air, oxygen, and nitrogen supplies which can perform standard and R&D tests in quiescent environments up to 50% oxygen concentrations for the understanding of materials' flammability behaviour.

Furthermore, the facility's total pressure ranges from 0.7 bar to 2.0 bar and there is also the possibility to choose different ignition sources (chemical or hot wire) and the ability to monitor and record all the relevant parameters (oxygen concentration, total pressure and electrical source). All these characteristics together make this flammability facility unique in Europe.

IOP Conf. Series: Materials Science and Engineering **1328** (2025) 012022 doi:10.1088/1757-899X/1328/1/012022

The FIAMMA is equipped with four (4) windows in order to observe the burning behaviour of the flame during testing. The versatile sample holder allows the execution of the test in non-standard conditions and thermal and video-recorder systems are used to evaluate and monitor the flame behaviour. An overview of the FIAMMA capabilities are reported in Table 1 and Figure 2 below as well as an overview of the facility (Figure 3).

The present work will present the results of the test campaigns conducted on various classes and forms of materials using different test environments as well as different ignition sources.

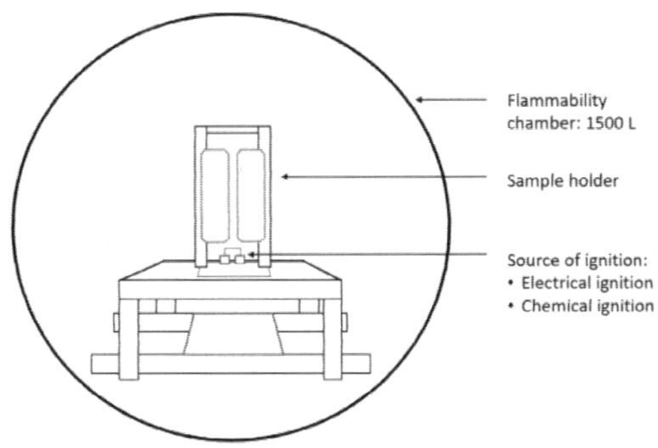

Figure 2: Schematic of the FIAMMA flammability chamber. Credits: TASI.

Results of performing the Upward limiting oxygen index (ULOI) and the Maximum oxygen concentration (MOC) tests are also presented.

Moreover, additional results arising from not-standard (configurational) test campaign are also reported.

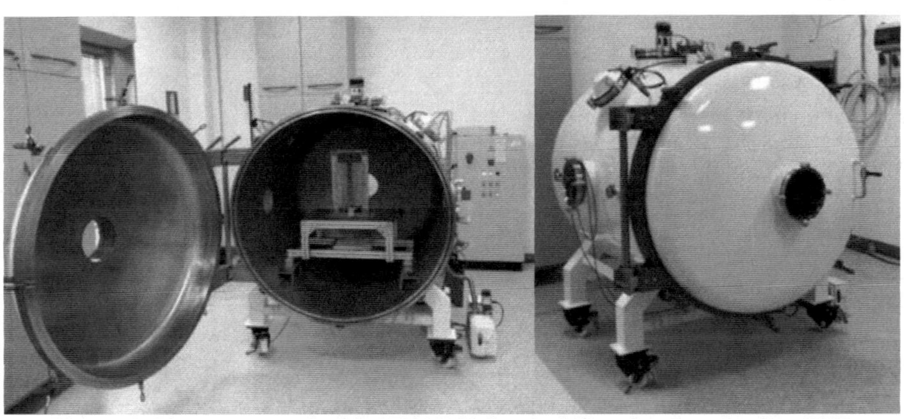

Figure 3: Photo overview of the FIAMMA flammability chamber. Credits: TASI.

Table 1: Flammability facility key figures. Credits: TASI.

Characteristics	Data
Chamber volume	Approximately. 1500 litters
Chamber size	Approximately. Φ1200mm X 1600mm (inner size)
Relative depletion of oxygen concentration	No more than a 5% relative depletion
Evacuation capability	To 1 mbar within 40 minutes
Oxygen concentration	Tuneable from 0% to 50%
Evacuation ports	2
Inlet ports	2 (one for pure nitrogen and one for pure oxygen)
Pressure gauge resolution	±5 mbar
Oxygen analyser accuracy	0.5%
Maximum operating pressure	2 bar
Minimum operating pressure	0.7 bar
Viewports	4

3. UPT tests

The purpose of UPT is to determine the flammability characteristics of a standard format candidate materials when exposed to an ignition source applied at the bottom edge. This test in general is also applicable to electrical wire insulation with a dedicated set-up.

This test evaluates materials in a standardized test configuration for two failure criteria: self-extinguishment within 15 cm (6 in.) and the propensity of flame propagation by means of flaming material transfer. [7-8]

Figure 4 shows an example of an UPT (Test 1) in which the test has failed.

Figure 4: UPT test of a PMMA thin sheet. Credits: TASI.

The configuration test, however, is designed to determine the flammability characteristics of materials configured as in qualification and subsequent spacecraft models to determine whether a flammability hazard exists when a material, which fails the basic screening test, is desired to be used. If the test is deemed successful, these materials can be accepted for restricted application, within the constraints of the test conditions used.

The standard ignition source currently used in all the applied standards is a chemical igniter, *i.e.* a mixture of flammable powders that is extruded to form sticks of white or yellowish colour.

These chemical igniters should provide approximately 3000 J for a duration of 25 ± 5 s and a certification campaign is performed to evaluate the quality of each batch and guarantee the reproducibility of the measurements. Thales Alenia Space in Italy realized the chemical igniters for the test campaign according to [7] with the relevant certification procedure.

In recent years, an alternative ignition source has been evaluated which allows to disengage from the synthesis process of chemical igniters. This hot-wires ignition source involves a 20 AWG KANTHAL wire set up in a flat coil configuration resembling a sine curve placed on the bottom of the sample in close contact. The test pressure and atmosphere shall represent the most hazardous atmosphere anticipated in the spacecraft within the confines of the scope of this test. The test condition (Table 2) are usually the maximum Oxygen Concentrations and Pressures for Crewed Spacecraft [9].

Tests are usually performed at the worst-case environmental conditions consisting of the highest anticipated pressure and oxygen concentration, commonly 30 percent oxygen and 70.3 kpa (10.2 psia). As all materials used in the ISS must be compliant to 24.1 percent oxygen and 14.5 psia.

Table 2: Flammability test conditions. Credits: NASA [9].

Vehicle	Maximum Oxygen Concentration (percent)	Cabin Pressure at Maximum Oxygen Content (psia)
Space Shuttle Orbiter Cabin	30	10.2
Space Shuttle Orbiter Payload Bay	20.9	14.7
ISS Internal	24.1	14.5
ISS Airlock	30	10.2
ISS External	20.9	14.7
MPCV Crew Module Internal	30	10.2
MPCV Crew Module External	20.9	14.7
Gateway	30	10.2
CCP Crew Module Internal	30	10.2
Habitable Exploration Elements on Lunar or Martian Surface	36-40	8.2
Habitable Exploration Elements Operating in Cislunar Vicinity or Deep Space When Docked to Other Vehicles	30	10.2
SLS	20.9	14.7

IOP Conf. Series: Materials Science and Engineering **1328** (2025) 012022 doi:10.1088/1757-899X/1328/1/012022

3.1. UPT-Driving parameters

When the sample is exposed to a standard configuration and ignition source, different parameters can influence the flame propagation, namely:

- Oxygen concentration
- System pressure (*aka* MTP-Maximum Total Pressure)
- Material thickness
- Ignition source

From literature, it is possible to deduce that oxygen concentration is the major driver for material flammability, with little effect from total pressure in standard pressure ranges above 410 mbar [10-14]. Figure 5 below shows, for example, two different oxygen concentrations (*i.e.* 24.1 % and 30% of oxygen at 1 bar) on a similar material (an aramid fabric) currently in use in the aerospace environment.

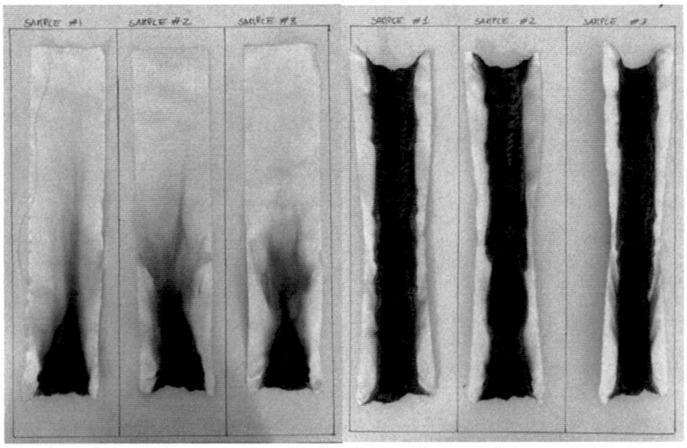

Figure 5: UPT test of aramid fibre at 24.1% O2 (left) and 30% O2 (right). Credits: TASI.

The effect identified is related to an increasing of burn length from 120 to 300 mm, respectively considering the increasing oxygen concentration.
In the first case the test is a PASS (i.e. the burn length < 150 mm) while it is a FAIL in the second one. In both no dripping, sparking phenomena are observed.

Material thickness can also have a marked effect on the UPT test results. The effect of thickness was evaluated on standard sized polycarbonate samples with different thicknesses:

- 12.7 mm
- 4.0 mm

Our study showed that the test result passed for the first two cases, samples which have been tested at both 24.1% oxygen (1 bar) and 30% oxygen (0.7 bar) showed consistent results (Table 3) which are in line with MAPTIS.

Table 3: Effect of the thickness study: Test results on PC (polycarbonate) samples. Credits: TASI.

Material	Oxygen concentration (%)	Total Pressure (bar)	Test result (Pass/fail)
PC (Thickness 12.7 mm)	24.1	1	Pass
	30	0.7	Pass
PC (Thickness 4 mm)	24.1	1	Pass
	30	0.7	Pass

3.2. Ignition source

The flammability test rating (PASS/FAIL) has been obtained by performing a similar test condition on similar material substrates using two different ignition sources specifically:

- Hot Wire (HW) based on an electrical wire;
- Chemical Igniter (CI) based on a chemical rod, certified in accordance with [7].

To evaluate the standard flammability and sensitivity of the ignition source an oxygen concentration was selected as a reference parameter considered for all types of flammability tests whilst keeping the MTP equal to 14.7 psi. This is because it is known that the total pressure has less effect on flame propagation as already mentioned [15].

The UPT test has been performed on 4 categories of materials, namely:

- fabric
- bulk
- film
- foam

For the sake of clarity, only the results for one representative material of the film category (Polymethyl methacrylate, PMMA) and one of the foam category (Polystyrene, PS) are reported in Table 4 below.

IOP Conf. Series: Materials Science and Engineering **1328** (2025) 012022 doi:10.1088/1757-899X/1328/1/012022

Table 4: Comparison test result between hotwire igniter and chemical igniter. Credits: TASI [16].

Material	Ignition source	Oxygen concentration	
		24.1%	30%
PMMA	Chemical Igniter	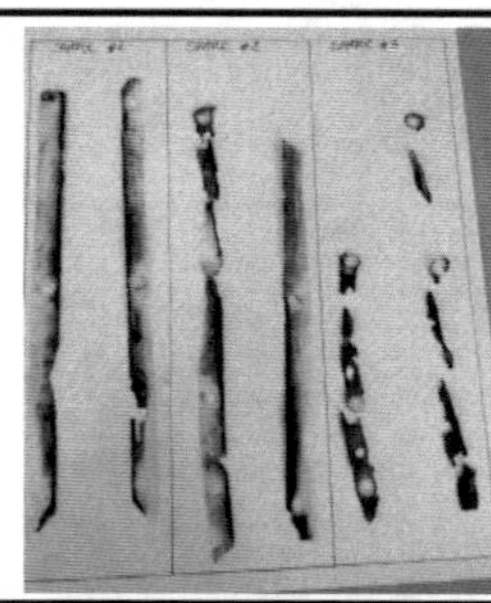	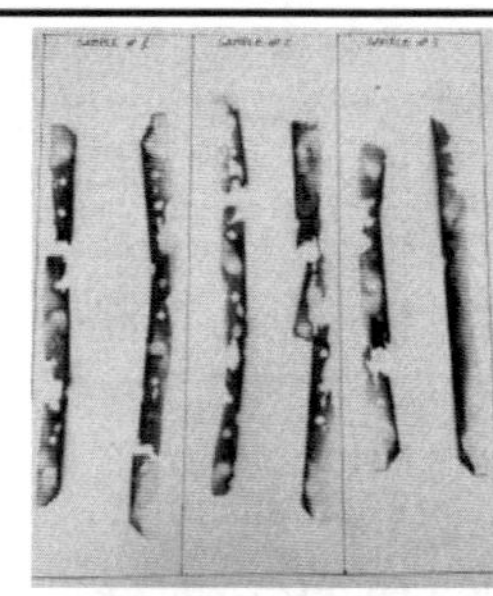
	Hotwire Igniter	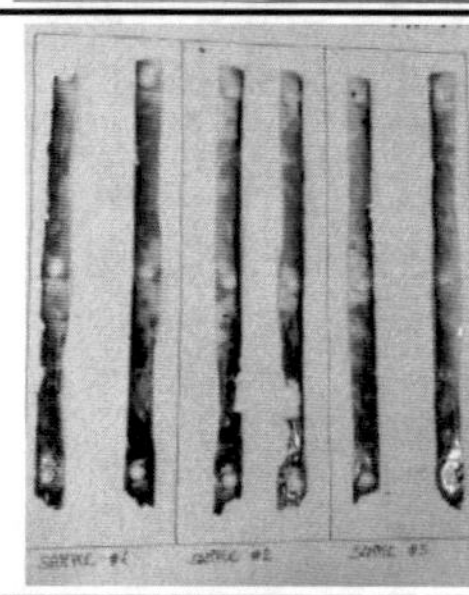	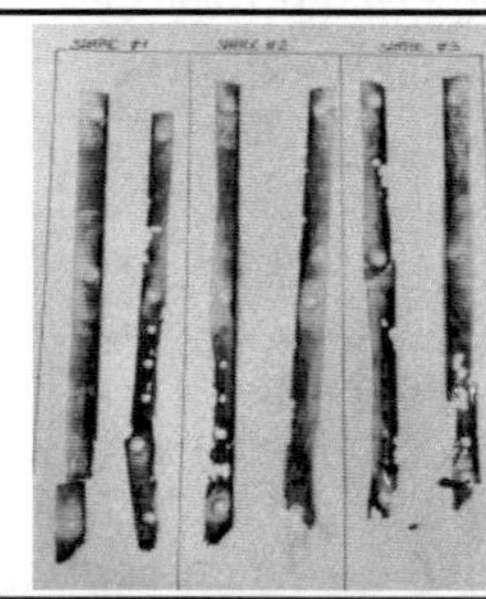
PS	Chemical Igniter	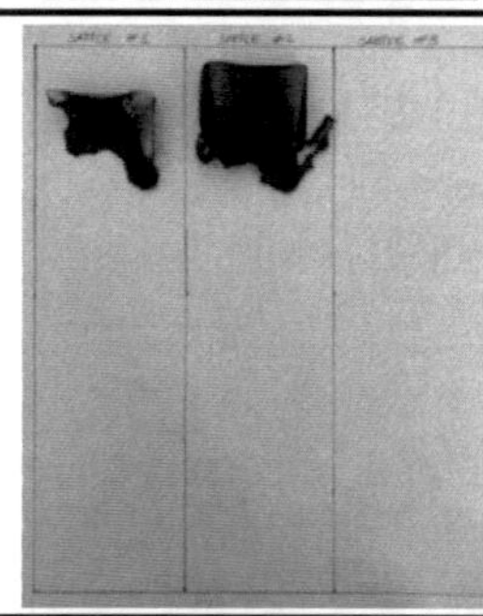	
	Hotwire Igniter		

The PMMA results were similar (300mm as burn length) for both igniters even if small differences were highlighted:

- Paper ignition occurred only with HW;
- Effect of oxygen concentration was more evident with HW in terms of propagation time and weight loss, where an expected behaviour was noted: mass loss was higher at low oxygen concentrations;

IOP Conf. Series: Materials Science and Engineering **1328** (2025) 012022 doi:10.1088/1757-899X/1328/1/012022

- Dripping was more evident with HW;
- At T=20 (arbitrary time from starting of the ignition, used as reference) the CI flame was usually still active while the sample flame was almost full extinguished.

In the case of PS, HW, the tests were a "PASS", the heat propagation of HW was "local" and in this case it seemed that the results depended on the thickness to an even greater extent. It is hypothesized that such high quantity of material can act as heat sink, stopping the propagation of the flame. Moreover, the placement of the hotwire igniter with respect to the lower side of the samples was also found to influence flammability, namely placing the igniter at the edge of the sample was more conservative (higher burn length) than placing it in the middle of the lower side of the sample.

4. ULOI/MOC

To evaluate the safety of materials under such "beyond-ISS" environments, an ULOI (Upward Limiting Oxygen Index) test has been proposed. This test is regulated under standard ISO-TS16697 which stands for the index to reveal the minimum oxygen concentration needed to achieve a successful upward flame spread. ULOI's were calculated with the "up-and-down method" [4-5], similar to LOI regulated under standard ISO 4589-2.

The definition is hereafter reported:

- Upward limiting oxygen index ULOI: oxygen concentration where approximately 50 % of samples fail the test;
- Maximum oxygen concentration MOC: highest oxygen concentration where all samples tested (at least five) pass the test.

In the frame of previous collaborations with ESA and JAXA, ULOI and MOC test have also been performed in FIAMMA on different materials, comparing the ULOI and MOC data obtained by different facility and ignition sources.

The test was performed in several steps, after selecting the oxygen concentration to be used for the next test as follows:

- the oxygen concentration is decreased if the burning behaviour of the preceding specimen gave an "X" response;
- the oxygen concentration is increased if the burning behaviour of the preceding specimen gave an "O" response.

The test procedure was repeated using oxygen concentration changes of any convenient step size until two oxygen concentrations, in percent volume, have been found that differ by 1 % oxygen and of which one gave an "O" response and the other an "X" response. An example of polycarbonate O and X responses is shown in Figure 6 .

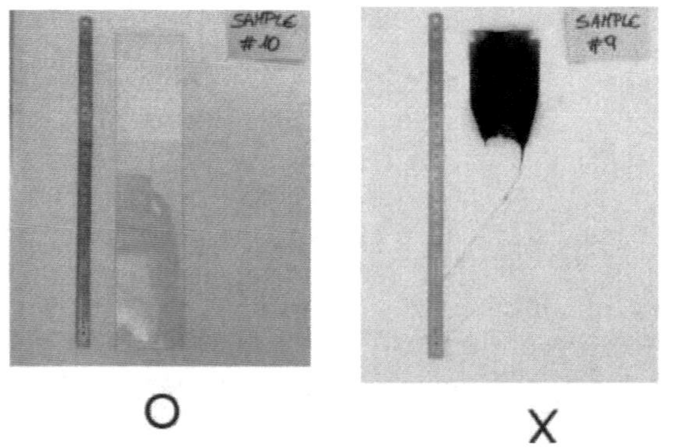

Figure 6: O- Concentration PASS, X-Concentration FAIL. Credits: TASI.

From this set of oxygen concentrations, the one that gave the "O" response is the preliminary oxygen concentration.

Using the preliminary oxygen concentration found initially, conduct the test again whilst recording the data in the data sheet. This is the first result of the N_L series. Then change the oxygen concentration for the next tests by 1 % by volume and conduct testing until a difference in response is obtained. Test four more specimens to complete the N_T series, maintaining d = 1 %, and note the C_o and the response of each specimen. Designate the oxygen concentration used for the last specimen as C_F, this value summed with the k value reported in Table A.1 [5] gives the ULOI data. The highest concentration where, at least five samples gave "O" is the MOC value.

Usually more than 15 samples must be tested to obtain ULOI values, for example, in case of thin sheets, data on polycarbonate is hereafter reported (Table 5):

Table 5: Polycarbonate ULOI and MOC O_2 %. Credits: TASI.

	ULOI (O_2%)	MOC (O_2%)
PC	19.7	17.6

5. Configuration tests

For general-purpose solid materials used in crew vehicle compartments, the fundamental flammability test is Test 1. For hardware configurations a special configurational test (variation on Test 1) is also considered. [7].

Often the results of the flammability material analysis are inconclusive so the hardware configuration (or a simulated configuration with acceptable fidelity) may be tested for flammability to determine acceptability.

In the frame of an ESA research project, we tested several components, examples of which are reported hereafter:

5.1. Action Camera

Given the geometry and size of the object, the sample was placed on two ceramic supports. The external case, usually used to protect the video camera from water or other atmospheric agents was removed, whilst the battery was left inside.
The sample began to burn immediately. Given the sample small dimensions, after about 20 seconds more than 50% of the plastic case was burned. After about 90 seconds, the sample fell on the paper sheet and continued to burn.

Despite this, the flame did not propagate on the paper, which remained almost intact until the end of the test. Only the paper perimeter around the item was consumed.
The flame reached the battery after approximately 6 minutes, causing an explosion and re-feeding the flame, which changed colour due to the combustion of the substances contained within the battery. The flame extinguished after more than 10 minutes. A close-up image of the post-test action camera is shown in Figure 7 below on the right.

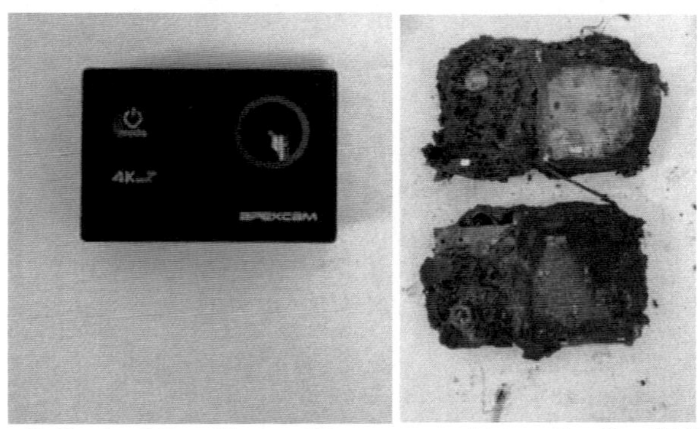

Figure 7: Pre and Post Test action camera photos. Credits: TASI [17].

5.2. PCB

PCBs used for the test have a dimension of 230 mm by 35 mm, with a thickness of 1.6 mm and a weight of 82.1 g. The PCB set-up reproduced a UPT set-up: the sample was placed on the metallic sheet of the sample holder and secured through metal grips, and the ignition point was placed at the bottom. The test was performed at 24.1% oxygen concentration and 100kPa (14.5 psi).
The PCB was self-extinguishing and burning effects detected up to 100 mm. Figure 8 shows pre and post-test photos.

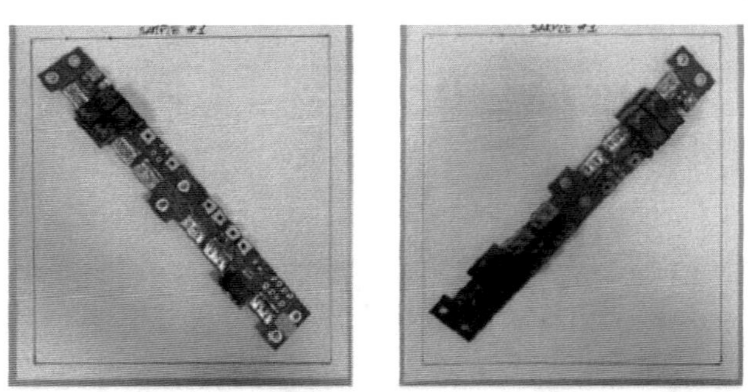

Figure 8: Pre and Post Test PCB photos. Credits: TASI [17].

5.3. ETFE-PVC wire

An ETFE insulated cable with a diameter of 3.5 mm and a PVC USB cable with a diameter 4.05 mm, each measuring 1 m in length have been tested by attaching them at a 15° ± 2° degree angle from vertical, fixed from one top corner of the sample holder to the opposite bottom corner using ceramic thread tensioner. The igniter was positioned 8 ± 3 mm below the sample and 3 ± 1 mm at the lower corner, where the sample was attached.
As expected, due to the presence of fluorinated groups it has a good resistance to flammability, the ETFE cable insulation allowed no flame propagation, and the material therefore is then self-extinguishing (Figure 9, left). On the other hand, the USB cable during the test, the flame propagated along the entire exposed length of the cable, as shown in a close-up photo of the burned cable (Figure 9, right). Even if the flame propagated and all the exposed area was burnt, no dripping or charred debris were observed.

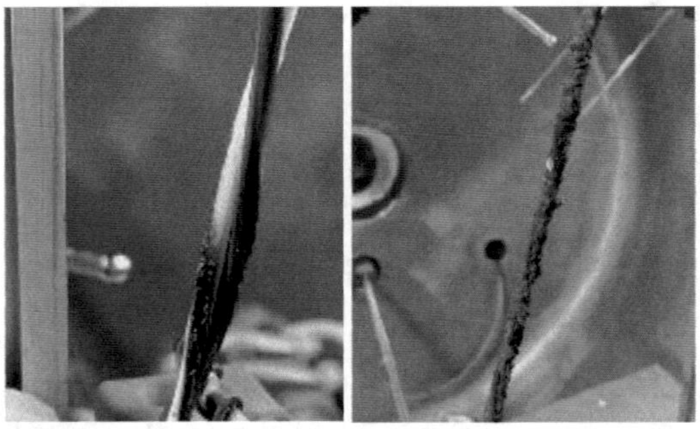

Figure 9: Detail of the EFTE (L) and USB cable (R) after test. Credits: TASI [17].

5.4. Mitigation strategies

The primary methods to reduce flammability hazards are the limitation of use of flammable materials by replacement with non-flammable materials, removal of ignition sources and the restriction of propagation paths, either by covering flammable materials with a non-flammable material or by separation of flammable materials within acceptable boundaries.[9]
For example, a foam material experienced huge dripping during Test 1 for flammability in 24.1 percent oxygen at 14.7 psia but using an Al-foil (0.075 mm) overwrapping the entire foam's surface passed the flammability test (Figure 10).

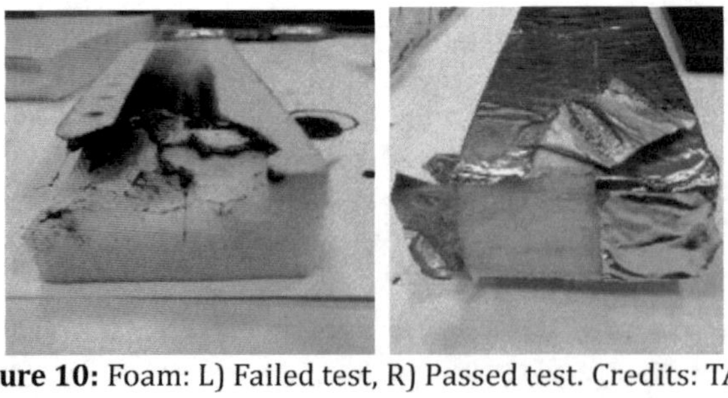

Figure 10: Foam: L) Failed test, R) Passed test. Credits: TASI.

5.5. FEP wrapped monitor

The monitor tested had a maximum height of 37.8 cm, a width of 41.4 cm, a thickness of 3 cm and a total weight of 7.3 kg. The FEP thin sheet, with a thickness of 0.3 mm, was used to wrap the entire monitor, LCD screen included, which turned out to be the most delicate exposed part.
The item was lit from the bottom up and left to burn. Given the dimensions and design of the monitor, it was interesting to observe how the flame behaved, whether reaching the LCD monitor, and how the FEP sheet responded on the surface and along the edges. It was also interesting to find out if the base and junction of the casing are areas where the flame tends to spread faster. A video recording was made from several points of view to collect data on the propagation of the flame and pre and post-test have been taken (Figure 11).

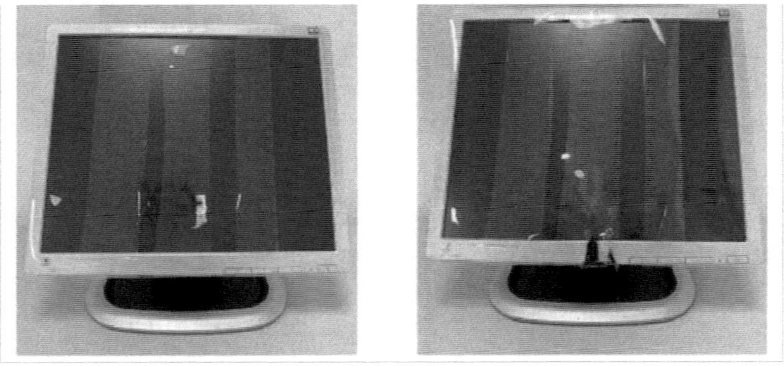

Figure 11: Pre and Post Test FEP-wrapped monitor. Credits: TASI [17].

The FEP sheet melted, but the flame did not propagate. The monitor case did not burn. A close up is shown in Figure 12 below.

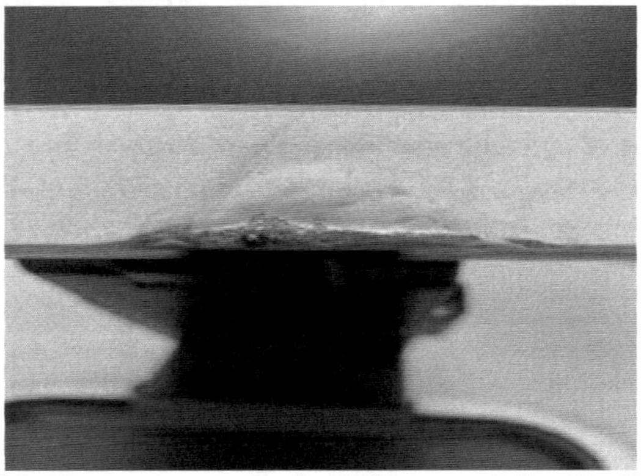

Figure 12: Monitor-post-test after FEP removal. Credits: TASI [17].

IOP Conf. Series: Materials Science and Engineering **1328** (2025) 012022 doi:10.1088/1757-899X/1328/1/012022

6. Conclusions

The FIAMMA facility conceived as a multipurpose test chamber can perform standard and R&D flammability tests in quiescent environments up to 50% oxygen concentrations for the understanding of materials' flammability behaviours. Furthermore, the facility's total pressure range from 0.7 bar to 2.0 bar gives unique flammability testing conditions.

In the FIAMMA facility the following tests have been performed:

- UPT test at materials level on different materials classes. Results are in line with MAPTIS one, when available;
- UPT using different ignition sources (chemical rod and hot wire). Fabrics and bulk materials, especially an aramid fabric and PC tested under HW and CI igniter sources, show similar test results. The ignition mode did not result in significant differences in the oxygen concentration flammability thresholds. The UPT test of polystyrene foam was strongly influenced by the effect of materials thickness. This behaviour is under further investigation;
- UPT test using different test conditions and specimens' thicknesses. The test results are in line with the MAPTIS used as reference;
- UPT tests at components level also in active conditions (i.e. under powered electrical configurations). The effect of mitigation actions has been also assessed;
- Results of ULOI and MOC of different materials classes are comparable with the reference samples.

Flammability investigations under total pressures outside of the 48.2 to 101.3 kPa range (7 to 14.7 psia) will be invaluable to the development and understanding of new spacecraft and extraterrestrial habitation module designs in the future, which may involve spacecraft depressurization. For this reason, the FIAMMA facility has been conceived, and its performances have been demonstrated through several R&D projects by performing UPT, MOC and ULOI testing campaigns.

A feasibility study is on-going to implement the in-situ monitoring of parameters such as temperature and to identify combustion reaction gases produced during the flammability test, as well as to reduce the pressure operative up to 0.4 bar.
The participation to a round robin study with other flammability test facilities is also envisaged for future accreditation.

IOP Conf. Series: Materials Science and Engineering **1328** (2025) 012022 doi:10.1088/1757-899X/1328/1/012022

References

[1] Olson S *et al* 2014 *Applying Flammability Limit Probabilities and the Normoxic Upward Limiting Pressure Concept to NASA STD-6001 Test 1* ICES conference

[2] SLS-RQMT-287 *SLS Payload Integration Process Cross-Program Post-Artemis I Payload Safety Requirements Document* Rev A, 2023

[3] https://www.esa.int/Science_Exploration/Human_and_Robotic_Exploration/Exploration/Gateway "Accessed on 14.04.2025"

[4] ISO 14624-1:2003 Space systems *Safety and compatibility of materials* Part 1

[5] ISO/TS 16697:2012 Space systems Safety and compatibility of materials *Method to determine the flammability thresholds of materials*

[6] Simone A, Propato L, Grizzaffi L, Portaluppi M, Mooney C 2022 *FIAMMA -Flammability facility for Future Human Spaceflight Missions 15th ISMSE/ 13th ICPMSE*

[7] ECSS-Q-ST-70-21C *Flammability testing for the screening of space materials, 2010*

[8] NASA STD-6001B *Flammability, Odor, Offgassing, and Compatibility Requirements and Test Procedures for Materials in Environments that Support Combustion Test 1 Upward Flame Propagation, 2016*

[9] JSC 29353D *Flammability Configuration Analysis for Spacecraft Applications, 2024*

[10] Harper S A, Juarez A, Iii H P, Hirsch D B, Beeson H D 2016 *Oxygen Partial Pressure and Oxygen Concentration Flammability: Can They Be Correlated?* **2** *JOURNAL OF ASTM INTERNATIONAL.*

[11] Hirsch D B *et al* 2011 *Selected Parametric Effects on Materials Flammability Limits 41st ICES*, Portland

[12] ISO-TS1669 2016 *Effect of Ambient Oxygen Concentration on Heating Performance of Chemical Igniter used in NASA-STD-6001B* Trans *JSASS Aerospace Tech. Japan* **14** 30

[13] Hirsch D *et al* 2006 *Pressure Effects on Oxygen Concentration Flammability Thresholds of Materials for Aerospace Applications* Journal of Testing and Evaluation

[14] https://ntrs.nasa.gov/api/citations/20070018274/downloads/20070018274.pdf "Accessed on 14.04.2025"

[15] Harper S A 2017 *NASA-STD-6001B Test 1 Upward Flame Propagation: Sample Length Impact on MOC Investigation* ICES conference

[16] TASI-SD-GEN-TNO-0578 CECILY TN4 Final report, 2021

[17] TASI-SD-GEN-TNO-0624 STEFANIA Non -standard evaluation flame propagation & mitigation TN4 Final report, 2022

AUTHOR INDEX

Institute of Physics
Dirac House, Temple Back
Bristol BS1 6BE UK

ISSN: 1757-8981
ISBN 979-8-3313-2245-8
9 798331 322458

CURRAN ASSOCIATES INC.
proceedings.com